1*95

Polyamine-Chelated Alkali M Compounds

Polyamine-Chelated Alkali Metal Compounds

Arthur W. Langer, *Editor*

A symposium co-sponsored by the Division of Polymer Chemistry and the Organometallic Subdivision of the Division of Inorganic Chemistry at the 164th Meeting of the American Chemical Society, New York, N.Y., Aug. 28, 1972.

ADVANCES IN CHEMISTRY SERIES **130**

AMERICAN CHEMICAL SOCIETY
WASHINGTON, D. C. 1974

ADCSAJ 130 1-290 (1974)

Copyright © 1974

American Chemical Society

All Rights Reserved

Library of Congress Catalog Card 74-75334

ISBN 8412-0194-3

PRINTED IN THE UNITED STATES OF AMERICA

Advances in Chemistry Series
Robert F. Gould, *Editor*

Advisory Board

Kenneth B. Bischoff

Bernard D. Blaustein

Ellis K. Fields

Edith M. Flanigen

Jesse C. H. Hwa

Phillip C. Kearney

Egon Matijević

Thomas J. Murphy

Robert W. Parry

FOREWORD

ADVANCES IN CHEMISTRY SERIES was founded in 1949 by the American Chemical Society as an outlet for symposia and collections of data in special areas of topical interest that could not be accommodated in the Society's journals. It provides a medium for symposia that would otherwise be fragmented, their papers distributed among several journals or not published at all. Papers are refereed critically according to ACS editorial standards and receive the careful attention and processing characteristic of ACS publications. Papers published in ADVANCES IN CHEMISTRY SERIES are original contributions not published elsewhere in whole or major part and include reports of research as well as reviews since symposia may embrace both types of presentation.

CONTENTS

Preface ix

1. Some Mechanistic Aspects of N-Chelated Organolithium Catalysis   1
   A. W. Langer, Jr.

2. Synthetic Aspects of Tertiary Diamine Organolithium Complexes ..   23
   W. Novis Smith

3. Stereochemical Properties of N-Chelated Alkali Metal Complexes ..   56
   Galen Stucky

4. Magnetic Resonance Studies of Polytertiary Amine Chelated Alkali Metal Compounds 113
   M. T. Melchior, L. P. Klemann, and A. W. Langer, Jr.

5. Ac Conductivity of Some Organolithium Complexes in Aromatic Solvents 131
   E. O. Forster and A. W. Langer, Jr.

6. Inorganic Complexes and Separation Processes 142
   Lawrence P. Klemann, Thomas A. Whitney, and Arthur W. Langer, Jr.

7. Polymerizations Using N-Chelated Alkali Metal Catalysts 163
   C. W. Kamienski

8. Metalation and Grafting by Anionic Techniques 177
   Adel F. Halasa

9. Telomerization Reactions Involving Amine-Chelated Lithium Catalysts 186
   W. A. Butte and G. G. Eberhardt

10. Telomerization of Conjugated Diolefins with Aromatics and Olefins Using Chelated Organosodium Catalysts 201
    William Bunting and Arthur W. Langer, Jr.

11. Polylithiation of Hydrocarbons 211
    Robert West

12. Directed Metalation 222
    D. W. Slocum and D. I. Sugarman

13. Synthetic Applications of N-Chelated Organolithium Compounds .. 248
    M. D. Rausch and A. J. Sarnelli

14. Asymmetric Synthesis via Lithium Chelates 270
    Thomas A. Whitney and Arthur W. Langer, Jr.

Index 283

PREFACE

This volume is the first book to be published on polyamine-chelated alkali metal compounds and their uses. It combines papers from the first American Chemical Society symposium on N-Chelated Alkali Metal Compounds with five new papers (Chapters 1, 2, 10, and 12) to achieve broad coverage of this rapidly developing area.

The first two chapters provide general background, mechanistic aspects, and practical preparative procedures. Chapters 3, 4, and 5 deal with the relationships between structure and properties as determined by x-ray, magnetic resonance, and electrical conductivity. Chapter 6 examines inorganic complexes and their uses for separating mixtures of chelating agents or salts. Polymerization, polymer grafting, and telomerization are reviewed in the next three chapters followed by telomerization of dienes with aromatics using chelated organosodium catalysts. Chapters 11, 12, and 13 cover metalation reactions from the standpoints of polylithiation, directed metalation, and general synthetic applications. The final chapter presents a novel technique for asymmetric synthesis in which steric control is provided by asymmetric chelating agents.

The chelating agents are of the polyalkylene–polyamine type having two to six functional groups. Tertiary polyamines are required for alkali metal compounds having reactive carbanions whereas secondary amine, primary amine, or ether groups may be used with compounds having less reactive carbanions or inorganic anions. The primary emphasis to date has been on complexes of lithium and sodium compounds. Little has been done with the higher alkali metal compounds because the simpler chelating amines do not form strong complexes and the necessary tetra and higher polyamines, particularly cyclic types, are not yet readily available.

Alkali metal chelates possess extraordinary anion reactivity which is attributed to deaggregated, cation-solvated ion pairs in hydrocarbon solvents where anion solvation is negligible. In contrast, solvated anions must undergo desolvation before reaction, making them less reactive than the naked anions obtained in the chelate systems.

Polyamine chelation research was initiated in industrial laboratories where potential applications and patent activities delayed publication. The literature has begun to increase dramatically during the last five years, but even today the bulk of this research appears only in the patent literature. In contrast, there is extensive literature on the polyether systems although polyether research was initiated some four to five years later when the first polyamine chelate papers were published. Clearly the differences can be related to industrial academic interests.

The closely related research on polyether chelates by Michal Szwarc and his co-workers led to a detailed determination of the structure and properties of carbanions in ion pairs and free ions. The fundamental principles which were developed and clarified in their numerous publications contribute to an understanding and interpretation of much of the polyamine chelate work as well. More recently the crown ether chelates, pioneered by Pederson and co-workers at the Dupont Laboratories, have given additional impetus to research on chelated alkali metal compounds. Crown ethers and amines are cyclic variations which can provide greater stability and specificity in complexation of cations, particularly the heavier alkali metal ions.

The polyethers and crown ethers significantly broaden the range of properties attainable by chelation of alkali metal compounds. In practical applications the chelating tertiary amines are often preferred because of greater stability toward metalative decomposition and generally higher solubility of the complexes in hydrocarbon solvents. It is hoped that this book will provide the background and stimulation for continued rapid growth of this field.

ARTHUR W. LANGER

Linden, N.J.
November 1973

Some Mechanistic Aspects of *N*-Chelated Organolithium Catalysis

A. W. LANGER, JR.

Corporate Research Laboratories, Esso Research and Engineering Co., Linden, N.J. 07036

> *Chelating tertiary polyamines have a dramatic effect on the reactivity and properties of organolithium compounds. The unusual properties of the chelates have led to their extensive use as unique catalysts and chemical reagents. Some general aspects related to structure and properties are discussed including the nature of the chelating agent, chelate/LiR ratio, aggregation, and metalation. These are examined as they relate to ion pairing of the lithium–carbon bond, reactivity, and some aspects of polymerization. Metalations produce kinetically favored products with very slow rearrangement to the thermodynamically more stable products. The effects of ion pair structure and steric hindrance in the chelating agent are illustrated in butadiene polymerization. Chain transfer mechanisms in ethylene polymerization are presented to explain cyclopentane rings in the product.*

Before 1964 there was extensive literature on the effects of monofunctional Lewis bases on the reactivity of organolithium compounds, particularly in the polymerization field (*1*). However no information was available on the effects of chelating bases. Since chelating polyethers were in wide use, one can only speculate that they had been tried and were discarded because they rapidly decomposed the alkyllithium.

Chelating tertiary polyamines have a dramatic effect on the reactivity and properties of organolithium compounds. The chelates are new compositions which are rapidly finding wide use as unique catalysts and chemical reagents.

N-chelated alkali metal compounds were discovered in 1960 at Esso Research and Engineering Co. when butyllithium was complexed with TMED (*N,N,N′,N′*-tetramethylethylenediamine) for use in ethylene polymerization (*2*). At that time the objective was to make new nontransition metal polymerization catalysts, and the question was raised whether triethylaluminum could be simulated by complexing an alkyllithium with a chelating tertiary diamine. Ziegler (3) had shown that AlEt$_3$ dimer dissociates to an electron deficient monomeric species which adds to ethylene to form longer chain aluminum alkyls (the growth reaction). Presumably the ethylene coordinates weakly with the aluminum and is polarized by the partially ionic Al–C bond to facilitate addition, shown at the top of Figure 1.

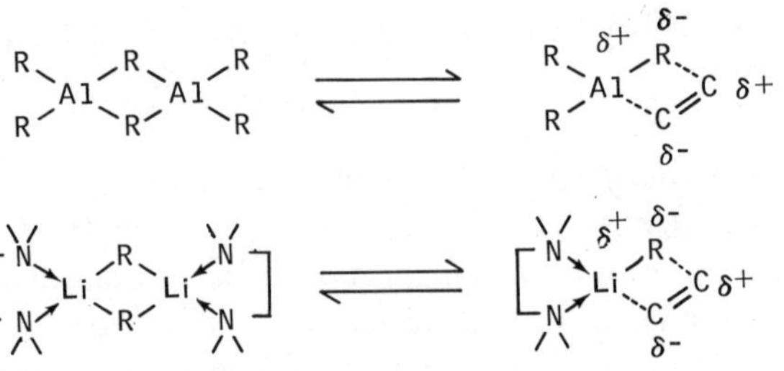

Figure 1. Trialkylaluminum simulation with an organolithium chelate

Butyllithium was also known to polymerize ethylene (4, 5, 6), but it was less active than triethylaluminum. Conceptually it was felt that BuLi should be more active because of the smaller cation and more ionic metal–carbon bond, but the low polymerization activity could be caused by greater difficulty in breaking down the strong aggregates. Solvation by TMED was visualized to give dimer and monomer structures which were directly related to AlEt$_3$, shown at the bottom of Figure 1. At the same time, solvation of lithium by TMED was expected to further increase the ionic character of the Li–C bond. TMED was used rather than ethers because it was expected to be less reactive toward BuLi.

These expectations were realized by the preparation of high molecular weight polyethylene at 25°–50°C whereas AlEt$_3$ activity is negligible below about 90°C. The increased ionic character of the catalyst caused an increase in metalation capability and led to the discovery of telomerization reactions (7) and new organometallic syntheses (8). The extraordinary reactivity of the N-chelated organolithium complexes stimulated extensive research directed toward defining the scope of the catalyst system, finding new applications, and examining the relationship of structure to chelate properties (9, 10, 11).

In 1962 and 1963 the chelated organolithium complexes and their application in ethylene telomerization were discovered independently of each other by Eberhardt (12) at Sun Oil Co. This research led to two patents (13, 14) in which the claims involving chelated catalysts were overturned by interference procedures (2, 7, 15, 16). Several papers were published (12, 17, 18) which helped stimulate interest in the catalysts. In addition Eberhardt discovered their use for the telomerization of ethylene with higher olefins containing allylic hydrogens (19). Eberhardt's and his co-workers' research resulted in patents of chelated organolithium catalysts used for isomerization of olefins (20), telomerization of ethylene with benzyl dialkylamines (21), ethylene polymerization above 100°C (22), and preparation of telomer waxes (23).

The scope of the chelated alkali metal system and the many uses for these new complexes are covered thoroughly in this volume. There are several general aspects related to structure and properties which are important in all preparations and applications. These include the ratio of chelating agent to the organolithium compound, the nature of the chelating agent, aggregation of the complexes, and the sharply increased metalation reactivity compared with uncomplexed organolithium compounds. This paper discusses these features in relation to the lithium–carbon bond, the chemical reactivity of the complexes, and some aspects of polymerization.

Structure of Lithium Chelates

Alkyllithium compounds exist as tetramers or hexamers in hydrocarbon solvents depending primarily upon the bulkiness and size of the alkyl groups (24, 25). Other largely covalent organolithium compounds probably have similar degrees of association although one might expect higher aggregates from dipole interactions for those having more ionic Li–C bonds. Support for this idea can be found in work by Makowski and Lynn (26) who showed that low molecular weight polybutadienyllithium has a degree of association considerably greater than six in the absence of solvent. They also showed that association then decreases with increasing molecular weight to a limiting value of two. In tetrahydrofuran both polyisoprenyllithium (27) and polystyryllithium (28) were reported to be monomeric although they are dimeric in hydrocarbon solvents (29).

n-Butyllithium is hexameric in benzene and cyclohexane at 25°C (30) and at their freezing points (31), but no cryoscopic measurements have been reported in paraffinic solvents. In our laboratory cryoscopic studies in benzene, toluene, and n-heptane showed that it was a hexamer in benzene and toluene but a tetramer in n-heptane at concentrations between $0.2M$ and $1.0M$ (32). The heptane result might be attributed to stronger solvation of the butyl groups at the freezing point ($-90.7°C$), thereby stabilizing a smaller aggregate.

Many chelated organolithium compounds can be obtained as 1:1 complexes, but only certain lithium aggregates appear to form insoluble complexes (8, 10). TMED, *trans*-N,N,N',N'-tetramethyl-1,2-cyclohexanediamine (*trans*-TMCHD) and tetramethyl-1,3-propanediamine (TMPD) formed crystalline complexes with butyllithium dimer and tetramer. Attempts to isolate complexes having 3:1, 5:1, or 6:1 BuLi:chelate ratios were unsuccessful. Thus the stable aggregate structures in butyllithium persist in complexes formed with bidentate chelating agents. However an equimolar amount of a strong chelating agent will produce 1:1 complexes in dilute solutions. BuLi:TMED is monomeric at low concentrations at both 1:1 (10, 32, 33) and 1:2 (33) BuLi/TMED ratios and dimeric at 1:1 ratio at high concentrations.

TMED produces a crystalline complex with methyllithium tetramer, but even a large excess of TMED does not break down the tetramer to dimer or monomer (8). This result shows that methyl bridges are stronger than tertiary amine complexation to lithium.

Various other organolithium complexes with bidentate chelating agents also contained lithium dimer and tetramer species (8). However preliminary attempts to prepare $(C_6H_5Li)_4 \cdot$ TMED from solutions containing 4:1 and 6:1 molar ratios yielded only crystals having the unusual ratio of three phenyllithium per TMED (8). A trimeric lithium species has been reported for lithium bis(trimethylsilyl)amide in x-ray studies (34) although Kimura and Brown (35) interpreted the solution behavior as an equilibrium between dimer and tetramer. It would be interesting to determine whether the phenyl anions bridge in a similar manner to amide ions or whether a chelating agent produces a different aggregate structure. Molecular weight data are needed to distinguish between $(C_6H_5Li)_3 \cdot$ TMED and $(C_6H_5Li)_6 \cdot$ 2TMED. With the unusual ratio, this work must be considered tentative until it is repeated. There was also evidence for complex formation with phenyllithium dimer which would be analogous to the phenyllithium dimer complexes in ether solvents reported by West and Waack (36).

With the exception of methyllithium, all types of organolithium compounds have yielded 1:1 complexes with bidentate, tridentate, or tetradentate chelating tertiary polyamines (16). Most are crystalline but BuLi · TMED is an oil.

Stable complexes containing two molecules of chelating agent per organolithium are obtained only when solvent separated ion pair formation is favorable. This occurs with ionic lithium compounds having a large, soft anion. For example crystalline complexes have been isolated having the compositions $(C_6H_5)_2CHLi \cdot (TMED)_2$, $(C_6H_5)_3CLi \cdot (TMED)_2$, and $(C_6H_5)_2CHLi \cdot HMTT$, where HMTT is hexamethyltriethylenetetramine (8). Such structures should be general for all lithium compounds in which the negative charge on the anion is more diffuse than in the benzyl anion.

In order to avoid chelate decomposition the carbanion must be less reactive than benzyl anion, or it will metalate the chelating agent with subsequent decomposition. Figure 2 illustrates crystalline complexes of benzyllithium with bi-, tri-, and tetradentate chelating agents. While the first two are stable, the tetramine complex decomposes extensively in 24 hours. In contrast, all three complexes with diphenylmethyllithium are stable. For the first two complexes with benzyllithium, the UV spectrum in benzene shows only one absorption at 330nm for the contact ion pair. The tetramine complex also shows a small absorption at 367nm, presumably from a loose or separated ion pair structure. In the crystal lattice where ion pair separation should be more favorable, decomposition occurred rapidly *via* benzyl ion attack on an N—CH_3 group. For $(C_6H_5)_2CHLi \cdot HMTT$ complex the UV absorption at 460nm for the separated ion pair is larger than that at 420nm for the contact ion pair, indicating a higher proportion of separated ion pairs than in the benzyllithium complex. This is consistent with the greater resonance stabilization of diphenylmethyl compared with benzyl anion. In this case, however, the complex is stable to decomposition because the less reactive carbanion cannot metalate the chelating agent despite the greater concentration of separated ion pairs. These results indicate that the tetramine complexes exist in two forms: (a) a contact ion pair in which only three nitrogens solvate lithium and the fourth is uncoordinated and (b) a separated ion pair in which the fourth nitrogen displaces the carbanion to the outer sphere.

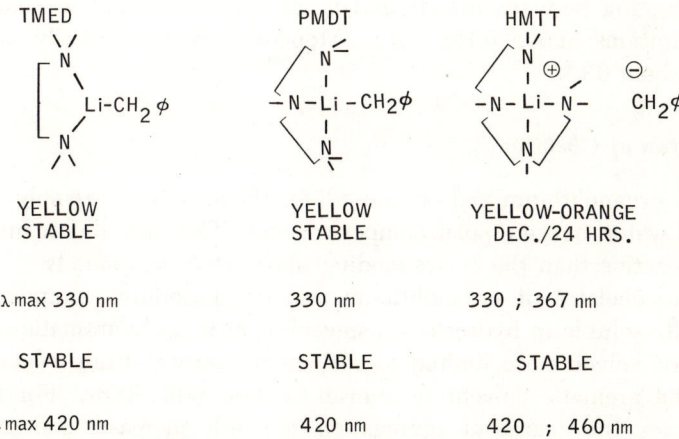

Figure 2. Organolithium chelates.

Top: Crystalline benzyllithium complexes. Bottom: Crystalline diphenylmethyllithium complexes.

X-ray structures for various crystalline organolithium complexes are covered in Chapter 3 which provides proof for chelate ring structures as opposed to open chain polymeric structures. It also supports the NMR finding of increased ionic character in the Li–C bond upon solvation of the lithium (*10*).

The resistance of organolithium aggregates to dissociation by donor solvents or chelating Lewis bases is undoubtedly related to the lattice energy of the aggregates just as was found for the complexation of inorganic salts (*see* Chapter 6, Klemann *et al.*). As stability increases, more powerful solvating agents are required to overcome this lattice energy. If stability is too high for 1:1 complexation by permethylated polyethylenetetramines, the lattice energy of the aggregate is above about 200 kcals/mole at 25°C. Methyllithium tetramer appears to be in this category.

All of the above discussion deals with the ratio of chelating agent to organolithium compound in various complexes. In most cases molecular weight has either not been determined or it was determined at a single concentration in a single solvent. It should be understood that all of the chelates have a strong tendency to aggregate because of their increased ionic character. The dipole–dipole interactions are dependent on temperature, concentration, solvent, etc. Thus the degree of aggregation for any specific system should be determined whenever it is suspected of influencing properties. In dilute solutions, as in catalysis, the chelates are believed to be monomeric. Where concentration studies were made (TMED · LiCH$_2\phi$ and ϕ_2CHLi with di-, tetra- and pentamines), aggregates ranging between dimers and hexamers were found in benzene at concentrations above $0.1M$ (*37*). Monomers were generally observed below about $0.1M$.

Properties of Chelates

All organolithium and organosodium chelates are extremely reactive toward water, oxygen, polar compounds, etc. They are almost invariably more reactive than the corresponding unchelated compounds.

The chelates of organolithium and organosodium compounds are generally soluble in hydrocarbon solvents, particularly aromatic solvents. Aromatic solvents are limited to carbanions derived from weaker acids than the aromatic solvent or transmetalation will occur. For the 1:1 complexes, the degree of aggregation generally increases and solubility decreases with increasing ionic character of the M–C bond. This is a consequence of stronger dipole interactions. Interestingly, in cryoscopic studies with highly soluble ionic complexes in benzene, deaggregation appears to occur at high concentrations probably because of non-ideal solutions arising from the strong solvation of the complex by benzene (*37*). The formation of ion pair aggregates and multiple ions in aromatic solvents gives these solutions unusual electrical properties which may be useful in batteries (*38*). There is NMR evidence that chelating agent separated ion pair formation in the 2 to 1 complex of TMED with LiCHϕ_2 increases with increasing concentration of the complex (*39*). Apparently the high local dielectric constant in the aggregates facilitates ion separation. One suspects that in the crystalline state this separation would be still more extensive, if not complete.

Metalation of Weak Acids. The ability to metalate very weak acids is a property which has been used widely for the direct synthesis of organolithium compounds, for new or improved organic syntheses, for polymer grafting, and for telomerizations involving chain transfer by transmetalation. These major topics are covered in seven chapters of this volume, indicating both the scope and utility of this property of N-chelated organoalkali metal compounds.

Increased reactivity in hydrogenolysis has been correlated to increased ionic character of the Li–C bond caused by strong solvation of lithium by chelating agent (*10*). The upfield chemical shift of the alpha-methylene protons of the butyl group caused by chelation was attributed to increased negative charge on the carbanion which is related to the amount of ionic contribution to the bond. As the TMED:BuLi ratio was increased from 0:1 to 1:1, both the chemical shift and the hydrogenolysis rate increased. No further change occurred above a 1:1 mole ratio, indicating that no 2TMED/BuLi complex forms (Figure 3). At 820 mm

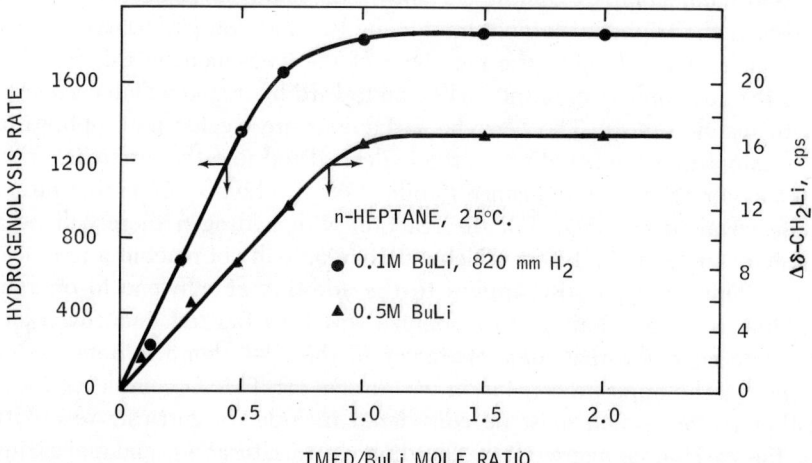

Figure 3. Parallel between BuLi hydrogenolysis rate and α-methylene proton chemical shift

hydrogen and 25°C in *n*-heptane, the 1:1 complex reacted approximately 7000 times faster than uncomplexed BuLi to produce butane and lithium hydride. Gilman reported that butyllithium reacted completely with hydrogen only after 61 hours at 100 psig (*40*). Furthermore the reactivities of phenyllithium and phenylpotassium toward hydrogenolysis in benzene were compared. Phenyllithium completely reacted in 32.2 hrs at 100 psig whereas the reaction with phenylpotassium was 90% complete in 0.54 hrs at slightly above atmospheric pressure (*40*). An increased rate of hydrogenolysis of phenyllithium has also been reported upon addition of ether solvents (*41*). The above facts taken together provide additional evidence that chelation increases the ionic character of the M–C bond.

Reactivity of butyllithium further increases when TMED is replaced by a tridentate (PMDT) or a tetradentate (HMTT) chelating agent. In these complexes reactivity is excessively high and leads to rapid metalation of the chelating agent. Thus, more effective or more extensive solvation of lithium by nitrogen bases gives the Li–C bond ion pair properties similar to uncomplexed alkylsodium or alkylpotassium. However the lithium chelates have advantages in solubility and much lower aggregation in hydrocarbon solvents.

This relationship to ionic character applies only to a given organolithium compound, however, and not to a comparison of different carbanions. For example hydrogenolysis of 1:1 complexes of TMED with butyllithium, benzyllithium, diphenylmethyllithium, and triphenylmethyllithium gives sharply decreasing rates in the order shown (*42*). Yet the butyllithium complex almost certainly has some covalent contribution whereas the triphenylmethyllithium complex is an ion pair. Here we are concerned with the intrinsic reactivity of the carbanion; in this reaction it is the nucleophilicity of the carbanion toward hydrogen which correlates with reaction rates. The benzylic carbanions are weaker nucleophiles to the extent that the negative charge is delocalized into the aromatic rings. At the same time this resonance stabilization also favors ion pairing rather than covalent bonding. For the reaction with hydrogen there will be a basicity cut-off point below which a carbanion will not react at a practical rate. This principle also applies to the addition of ethylene to organolithium reagents. Failure to recognize this lead Bartlett and coworkers (*43*) to conclude that ionic character in the Li–C bond is unfavorable although the opposite conclusion was drawn for TMED complexes (*44*). Other factors which must be considered include the intrinsic reactivity of the carbanion, aggregation, coordinative unsaturation, and polarizing ability of the metal cation. Qualitatively this cut-off point appears to be slightly below triphenylmethyl carbanion, at about a pK_b of 30-31 on the MSAD scale (*45*). This places the pK_a of hydrogen at about 29-31 (same scale). Chelated alkyllithium can metalate any compound having an acidity greater than methane (less than about 40 pK_a).

Aggregation adversely affects metalation reactivity. With ϕ_2CHLi · TMED in toluene at 25°C and 820 mm hydrogen the rate at 0.4M was only one-fifth that at 0.025M (46). One possible explanation is that this is related to the increased formation of separated ion pairs in the aggregates. One could rationalize more facile hydrogenolysis in a contact ion pair *via* a four center reaction mechanism with lithium participation than in a simple anionic attack on unpolarized hydrogen.

Kinetic *vs*. Thermodynamic Metalations. Metalations using 1:1 complexes in hydrocarbon solvents initially yield the products determined by kinetic acidities rather than thermodynamic acidities. We attribute this to contact ion pair structures which equilibrate only very slowly in nonpolar solvents. This feature has considerable synthetic value since it allows direct preparation and isolation of organolithium compounds by metalation of weaker acids than the chelating agent. Table I illustrates this point for the reaction of TMED · *sec*-BuLi with tetramethylsilane (TMS). The first reaction in *n*-heptane gave over 95% yield of lithiomethyl trimethylsilane within minutes at 30°C. After standing about two months the mixture contained 30% metalated TMED and it probably had not attained thermodynamic equilibrium. From this we see that tetramethylsilane has a very high kinetic acidity relative to TMED, but its equilibrium or thermodynamic acidity is comparable with TMED. The second reaction shows a competitive metalation of equimolar amounts of tetramethylsilane and benzene. Initially 35% TMED · Li–TMS and 65% TMED · Liϕ formed compared with 90% Liϕ at equilibrium. These results show the remarkably high kinetic acidity of tetramethylsilane and also indicate that its thermodynamic acidity falls between that of benzene and TMED. This latter observation is contrary to predictions based on electronegativity and suggest carbanion stabilization by pπ–dπ overlap.

Table I. Kinetic and Thermodynamic Product Yields from Metalation Reactions With TMED · *Sec*-BuLi

		Chelated Products		
Reactants		Li-TMS	Li-TMED	Li—C_6H_5
TMED·*sec*-BuLi + TMS	Initial	>95%	<5%	—
	Equilib.	70%	30%	—
TMED·*sec*-BuLi + TMS + Benzene	Initial	35%	—	65%
	Equilib.	10%	—	90%

The metalation of tetramethylsilane is another dramatic example of the effect of chelation on reactivity since both *n*-BuLi and *sec*-BuLi are nearly inert under the same conditions. Chelated *sec*-BuLi is the most reactive soluble metalating agent we have found. TMED · Li-*sec*-Bu reacts with tetramethylsilane about 1000 times faster than TMED · Li-*n*-Bu and yields purer product (*47*). Broaddus (*48, 49*) has discussed kinetic metalation of olefins and alkyl aromatic compounds using TMED · LiBu, and he also observed the slow equilibration to the thermodynamically favored isomers. The extent of ring metalation in toluene and the conditions for isomerization to benzyllithium are discussed in Chapter 2, Smith.

Metalation of the Chelating Agent. A major consequence of the metalation reactivity is the competing metalation of the chelating agent. Although it is often an undesirable by-product, the metalated chelating agent has its own utility in organometallic syntheses (*see* Chapter 2), in the preparation of new chelating agents (*50*), and as polymerization initiators (*51*). The attack on chelating agent could be facilitated by complexation with the lithium cation which could enhance the acidity of the N–CH$_3$ protons. This is consistent with the 0.1 ppm ^{1}H NMR downfield chemical shift upon chelation of TMED with BuLi in methylcyclohexane solvent (*37*). Metalation can occur with any organolithium derived from a compound having a pK_a greater than about 35 (toluene) (MSAD scale). It occurs more readily with alkyllithium and with increasing activation of the organolithium by increasing solvation as the number of chelating nitrogens is increased from two to four. Both tetramines and excess diamines with BuLi lead to rapid metalation and decomposition. For this reason complexes should always be made by slow addition of the chelating agent to the organolithium (or sodium) and not vice-versa. When a chelated alkyllithium, especially *sec*-alkyllithium, is to be used for metalation, the complex should be prepared in the presence of excess substrate preferably using the substrate as solvent when possible to minimize metalation of chelating agent. If some metalated chelating agent forms under these conditions, it will often be sufficiently reactive to attack the substrate before decomposing to inactive products.

Lithiation of the chelating agent was first noted from aging TMED · LiBu catalyst before it was used for ethylene polymerization (*2, 9*). This led to polymer containing an end group of chelating agent since the lithiated chelating agent is a new, active organolithium compound.

An extensive ^{1}H and ^{7}Li NMR study of the aging of this system has shown that a complex series of reactions occurs (*47*). The first step in the aging of a TMED · LiBu solution is metalation of the chelating agent *via* a second order reaction to form presumably monomeric LiCH$_2$N(CH$_3$)CH$_2$CH$_2$N(CH$_3$)$_2$. The metalation rate was followed by the disappearance of the high field ^{1}H NMR peak caused by the alpha protons of the butyl group. The freshly metalated species (Li–TMED) can either decompose to lithium dimethylamide plus vinyl dimethylamine or it can aggregate to form a more stable metalated species. In the n-BuLi complex metalation, decomposition, and aggregation have comparable rates.

However with sec-BuLi · TMED, metalation was about 100 times faster, and it was complete before the subsequent reactions proceeded very far. The 100-MHz ^{1}H spectrum of the final (Li–TMED)$_n$ aggregate suggested a repeating unit with two CH_2 groups in an A_2B_2 pattern, three CH_3 groups two of which are nearly equivalent, and a CH_2 peak at much higher field. This spectrum is consistent with a structure which has the Li–TMED unit in a rigid cyclic arrangement such as the dimer (Structure I) and higher aggregates. The initial metalation product most likely has a bicyclic structure (Structure II) which undergoes rearrangement to a less strained ring structure after aggregation.

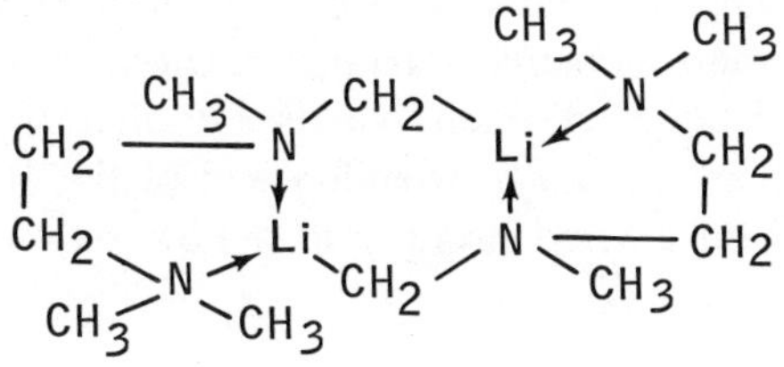

(I) 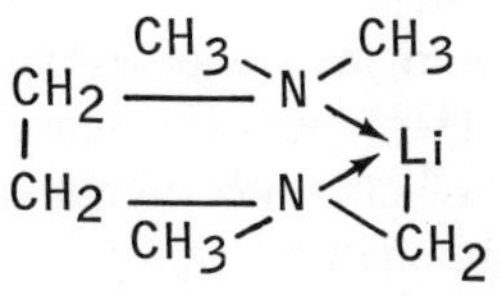   (II)

Both freshly metalated TMED and aged product contain an active Li–CH$_2$N structure shown by ethylene polymerization activity and transmetalation reactions. For example metalated TMED reacts slowly with toluene to produce TMED · LiCH$_2\phi$. Hydrogenation of 0.5M TMED · LiBu complex in n-heptane aged one week at 25°C gave an 82% recovery of TMED and 18% decomposition products. These experiments also demonstrated that the rearrangement observed by NMR did not involve the skeletal structure of TMED.

Some General Factors Affecting Polymerizations

Four chapters in this volume are addressed to the uses of chelated alkali metal complexes in various polymerizations, telomerizations, and polymer grafting applications. They fully cover all of the published work in these areas. There are, however, several general features based on our unpublished results which warrant a general discussion. These include the effect of catalyst ion pair structure on polymerization activity and polybutadiene microstructure, the effect of steric hindrance on catalyst activity, and the mechanisms for chain transfer.

Effect of Ion Pair Structure on Butadiene Polymerization. Several different studies of the effects of chelating agent structure on organolithium catalysts show that the proportion of 1,2-polybutadiene microstructure is directly related to the stability of the lithium chelate. For example in the $(CH_2)_n(NMe_2)_2$ series where $n = 1$ to 6, the highest 1,2 microstructure was obtained when $n = 2$ or 3 (*10*). These diamines produce the most stable 5 and 6 membered chelate structures with lithium and therefore provide the greatest driving force for ion pair formation in the series. Similarly polymerization at lower temperatures with any given chelated lithium catalyst produced a higher per cent of 1,2 microstructure because of the stronger solvation of lithium by the chelating agent (*10*). Lower temperature shifts the equilibrium between covalent bonding (localized) and ion pair bonding (delocalized) to a higher proportion of ion pairs. With a very weak solvating agent like diethyl ether, NMR evidence for such an equilibrium has been observed at $-45°C$ (*52*). We have also found that decreasing the steric hindrance to chelation (TMED *vs.* TEED) shows this same effect at constant temperature (*42*). In all these instances the more effective solvation favors ion pair formation in which propagation from allyl anions yields 1,2-microstructure. The kinetically more reactive secondary carbanion end of the allyl anion reacts with monomer to form a vinyl group and a new terminal allyl anion:

$$RCH_2CH\overset{\ominus}{\cdots\cdots}CH\cdots\cdots CH_2 \;+\; CH_2=CH-CH=CH_2 \;\longrightarrow\; RCH_2\overset{CH=CH_2}{CH}-CH_2-CH\overset{\ominus}{\cdots\cdots}CH\cdots\cdots CH_2$$

If solvent-separated ion pairs or free ions were present, they should produce similar polymer microstructure to that obtained from contact ion pairs since propagation will involve only the allyl anion. There is no evidence for anything other than contact ion pairs in 1:1 lithium complexes with chelating diamines or triamines in hydrocarbon solvents. Only by using excess diamine or the more powerful chelating tetramines can we test the idea. As mentioned previously these are capable of producing some separated ion pairs when the anion is a sufficiently weak nucleophile to be displaced from the lithium by a neutral tertiary amine. With a benzyllithium · tetramine complex, both contact and separated ion pair structures were observed spectroscopically. Since allyl and benzyl anions have rather similar charge delocalization, it is reasonable to expect that a tetramine complex of polybutadienyllithium would have similar proportions of contact and separated ion pairs.

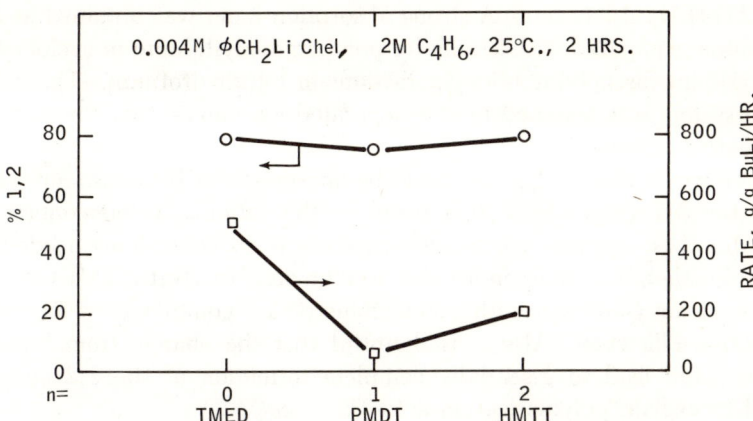

Figure 4. Effect of polydentate ligand $Me_2N(CH_2CH_2N\overset{CH_3}{})_n CH_2CH_2NMe_2$ on butadiene polymerization

Figure 4 shows polymerization rate and polymer microstructure found using benzyllithium initiator combined with diamine, triamine, or tetramine chelating agents (53). In all three cases the 1,2-microstructure was about 80% indicating that propagation occurred from allyl anions. (In the absence of solvating agents lithium catalysts make polybutadiene having nearly 90% 1,4-enchainments). Thus microstructure alone does not provide evidence for polymerization involving separated ion pairs. However polymerization rates are more informative. The high activity with TMED can be attributed to a 1:1 chelate with a contact ion pair structure having low steric hindrance to an incoming monomer. On the other hand the triamine PMDT produces a contact ion pair in which lithium is fully coordinated. The lithium allyl ion pair is highly hindered by the methyl groups in the chelating agent and only negligible activity was obtained. Increased activity obtained with the tetramine HMTT could be attributed to propagation from a small fraction of separated ion pairs in which all four nitrogens are coordinated to lithium and the allyl anion is displaced to the outer sphere where it is unhindered. Propagation from a contact ion pair structure having only three of the four nitrogens coordinated to lithium is ruled out because it would be even more hindered than the triamine complex. Increased activity was also obtained using excess TMED.

Analogous results were reported by Hay and co-workers (33, 54). In the polymerization of butadiene initiated by BuLi/TMED, the rate was found to increase with increasing TMED to 1Li:2TMED, but additional TMED had no further effect. The ultraviolet spectrum of the polybutadienyl anion suggested chelating agent separated ion pairs in the 2TMED/1Li system. A strong absorption band was obtained at λ_{max} = 314nm compared with 275nm for polybutadienyllithium in cyclohexane and 312nm for polybutadienylpotassium in tetrahydrofuran. The potassium system was assumed to give separated ion pairs rather than contact ion pairs.

However this assignment could be questionable. It seems more likely that the 312-314nm band is a result of the cation-solvated contact ion pair for these systems particularly at the low concentrations needed for UV. Furthermore their molecular weight data for BuLi/TMEDA mixtures are not consistent with a stoichiometric 1:2 complex used to explain initiation efficiency. Also it is doubtful that the change from butyl to allyl would lead to essentially complete formation of the 1:2 complex used to explain polymerization activity.

Butadiene polymerization studies with HMTT/ϕCH$_2$Li catalysts have given results which are directly contrary to those expected from the Hay mechanism. Activity at 0.5 HMTT/ϕCH$_2$Li was double that at 1:1 ratio whereas the reverse should have been obtained if the tetramine solvated lithium compound were the active species. Our lithium catalyst studies suggest that all of the known TMED complexes are active for butadiene polymerization with activity increasing roughly in the order (RLi)$_4$ · TMED < (RLi)$_2$ · TMED < RLi · TMED << RLi(TMED)$_2$. The equilibrium to form RLi(TMED)$_2$ is believed to be unfavorable except when R$^-$ is a highly delocalized carbanion.

DIAMINE	G./G. BuLi	
	C$_4$H$_6$, 2 HRS.	STYRENE, 1 HR.
C\N-C-C-N/C (C,C / C,C)	> 750	300
C\N-C-C-N/C-C (C,C / C,C-C)	550	230
C\N-C-C-N/C-C (C,C / C-C,C-C)	420	145
C\N-C-C-N/C-C (C,C-C / C-C,C-C)	290	80
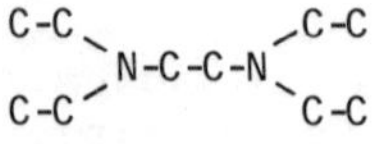 C-C\N-C-C-N/C-C (C-C,C-C / C-C,C-C)	260	40
NONE	65	30

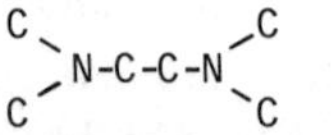

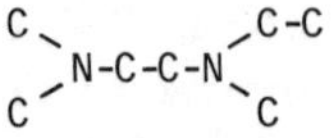

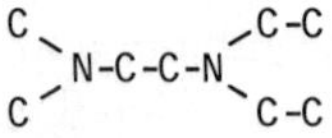

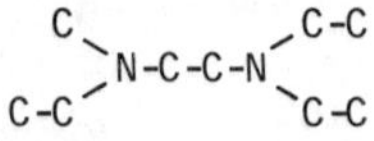

Figure 5. *Effect of diamine substituents on polymerization rates, 0.002M BuLi-Diamine, n-C$_7$, 25°C*

Steric Effects of the Chelating Agent. In the preceding section a triamine chelate was shown to have low activity in butadiene polymerization compared with the diamine chelate. The effects of more gradual changes in chelate structure on the rates of polymerization for butadiene and styrene are shown in Figure 5. The steric bulk of tetramethylethylenediamine was systematically increased by successive replacement of methyl groups by ethyl groups, and activity decreased accordingly. With both monomers, however, the activity of butyllithium alone was lower than that with the most hindered chelating agent, tetraethylethylenediamine (TEED).

Instead of steric hindrance to propagation, an alternative explanation is dissociation of the chelating agent from the lithium caused by the increasing hindrance to complexation. Dissociation definitely is a contributing factor with chelating agents having larger alkyl substituents like tetrapropylethylenediamine, particularly at higher polymerization temperatures because the per cent 1,2-polybutadiene structure decreases substantially. However TEED and TMED produced nearly the same per cent 1,2-polybutadiene structure at 25°C indicating that the polymer was produced at a chelated lithium site. These data and the data from PMDT (a triamine) suggest that monomer approach to the ion pair site is hindered by substituents on the chelating agents. A quantitative measure of the steric factor on polymerization rates can be obtained without a study of chelate dissociation equilibria under polymerization conditions and a knowledge of the specific propagation rate constants for the covalent, contact ion pair and separated ion pair species.

The results are easily rationalized by assuming propagation takes place at a TMED · Li$^+$R$^-$ contact ion pair site where the bulkier chelating agents could hinder monomer approach to the ion pair. However Hay concluded that the active site is (TMED)$_2$Li$^+$R$^-$ even at a 1:1 mole ratio of BuLi/TMED (33). One would not expect steric hindrance to monomer judging from the bulkiness of the chelating agents in such a chelate separated ion pair unless the allyl anion is buried in the chelate hydrocarbon shell or strongly hydrogen bonded to the shell. More likely, the hindered chelating agents would have greater difficulty forming the active dichelated lithium species. Lower activity with hindered chelating agents is also consistent with poorer ability to break the lithium dimer and form the active 1:1 complex.

Increased steric bulk in the chelating agent adversely affects catalyst activity. However there are several mechanisms by which steric hindrance could affect activity, but we cannot distinguish them at present.

Chain Transfer Mechanisms. Chelated alkali metal catalysts can undergo a variety of chain transfer reactions depending primarily upon the ionic character of the metal–carbon bond and the nature of the carbanion. In general, chain transfer reactions increase with catalyst structural changes in the order covalent < contact ion pair < separated ion pair, which is the usual order for anionic polymerization activity (55). Also some types of chain transfer reactions, like metalation, increase with increasing nucleophilicity of the carbanion.

For ethylene polymerization with lithium catalysts the most probable chain transfer reactions include (1) monomolecular elimination of LiH, (2) bimolecular displacement of the polymer by monomer, (3) metalation of monomer, and (4) metalation of solvent as shown below.

$$LiCH_2CH_2R \longrightarrow LiH + CH_2 = CHR \tag{1}$$

$$LiCH_2CH_2R + C_2H_4 \longrightarrow LiC_2H_5 + CH_2 = CHR \tag{2}$$

$$LiR + C_2H_4 \longrightarrow LiCH = CH_2 + RH \tag{3}$$

$$LiR + R'H \longrightarrow LiR' + RH \tag{4}$$

Both Reaction 1 and Reaction 2 are well known with other alkyl metals like AlR_3 whereas Reaction 3 and Reaction 4 are expected of strong carbanion catalysts.

Reaction 1 appears to result solely in termination. In hydrogenolysis experiments with various chelates we have observed precipitation of lithium hydride in all cases at room temperature. Attempts to generate chelated LiH *in situ* by adding hydrogen during ethylene polymerization also caused a rapid, irreversible loss of activity. Since there is no evidence that lithium hydride can add to ethylene under moderate polymerization conditions, it is unlikely that any significant chain transfer occurs *via* this mechanism. Potassium alkyls readily eliminate olefin with the formation of metal hydride, and sodium alkyls do so at elevated temperatures (56). It was noted earlier that chelation of lithium alkyls makes them more like sodium or potassium compounds, so it is quite probable that some termination occurs by eliminating LiH. It is conceivable that this could be a chain transfer mechanism with more reactive monomers than ethylene because addition to lithium hydride would be more favorable.

Eberhardt (57) reported that olefins were obtained using chelated butyllithium catalysts and assumed that they were produced by Reaction 2. However olefins are potentially available from three different reactions. The amount of olefin produced by Eberhardt is not known to be in excess of that from the catalyst so Reaction 1 cannot be excluded. Reaction 2 must be considered plausible by analogy with aluminum alkyls.

Broaddus (49) has shown that BuLi · TMED can metalate an alpha olefin at the vinylic hydrogens as well as at the allylic positions. This reaction must also take place with ethylene as shown in Reaction 3. If the vinyllithium added to ethylene and then chain transferred *via* transmetalation, one would again obtain linear alpha olefins. However we have identified alkylcyclopentanes as the major product type (over 95%) in ethylene polymerization studies with BuLi/2TMED in *n*-heptane at 100°C and 1000 psig ethylene. These cyclic structures are produced at the hexenyllithium stage during polymerization initiated by vinyllithium:

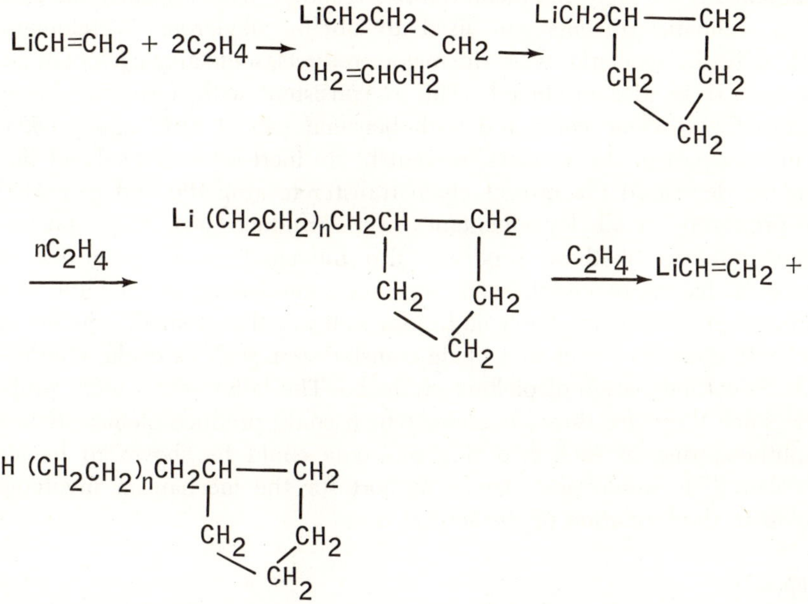

$$LiCH=CH_2 + 2C_2H_4 \rightarrow \begin{matrix} LiCH_2CH_2 \\ \\ CH_2=CHCH_2 \end{matrix} CH_2 \rightarrow \begin{matrix} LiCH_2CH-CH_2 \\ | \quad\quad | \\ CH_2 \quad CH_2 \\ \diagdown \diagup \\ CH_2 \end{matrix}$$

$$\xrightarrow{nC_2H_4} \begin{matrix} Li(CH_2CH_2)_nCH_2CH-CH_2 \\ | \quad\quad | \\ CH_2 \quad CH_2 \\ \diagdown \diagup \\ CH_2 \end{matrix} \xrightarrow{C_2H_4} LiCH=CH_2 +$$

$$\begin{matrix} H(CH_2CH_2)_nCH_2CH-CH_2 \\ | \quad\quad | \\ CH_2 \quad CH_2 \\ \diagdown \diagup \\ CH_2 \end{matrix}$$

If the ring structure were produced by metalation of olefinic products and cyclization after addition of two molecules of ethylene, they would be cyclohexane derivatives:

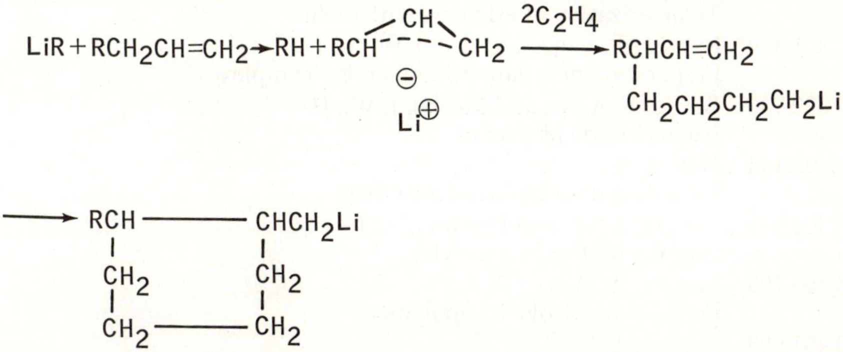

$$LiR + RCH_2CH=CH_2 \rightarrow RH + RCH\begin{matrix}\diagup CH \diagdown \\ \\ \ominus \quad\quad CH_2 \\ Li^\oplus \end{matrix} \xrightarrow{2C_2H_4} \begin{matrix} RCHCH=CH_2 \\ | \\ CH_2CH_2CH_2CH_2Li \end{matrix}$$

$$\longrightarrow \begin{matrix} RCH-CHCH_2Li \\ | \quad\quad | \\ CH_2 \quad CH_2 \\ | \quad\quad | \\ CH_2-CH_2 \end{matrix}$$

Mass spectroscopic analysis of separated products in the C_{6-50} range showed only cyclopentane rings. Therefore the major cyclization reaction must proceed *via* vinyllithium and the major chain transfer process in the absence of acidic solvents must be Reaction 3. A previously proposed cyclic elimination mechanism (53, 58) must be considered less probable because one would expect both five and six membered rings to be produced.

When compounds are present which are more acidic than ethylene, Reaction 4 is the preferred chain transfer reaction. This is clearly the case when aromatic solvents are used to obtain alkylaromatic telomers. Although alkylaromatics were the major products, some alkylcyclopentane products were also produced. This is consistent with a slightly lower acidity for ethylene compared with benzene (about $1pK_a$ unit) (59). Replacing part of the aromatic solvent by an inert saturated solvent like heptane decreased the rate of chain transfer to aromatic and increased the proportion of alkylcyclopentanes in the product (60). Mass spectroscopic analysis of these products also indicated small quantities of molecules having two naphthenic rings and some having an aromatic ring and a naphthene ring. Such molecules indicate that a small amount of cyclization occurs either in a chain transfer step, such as cyclic elimination, or *via* metalation of olefinic products. The latter seems more probable since there are three reactions which could produce olefins. If one naphthene ring in each two ring molecule could be shown to be six membered, it would add strong support for the mechanism involving olefins in the formation of the second ring.

Appendix

Bibliography of U.S. Patents on *N*-Chelated Alkali Metal Compounds

3,206,519[a] Eberhardt, G. G.
              Telomerization of ethylene with aromatics
3,257,364   Eberhardt, G. G.
              Telomerization of ethylene with olefins
3,321,479[b] Eberhardt, G. G. and Butte, W. A., Jr.
              Preparation of organolithium amine complexes
3,329,736   Butte, W. A., Jr. and Morris, J. W., III
              Isomerization of Olefins
3,402,144   Hay, A. S.
              Lithiation of polyphenylene ethers
3,404,042   Langer, A. W. and Forster, E. O.
              Organic battery electrolytes
3,450,795   Langer, A. W.
              Preparation of block copolymers
3,451,988   Langer, A. W.
              Compositions and uses for polymerization
3,458,586   Langer, A. W.
              Telomerization of ethylene with aromatics

3,474,143	Butte, W. A., Jr. Telomerization of ethylene with $C_6H_5CH_2NR_2$
3,498,960	Wofford, C. F. Random copolymerization of dienes and styrenes
3,502,731	Peterson, D. J. Lithiation of methylsulfides
3,509,188ᶜ	Halasa, A. F. and Tate, D. P. Polylithiation of ferrocenes
3,511,865	Peterson, D. J. Lithiation of methylsilanes
3,517,042	Peterson, D. J. Lithiated methylsilanes
3,522,326	Bostick, E. E., Hay, A. S. and Chalk, A. J. Grafting on lithiated polyphenylene ethers
3,532,772	de la Mare, H. and Neumann, F. E. Polymerizations using chelated $LiPR_2$
3,536,679	Langer, A. W. Lithiated chelating agents
3,541,149	Langer, A. W. Crystalline organolithium chelates
3,567,703	Eberhardt, G. G. Ethylene polymerization at 101-150°C
3,579,492	Smith, W. N., Jr. Narrow molecular weight distribution polyethylene at 0-75°C
3,584,056	Peterson, D. J. Uses for lithiated methylsulfides
3,594,396	Langer, A. W. Lithiated methylsilane chelates
3,597,463	Peterson, D. J. Uses for lithiated methylsulfides
3,598,793	Koch, R. W. Preparation of carboxylated rubbers
3,624,260	Peterson, D. J. Uses for lithiated methylsulfides
3,626,019	Black, E. P. Ethylene-pseudocumene telomer wax
3,627,837	Webb, F. J. Preparation of graft copolymers
3,632,658	Halasa, A. F. Polylithiation of aromatics
3,634,548	Harwell, K. E. and Galiano, F. R. Preparation of graft copolymers
3,646,219	Peterson, D. J. Uses for lithiated methylsulfides
3,647,803	Schlott, R. F., Hoeg, D. W. and Pendleton, J. F. Preparation of organosodium chelates

3,652,696　Honeycutt, S. C.
　　　　　　Ethylene-aromatic telomer wax
3,657,373　Peterson, D. J.
　　　　　　Olefin preparation using lithiated methylsilanes
3,658,925　Erman, W. F. and Broaddus, C. D.
　　　　　　Lithiation of limonene
3,663,585　Langer, A. W.
　　　　　　Lithiation of ferrocene
3,666,618　Black, E. P.
　　　　　　Uses for telomer laminating wax
3,666,744　Black, E. P.
　　　　　　Process for preparing telomer laminating wax
3,674,895　Gaeth, R. H. and Farrar, R. C.
　　　　　　Polymerizations using multilithium initiators
3,678,088　Hedberg, F. L. and Rosenberg, H.
　　　　　　Lithiation of metallocenes
3,678,121　McElroy, B. J. and Merkley, H.
　　　　　　Narrow molecular weight distribution polydienes using dilithium initiators
3,679,650　Schott, H. and Herwig, W.
　　　　　　Polymerization of alpha-methylstyrene
3,691,241　Kamienski, C. W. and Merkley, J. H.
　　　　　　Uses for alkali metal magnesium organohydrides
3,734,963　Langer, A. W. and Whitney, T. A.
　　　　　　Inorganic lithium chelates
3,737,458　Langer, A. W.
　　　　　　Grignard reactions of lithiated chelating agents
3,742,057　Bunting, W. M. and Langer, A. W.
　　　　　　Chelated organosodium compositions
3,755,484　Langer, A.W.
　　　　　　Telomer wax finishing process
3,755,447　Klemann, L. P., Whitney, T. A. and Langer, A. W.
　　　　　　Separation and purification of chelating polyamines
3,755,533　Langer, A. W. and Whitney, T. A.
　　　　　　Separation and recovery of lithium salts
3,751,384　Langer, A. W.
　　　　　　Chelated organolithium compositions
3,758,580　Langer, A. W., Whitney, T. A. and Klemann, L. P.
　　　　　　Separation and purification of tertiary polyamines
3,758,585　Bunting, W. M. and Langer, A. W.
　　　　　　Inorganic sodium chelates
3,763,131　Langer, A. W.
　　　　　　Telomer wax process and composition
3,764,385　Langer, A. W., Whitney, T. A.
　　　　　　Electric battery using chelated inorganic lithium salts
3,767,763　Bunting, W. B., Langer, A. W.
　　　　　　Separation and recovery of sodium compounds

3,769,345 Langer, A. W.
           Preparation of organolithium amine complexes
3,770,827 Langer, A. W., Whitney, T. A.
           Separation of chelating tertiary polyamines

a Claims lost to 3,451,988 and 3,458,586.
b Claims lost to 3,769,345.
c Claims lost to 3,663,585.

Literature Cited

1. Bywater, S., *Advan. Polym. Sci.* (1965) **4**, 66.
2. Langer, A. W., Jr., U.S. Patent **3,451,988**.
3. Ziegler, K., *Angew. Chem.* (1959) **71**, 623.
4. Ziegler, K., Gellert, H., *Ann.* (1950) **567**, 195.
5. Bartlett, P. D., Friedman, S., Stiles, M., *J. Amer. Chem. Soc.* (1953) **75**, 1771.
6. Hanford, W. E., Roland, J. R., Young, H. S., U.S. Patent **2,377,779**.
7. Langer, A. W., Jr., U.S. Patent **3,458,586**.
8. Langer, A. W., Jr., U.S. Patent **3,541,149**.
9. Langer, A. W., Jr., *Trans. N.Y. Acad. Sci.* (1965) **27**, 741.
10. Langer, A. W., Jr., *Polym. Prepr., Amer. Chem. Soc., Div. Polym. Chem.* (1966) **7** (1), 132.
11. Langer, A. W., Jr., *Gordon Conf. (Polym.), July, 1965.*
12. Eberhardt, G. G., Butte, W. A., *J. Org. Chem.* (1964) **29**, 2928.
13. Eberhardt, G. G., U.S. Patent **3,206,519**.
14. Eberhardt, G. G., U.S. Patent **3,321,479**.
15. Langer, A. W., Jr., U.S. Patent **3,541,149**.
16. Langer, A. W., Jr., U.S. Patent **3,769,345**.
17. Eberhardt, G. G., Davis, W. R., *J. Polym. Sci.* (1965) **A3**, 3753.
18. Butte, W. A., *Hydrocarbon Process.* (1966) **45** (9), 277.
19. Eberhardt, G. G., U.S. Patent **3,257,364**.
20. Butte, W. A., Morris, J. W., III, U.S. Patent **3,329,736**.
21. Butte, W. A., U.S. Patent **3,474,143**.
22. Eberhardt, G. G., U.S. Patent **3,567,703**.
23. Black, E. P., U.S. Patents **3,626,019**; **3,666,618**; **3,666,744**.
24. Brown, T. L., *Pure Appl. Chem.* (1971) **23**, 447.
25. Margerison, D., Pont, J. D., *Trans. Faraday Soc.* (1971) **67**, 353.
26. Makowski, H. S., Lynn, M., *J. Macromol. Chem.* (1966) **1** (3), 443.
27. Morton, M., Bostick, E. E., Livigni, R. A., *J. Polym. Sci., Part A-1* (1963) **1**, 1735.
28. Bywater, S., Worsfold, D. J., *Can. J. Chem.* (1962) **40**, 1564.
29. Morton, M., Fetters, L. J., *J. Polym. Sci., Part A-2* (1964) **2**, 3311.
30. Margerison, D., Newport, J. P., *Trans. Faraday Soc.* (1963) **59**, 2058.
31. Lewis, H. L., Brown, T. L., *J. Amer. Chem. Soc.* (1970) **92**, 4664.
32. Findeis, A. F., Langer, A. W., Jr., unpublished data.
33. Hay, J. N., McCabe, J. F., Robb, J. C., *Trans. Faraday Soc.* (1972) **68**, 1.
34. Mootz, V. D., Zinnius, A., Boettcher, B., *Angew. Chem.* (1969) **81**, 398.
35. Kimura, B. Y., Brown, T. L., *J. Organometal. Chem.* (1971) **26**, 57.
36. West, P., Waack, R., *J. Amer. Chem. Soc.* (1967) **89**, 4395.
37. Melchior, M. T., Klemann, L. P., Langer, A. W., Jr., *Int. Congr. Organometal. Chem., 6th, Aug. 1973, Prepr.*
38. Forster, E. O., Langer, A. W., Jr., U.S. Patent **3,404,042**.
39. Klemann, L. P., Melchior, M. T., Langer, A. W., Jr., *Int. Congr. Organometal. Chem., 6th, Aug. 1973, Prepr.*

40. Gilman, H., Jacoby, A. L., Ludeman, H., *J. Amer. Chem. Soc.* (1938) **60**, 2336.
41. Clauss, V. K., Bestian, H., *Ann.* (1962) **654**, 8.
42. Langer, A. W., Jr., unpublished data.
43. Bartlett, P. D., Tauber, S. J., Weber, W. P., *J. Amer. Chem. Soc.* (1969) **91**, 6362.
44. Bartlett, P. D., Goebel, C. V., Weber, W. P., *J. Amer. Chem. Soc.* (1969) **91**, 7425.
45. Cram, D. J., "Fundamentals of Carbanion Chemistry," p. 19, Academic, New York, 1965.
46. Langer, A. W., Jr., unpublished data.
47. Melchior, M. T., Klemann, L. P., Whitney, T. A., Langer, A. W., Jr., *Polym. Prepr., Amer. Chem. Soc., Div. Polym. Chem.* (1972) **13** (2), 649.
48. Broaddus, C. D., *J. Org. Chem.* (1970) **35**, 10.
49. Broaddus, C. D., *Amer. Chem. Soc., Div. Petrol. Chem. Prepr.* (1970) **15** (2), E82.
50. Langer, A. W., Jr., U.S. Patent **3,737,458**.
51. Langer, A. W., Jr., U.S. Patent **3,536,679**.
52. Morton, M., Sanderson, R. D., Sakata, R., *J. Polym. Sci., Part C* (1971) **9**, 61.
53. Langer, A. W., Jr., *Akron Summit Polym. Conf., 1st, June 1970.*
54. Hay, J. N., McCabe, J. F., *J. Polym. Sci., Part A-1* (1972) **10**, 3451.
55. Shimomura, T., Tolle, K. J., Smid, J., Szwarc, M., *J. Amer. Chem. Soc.* (1967) **89**, 796.
56. Finnegan, R. A., *Trans. N.Y. Acad. Sci.* (1965) **27** (7), 730.
57. Eberhardt, G. G., *Organometal. Chem. Rev.* (1966) **1**, 491.
58. Langer, A. W., Jr., Sixth Middle Atlantic Meeting, *Amer. Chem. Soc.*, Feb., 1971.
59. Maskornick, M. J., Streitweiser, A., *Tetrahedron Lett.* (1972) (17), 1625.
60. Langer, A. W., Jr., U.S. Patent **3,763,131**.

RECEIVED September 4, 1973.

Synthetic Aspects of Tertiary Diamine Organolithium Complexes

W. NOVIS SMITH

Foote Mineral Co., Exton, Pa. 19341

The metalations of benzene, toluene, propene, and 1-butene with butyllithium–TMEDA complexes were studied. Factors controlling the metalation rate are discussed including temperature, structure of butyllithium isomer, amount of TMEDA, and solvent. Optimum conditions are described for the formation of TMEDA complexes of benzyl-, phenyl-, allyl-, and crotyllithium and benzyllithium · triethylenediamine (TED). Synthetic uses of these complexes demonstrate that they react in the same way as other organolithium compounds. Ring-isomer formation during metalation of toluene using n-butyllithium–TMEDA complexes is also examined. The conditions for the lithiation of TMEDA and other tertiary diamines and monoamines (including trimethylamine) are given. The reaction of lithiated TMEDA with 1-bromobutane produced N-n-pentyl-N,N′,N′-trimethylenediamine in 40% yield.

The literature describing tertiary diamine organolithium compounds has grown tremendously since the initial announcements of these versatile reagents (*1, 2*). Much of this literature has centered on polymerization or polymer studies using these complexes as catalysts. The synthetic potential of these complexes and their preparations have been discussed, but there has been no intensive study to determine the optimum conditions for their preparation for synthetic applications (*1–6*). Synthetic applications usually require an isolation step or some procedure to ensure purity of the complex. Even investigations of the general utility and conditions for use of these complexes are not extensive. The most significant factor in their reactivity from a synthetic point of view is the ability of these complexes to metalate or replace certain activated hydrogens in hydrocarbons with the lithium cation (*7, 8*). The new aromatic, benzylic, or allylic organolithium tertiary diamine complex is then reactive enough to react selectively with a variety of reagents:

$$>N \sim N< \cdot R\text{Li} + R'H \rightarrow RH + >N \sim N< \cdot R'\text{Li} \qquad (1)$$

The results reported here describe an investigation of the optimum conditions of preparation of the tertiary diamine phenyllithium, benzyllithium, allyllithium, and lithiated TMEDA complexes. These reagents were allowed to react with some common reagents to help delineate the synthetic usefulness of the complexes.

The accompanying papers in this volume, the literature cited in them, and this paper discuss reasons for the enhanced reactivity of these complexes. Mechanistic discussion has been left to a minimum in this paper to emphasize preparative and synthetic aspects. (A useful analogy as to what these complexes can do or how they will react is to consider their behavior to be similar to hydrocarbon soluble alkylsodium compounds if they were to exist.)

Results and Discussion

Reactivity of Various Tertiary Diamines in the Preparation of Tertiary Diamine Organolithium Complexes by Metalation. Previous work has shown that the reactivity and the rate of metalation for Reaction 2 of the tertiary diamine organolithium complexes is a function of the tertiary diamine in the complex (1, 9). In the specific case of the metala-

$$R\text{Li} \cdot \genfrac{}{}{0pt}{}{R_3}{R_4}{>}N \sim N{<}\genfrac{}{}{0pt}{}{R_1}{R_2} + \left\{\genfrac{}{}{0pt}{}{\text{Aromatic}}{\genfrac{}{}{0pt}{}{\text{Benzylic}}{\text{Allylic}}}\right\} -\text{C}-\text{H} \rightarrow$$

$$R-H + \left\{\genfrac{}{}{0pt}{}{\text{Aromatic}}{\genfrac{}{}{0pt}{}{\text{Benzylic}}{\text{Allylic}}}\right\} -\text{C}-\text{Li} \cdot \genfrac{}{}{0pt}{}{R_3}{R_4}{>}N \sim N{<}\genfrac{}{}{0pt}{}{R_1}{R_2} \qquad (2)$$

tion of benzene as shown in Reaction 3, a tertiary diamine was added to n-butyllithium in benzene at room temperature in a 1:1 molar ratio. After the initial temperature rise, the heated solution began evolving n-butane at 45°-50°C (n-butane is very soluble in benzene, and usually

$$\text{C}_6\text{H}_5\text{-H} + \text{Tertiary diamine} \cdot n\text{-Butyllithium} \rightarrow$$

$$n\text{-Butane} + \text{Tertiary diamine} \cdot \text{C}_6\text{H}_5\text{-Li} \qquad (3)$$

no gas evolution occurs unless the temperature is raised). The rate of gas evolution was quite sensitive to the structure of the tertiary diamine. The order of reacivity observed using this procedure was N,N,N',N'-tetramethylethylenediamine (TMEDA) > N,N,N',N'-tetramethyl-1,3-butanediamine > N,N,N',N'-tetramethyl-1,3-propanediamine > triethylenediamine (TED or DABCO) and 2-methyltriethylenediamine > N,N,-N',N'-tetramethyl-1,4-butanediamine. The tertiary diamines, N,N,N',N'-tetramethylmethylenediamine and N,N'-dimethyl-1,4-piperazine, caused little or no gas evolution indicating essentially no metalation in the hour in which the other tertiary diamine complexes reacted.

The order of reactivity agrees with previous studies—TMEDA being by far the most reactive of the readily available tertiary diamines. [Sparteine appears to be the most activating of the tertiary diamines investigated (1, 10)]. Recent reports by Langer and others indicate that increased activation is expected from the methyl tertiary polyamines containing more than two basic nitrogens, all separated by two carbons (11). Preparative and synthetic work centered on TMEDA and TED (DABCO) because of their economic practicality.

Optimum Conditions for Preparing Phenyllithium from Benzene. The optimum procedure for metalation of an aromatic compound in high yield was to use the aromatic compound itself as the solvent. The alkyllithium compound used mainly in this study was n-butyllithium because it was convenient and economical.

The following general procedure was used to determine the completion of reaction. The calculated amount of TMEDA was added to the solution of n-butyllithium in benzene over one to two minutes. The stirred solution (under argon) was kept at the desired temperature; aliquots of this reacting solution and the final solution were then taken, allowed to react with trimethylsilyl chloride, and analyzed by GLC. These results may indicate the mechanism of the metalation reaction in addition to determining optimum metalation conditions for benzene. The results are listed in Table I.

A number of conclusions can be drawn. The effect of excess tertiary diamine above a 1-mole equivalent is not as significant for rapid metalation as is the initial amount of TMEDA. Practical metalation could only be obtained with a 1:1 molar ratio at 40°C and a 2:1 ratio at 60°C; the 4:1 ratio required reflux. Side reactions appeared to cause a problem only with the 4:1 ratio at the higher temperatures. Benzene as the solvent accelerated the reaction sixfold over hexane with 10% mole excess as the solvent. Some of this acceleration may be solvent effect. The use of sec-butyllithium increased the metalation rate eight times.

As expected, the metalation rate increased with temperature. The rate was relatively rapid over the initial 60% and then slowed dramati-

Table I. Study of the Metalation Rate of Benzene by Butyllithium and TMEDA

Run	Molar Ratio BuLi: TMEDA	T, °C	Elapsed Time, hrs	% Reaction[b]	% Yield	Comments
1	1:1[a]	22	1.0	34		
			2.5	60		
			7.8	79		
			21.8	100[c]	101	Almost gel initially, then became fluid. Some product ppt at end; sol. in benzene
2	2:1[a]	22	5.0	13		
			30.0	38		
			70.5	53	(100)[d]	
3	1:2[a]	22	1.0	32		
			2.0	50		
			4.5	82		
			20.0	100[c]	—	
4	1:1[a]	40	0.5	67		
			1.0	91		
			3.0	100[c]	97	Some product pptd; dissolved in added benzene, almost gel initially but became fluid rapidly; some product pptd, sol. in TMEDA.
5	2:1[a]	40	1.0	37		
			3.8	57		
			21.0	85	—	
6	4:1[a]	40	18.3	65	—	
7	1:1[a]	60	15 min	95		
			34 min	100[c]	100	Some product pptd; dissolved in added benzene.
8	2:1[a]	60	0.5	71		
			2.0	92		
			4.5	99	—	Some product pptd; dissolved in added benzene.
9	4:1[a]	60	2.0	60	—	Some product pptd; dissolved in added benzene.

Table 1. Continued

Run	Molar Ratio BuLi:TMEDA	T, °C	Elapsed Time, hrs	% Reaction[b]	% Yield	Comments
10	2:1[a]	60 (reflux)	0	—		
		76	12 min	—		
		81	24 min	95.0		
		82	39 min	99.1		
		82	44 min	100[c]	—	Haze, insol. in added benzene.
11	4:1[a]	68 (reflux)	0	—		
		81	38 min	99.5	93	Considerable product pptd; sol. in benzene.
12	1:1[e]	66	3.0	98.5		
			3.9	100.0[c]	92	Some product pptd.
13	2:1[e]	66	5.3	86	—	
14	4:1[e]	66	2.8	42	—	
15	1:1[f]	79	38 min	97.5		
			105 min	100.0[c]	84	
16	1:1[g]	67	32 min	99		
			35 min	100[c]	79	Considerable haze.

[a] Conditions unless otherwise noted: TMEDA added over 1 min to organolithium solution at temp. Used 24% n-butyllithium in benzene.
[b] Represents percent of butyllithium reacted. Assumed only metalation of benzene occurring.
[c] These numbers were very close to the exact time the reaction finished because gas evolution was observed up to the end of the reaction. A trace of butyllithium generally remained at this point (<0.1%).
[d] Complete accountability of butyllithium (phenyllithium plus butyllithium).
[e] Used 15.2% n-butyllithium in hexane and added 10% excess benzene over stoichiometric.
[f] Used 14.6% n-butyllithium in cyclohexane with 10% excess benzene.
[g] Used 11.9% sec-butyllithium in hexane with 10% excess benzene.

cally. (This complex type of behavior would be expected from previous mechanistic studies of organolithium reactions.) The metalation rate of aromatics by organolithium–tertiary diamine complexes is, therefore, a function of organolithium concentration, organolithium structure, concentration of aromatic, temperature of metalation, and amount and structure of tertiary diamine. Although none of these factors is surprising, they all must be kept in mind in optimizing a particular metalation using organolithium–tertiary diamine complexes.

In cases where the amount of TMEDA is to be minimized, the 2:1 ratio can be prepared at 60°C for five hours, but the reaction would have to be run slightly more dilute to maintain complete solubility. Another factor when preparing phenyllithium–TMEDA complexes with ratios greater than 1:1 is that these solutions tend to supersaturate and continue to throw down precipitate for hours—even days—after the reaction has been completed.

The metalation rate of TMEDA is much slower than that of benzene, and the phenyl anion is more stable than is the metalated TMEDA anion so that no interference by the TMEDA metalation was observed.

The 4:1 ratios were unsatisfactory; the excessive amount of precipitate that formed was insoluble in benzene. (This ratio might be satisfactory if the phenyllithium that formed reacted *in situ* but a somewhat lower yield would result.

Laboratory quantities of the dark, red-brown solutions although highly reactive, were not pyrophoric when in air. More dilute solutions of the complexes were bright yellow. When phenyllithium–TMEDA solutions of higher ratios were prepared, the precipitated complexes were also bright yellow. The odor of phenol was noticed after hydrolysis of the air-oxidized complexes.

Work with TED (triethylenedimaine) showed that reasonable metalation rates were obtained only at ratios of 1:1 in benzene. The reaction required 2.5 hours in refluxing benzene for completion. The bright yellow, crystalline product had a modest solubility in benzene and precipitated from solution. The solid complex may be pyrophoric in large amounts. The solubilities of various organolithium–tertiary diamine complexes are listed in Table II.

Table II. Solubility of Organolithium–Tertiary Diamine Complexes

Organolithium[a]	Tertiary Diamine	Ratio RLi:Diamine	Solvent	RLi, wt % of Solution
Phenyllithium	TMEDA	1:1	benzene	19
		2:3	hexane	8
		4:1[b]	benzene	7.9
Benzyllithium		2:1	toluene	36
		2:1	cyclohexane	Two liquid layers formed
		3:1	toluene	4.5, top layer (two layers)
	TED	1:1	hexane	0.13
Allyllithium	TMEDA		THF	10.7
			1,2-dimethoxyethane	Reacts
			hexane	0.3
			diethyl ether	3.5
			cyclohexane	0.4
		2:1	diethyl ether	1.2
Crotyllithium		1:1	hexane	11

[a] All solubilities at room temperature (22°C); analysis using Gilman double titration (17).
[b] Slurry present—only solution in equilibrium was analyzed.

Table III. Stability of Organolithum–TMEDA Complexes

Organolithium Compound	Ratio RLi: Diamine	T, °C	Conc wt %	Solvent	% Lost of Contained Active Material/Day[a]
Phenyllithium	1:1	35	15	benzene	0.67
Benzyllithium	1:1	35	14	toluene	0.31
	3:2	35	22		0.30
	2:1	30	38 (sat.)		0.046
	2:1	30	9		0.046
Lithiated N,N,N',N'-tetramethylethylenediamine	—	35	19	hexane	0.52

[a] Analysis of solutions using Gilman double titration with a time period of about three weeks (17)

Table III lists the decomposition rates of several organolithium–TMEDA complexes. The stability of the complexes is relatively good, but the presence of excess TMEDA above 1-mole equivalent significantly decreases stability. Solutions of the phenyllithium–TMEDA complex should be stored below 10°C for long shelf life.

It might be expected that in the presence of TMEDA or other tertiary diamines anomalous reaction products might be obtained with organolithium compounds such as benzyllithium. A number of reports in the literature disclose instances of the expected reaction products from reactions such as carbonation to the carboxylic acid and addition to benzophenone (1, 3, 4, 12). The phenyllithium–TMEDA (1:1) complex in benzene was allowed to react with benzophenone to give a 95% yield of triphenylcarbinol and with cyclohexanone to yield 59% of the 1-phenylcyclohexanol. The reaction with excess trimethylsilyl chloride is apparently quantitative. The main consideration in using these complexes is to use low temperatures for reaction and aqueous washes of ammonium chloride solution in the work-up to remove all of the tertiary diamine (the odor can be detected in low concentrations.)

Optimum Conditions for Preparing Benzyllithium from Toluene. Both the TMEDA and TED complexes of benzyllithium were investigated. Toluene metalation proceeds much faster than does benzene metalation under similar conditions. The benzyllithium complexes were more soluble in hydrocarbon solvents than were the corresponding phenyllithium complexes. This method of preparation of benzyllithium is the most convenient of the few literature procedures available. Other procedures described are the cleavage of benzyl methyl ether with lithium metal, the exchange of benzyltin compounds with an organolithium compound, and the dibenzylmercury reaction with lithium metal (*13, 15, 46*). The direct reaction of benzyl chloride and lithium metal in ether leads only to the coupling product.

Only a brief study was made of the metalation rate of toluene by *n*-butyllithium because the reaction proceeded readily with TMEDA under most conditions. The runs listed in Table IV indicate just how much faster toluene metalation proceeded compared with benzene metalation. This rate was about 6.5 times faster with hexane as the solvent and at least 10 times faster when the corresponding aromatics (toluene and benzene) were used as the solvents.

The crystalline yellow 1:1 benzyllithium–TMEDA complex has a lower solubility in toluene than does the (benzyllithium)$_2$–TMEDA complex which tends to form oils. Because of its higher solubility and to eliminate as much of the tertiary diamine as possible, the 2:1 complex was used for synthetic evaluations. It is probably the optimum ratio to use in this benzyllithium system. Solutions of up to 38 wt % benzyllithium in toluene or about 59 wt % of the (benzyllithium)$_2$–TMEDA complex have been prepared. At ratios of 4:1 or more in toluene, a dark-red tar or oil also separated from the solution. With hexane as the solvent, yellow crystals of the 1:1 benzyllithium–TMEDA complex precipitated; at a ratio of 2:1 or more, a dark-red oil separated instead.

Because toluene metalation is very fast, the best preparative procedure is to slowly add TMEDA to *n*-butyllithium in toluene at about 60°C because of the need to drive off butane as it forms and prevent its build-up in the reaction solution. The reaction is quite exothermic because of the initial solvation of TMEDA. In small-scale preparations (200-ml or less) this precaution is not necessary. An added benefit of adding the TMEDA at a steady rate to the warm solution is that the reaction rate is almost instantly controllable and readily followed by observing the rate of gas evolution. Concentrations of *n*-butyllithium in toluene to about 26 wt % could be used without having to add additional toluene to ensure product solubility.

The yield for the 26-mole Run 10, Table IV, was determined by GLC (using the trimethylsilyl derivative) and by the Gilman double titration technique, both of which gave 100% yield (*17*). The product solution was clear, dark red–brown, and had good thermal stability. The concentrated solutions were highly reactive to air but did not ignite. (The solution should be stored cool or frozen when stored for longer than four months.)

The possibility of using the solid, golden yellow complex benzyllithium–TED (1:1) as an alternative for some special applications prompted a study of the best conditions for its preparation. The benzyl-

Table IV. Toluene Metalation by *n*-Butyllithium–Tertiary Diamine Complexes

Run	Molar Ratio n-BuLi: Diamine	T, °C	Elapsed Time, min	% Reaction	% Yield	Comments
1	1:1[a] TMEDA	52	3	100[b]	98	Too fast to measure accurately
2	2:1[a] TMEDA	77	16	100[b]	100	
3	4:1[a] TMEDA	77	36	99[b]		
			45	100	96	Two layers formed; bottom was a red oil. Added TMEDA caused one phase to form
4	1:1[c] TMEDA	66	36	100[b]	100	Entire solution solidified to yellow crystals on cooling, holding the hexane
5	1:1[c] TED	68	135[b]	—	90	Isolated by filtration; bright yellow crystals
6	1:1[c] TED	68	135[b]	—	96	Isolated by evaporating solvent; bright yellow crystals
7	8:1[e] TMEDA	89	75	(100)[d]	—	Red-black tar
8	3:1[e] TMEDA	70	10	—	—	Two layers
9	1:1[f] TMEDA	70	120 (slow addition)	—	100	
10	2:1[g] TMEDA	60–65	60	—	100 (titration and glc)	Complete solution; large-scale run

[a] 14% *n*-butyllithium in toluene used for these runs.
[b] Evolution of gas ceased at this point, and trimethylsilyl derivative was made. Yields were based on GLC analysis.
[c] 15.2% *n*-butyllithium in hexane used with 10% excess toluene added.
[d] The reaction was run until gas evolution ceased. The product was too intractable to dissolve readily in solvents for analysis.
[e] 19.5% *n*-butyllithium in toluene used.
[f] 26.4% *n*-butyllithium in toluene used.
[g] 22.0% *n*-butyllithium in toluene used.

lithium–TED (1:1) is fairly soluble in toluene, but it has only a slight solubility in hexane. Therefore, the solvent of choice was an aliphatic hydrocarbon such as hexane. The metalation rate of toluene by n-butyllithium in hexane with TED was about a fourth of what it was with TMEDA, but it still proceeded at a convenient rate at 68°C. Greater than 1:1 ratios (less TED) seriously slowed the metalation reaction so that only the 1:1 ratio with TED is considered practical.

The optimum procedure for preparing the butyllithium–TED (1:1) complex was to add the TED solid directly to 15-20% n-butyllithium in hexane at 40°–50°C. (When the TED was added near room temperature, a gel stage formed.) The exotherm from the solution was moderate, and the temperature was allowed to rise to reflux. Only a 20% excess of toluene was added because large excesses solubilized the product. The yield of the dried, yellow crystals was from 82 to 90%. When the hexane solution was evaporated without filtration to dryness, a yield of 96% was obtained. No side products were observed. The crystals were very sensitive to air; they charred and smoked, but no flame was observed in laboratory quantities. The crystals were stable for at least five years at room temperature when protected from air and moisture. The TED could not be sublimed away (only 0.2% weight loss) from the 1:1 complex crystals at 0.14 torr before decomposition occurred at 125°C.

Isomer Formation during Metalation of Toluene Using n-Butyllithium and TMEDA. There have been several reports describing o-, m-, and p-tolyllithium from isomer formation stemming from ring metalation of toluene in addition to the normal benzyl metalation using n-butyllithium–TMEDA (1, 16, 18, 19). These previous works found about 10% ring metalation of which meta metalation comprised about 60% of the total. Analysis of the crystalline benzyllithium–TED complex showed no ring metalation isomers present.

Table V summarizes the results of both this investigation and those reported in the literature. The amount of ring metalation and relative ratios of isomers vary somewhat according to the procedure used and on standing. This is evidently a case of kinetic control $vs.$ thermodynamic control, as pointed out by Broaddus (18). Therefore the procedure for metalation affected the initial amount of ring-isomer formation directly as expected in a kinetically controlled reaction that involves temperature of metalation, amount of TMEDA, and subsequent treatment or storage of the product solution. A recent report discusses a similar system of anion equilibration between m- and p-xylyllithium with TMEDA, and anion equilibration between the meta and benzyl positions of ethylbenzene using TMEDA or N,N,N',N',N''-pentamethyldiethylenetriamine (PMDET) (20).

Ring-isomer formation during metalation was analyzed by GLC using the dimethyl sulfate derivative, which produced a mixture of xylenes and ethylbenzene in this system. (Dimethyl sulfate is a better reagent for organolithium derivative formation than is trimethylsilyl chloride or carbon dioxide.)

The toluene-metalation system in this study differs from that of Broaddus in that pure toluene was used as the solvent; a RLi to TMEDA ratio of two was used. Analyses were usually run after complete reaction of the n-butyllithium unless otherwise noted. Higher reaction temperatures were used, and the final concentration of benzyllithium in solution was higher.

The results reported by others as well as our own indicate that the main contaminating ring isomer is meta, the more acidic (reactive) position compared with the ortho or para positions when only electronic or resonance effects are considered. In addition, the statistical factor of two meta-hydrogens probably also contributes to the predominance of m-tolyllithium formation compared with formation of the para isomer. The combined meta and para isomer content is reported because of the difficulty in obtaining good separation of the xylene mixture formed by derivatization.

One of the specific effects influencing total ring-isomer content was the ratio of RLi to TMEDA. In Run 2, a ratio of 0.5 (large excess TMEDA) was used, and the total amount of ring isomer initially formed and present after one hour was small (less than 2.1%). When the amount of TMEDA was decreased to a RLi-to-TMEDA ratio of two, the amount of ring isomer present when the reaction was completed was 11.5%, agreeing with the work of previous workers. In Run 6, enough TMEDA was added to an already formed benzyllithium-in-toluene solution with a RLi:TMED ratio of two to change the ratio to 0.5. The amount of meta ring isomer decreased from 6.3 to 5.2% in about 15 minutes at 30°C. The effect of the large TMEDA excess was more pronounced when it was present during metalation, although still effective after the fact.

A number of experiments were carried out to determine the effect of temperature and time on the ring-isomer content of the (benzyllithium)$_2$ · TMEDA in toluene solutions. In all cases—0°C, room temperature, or 60°C—the amount of ring isomer decreased with time with the ortho isomer disappearing far more rapidly than the meta or para ring isomers. This change is caused either by equilibration of the tolyllithium isomers to the more thermodynamically stable benzyllithium, or by a more rapid decomposition reaction of these ring isomers with TMEDA. The measured stability of these solutions indicate that selective decomposition is not a good explanation. (Run 6, after being heated at

Table V. Isomer Content of

Run	Mole Ratio RLi: TMEDA	Addition T °C	Addition Time min	Heating T °C	Heating Time min	Elapsed Time at Temp. after Prep
1(a)	2	60	20	70	10	1 hr 2.5 days 60 days 90 days
2	0.5	60	20	70	10	1 hr
3	2	40	60	—	—	—
4	2	65	60	—	—	—
5	2	40	15	70	40	—
6	2	40	60	—	—	—
	2			70	15	—
				60	30	
				60	90	
				60	390	
7	2	55	90	70	20	180(0°) days
8(d)	1	30	20	69	15	1 hr 60 days 90 days
9(e, d, f)	1	30		—	—	15 min 2 hrs
10(d, g)	1	25		—	—	5 hrs

[a] TMEDA was added to 25% n-butyllithium in toluene while stirring at designated temp. The RLi compounds reacted with dimethyl sulfate to produce ethylbenzene and xylenes.
[b] Total resolution of m- and p-xylenes was not accomplished; thus the meta and para isomers are reported together.

Table VI. Reactions of (Benzyllithium)$_2$

Run	Reactant	T, °C	Time, min
1	Trimethylsilyl chloride	−5	45
2	1-Bromobutane	−20	90
3	Benzonitrile	−5	45
4	Ethyl acetate	5-10	60
5	Ethyl benzoate	5-10	60
6	Cyclopentanone	−40	30
7	Cyclohexanone	−5	60
8	2-Butanone	−15	60
9	Diisopropyl ketone	−5	60
10	Benzaldehyde	−35	20
11	Benzophenone	−15	20

[a] Yield based on recrystallized or distilled product unless otherwise noted; see

Benzyllithium–TMEDA Complexes

Ring-isomers			Comments
%(m+p)[b]	%o	% Benzyl	
8.5	3.0	88.5	
5.9	1.4	92.7	
1.6	—	98.4	
0.3	—	99.7	
1.1	—	98.9	
8.5	1.0	90.5	Addition of RLi to TMEDA (inverse)
9.1	1.6	89.3	
7.3	—	92.7	
6.5	9.1	84.4	Incomplete reaction at this point
6.6	5.0	88.4	Complete reaction
6.1	3.6	90.3	
6.3	0.9	92.8	Add TMEDA to produce 1:2 ratio at this point to give 5.2% *meta* and 94.8% benzyl.
3.8	—	96.2	Yield at this point is 98%
0.8	—	99.2	26-mole scale
6.3	1.0	92.7	
2.1	—	97.9	
—	—	100.0	
9	3	89.0	Incomplete reaction at this point
6	2	92.0	
7.6	1.8	90.6	

[c] No isomer is reported when less than 0.1%
[d] 15% *n*-butyllithium in hexane used in this run; hexane was solvent.
[e] Broaddus (*18*) used carbon dioxide for derivatization.
[f] Hexane-toluene mixture was the solvent for this run.
[g] Chalk and Hoogeboom (*16*) used trimethylsilyl chloride for derivatization.

–TMEDA in Toluene

Product	Product	
	% Yield	bp or mp, °C[a]
benzyltrimethylsilane	81	59-62 (5 mm)
n-amylbenzene	84	71-74 (5 mm)
benzylphenyl ketone	16	(Crude)
1,1-dibenzylethanol	43	122-125 (0.4 mm)
1,2,3-triphenyl-2-propanol	22	(Crude)
1-benzylcyclopentanol	35	55
1-benzylcyclohexanol	59	54
2-benzyl-2-butanol	53	92-94 (0.5 mm)
3-benzyl-2,4-dimethyl-3-pentanol	82	96-99 (0.1 mm)
1,2-diphenylethanol	55	64
1,1,2-triphenylethanol	92	86

experimental for literature physical constants and references.

60°C for 6.5 hours, still showed at least 98% yield, while ring-isomer content decreased to 2.8% with no ortho or para isomer.)

The equilibration of the ring anions to the benzyl anion is the probable explanation, especially considering the recent work of Gau and our observation of an equilibration between m-xylene and toluene in the presence of (benzyllithium)$_2$ · TMEDA (20). (In fact, this type of system might be the basis of another route for a hydrocarbon acidity scale.) The more rapid disappearance of the ortho isomer, compared with the meta, may be the result of a possible intramolecular route for conversion to the benzyl anion. The meta and para isomers probably change to the benzyl anion by an intermolecular route that would be slower and agree with what was observed. Although complete resolution by GLC of the para and meta isomers was not done in this study, the para disappeared faster than the meta, but much slower than the ortho. In other time studies on the disappearance of the meta isomer at room temperature, about half the initial amount of this isomer was gone in three days and all of it in two to three months.

Aging these benzyllithium solutions to remove the ring-isomer content is not practical in most cases, and use of large amounts of TMEDA to promote conversion can also cause problems later in synthetic use. Since heating the toluene product solution did not cause serious decomposition at 60°C, this could provide a solution to the problem for most

Table VII. Reactions

Run	Reactant	Solvent	T, °C	Time, min
1	1-Bromobutane	Et$_2$O	35	60
2	Bromobenzene	Et$_2$O	35	60
3	Iodobenzene	Et$_2$O	20	60
4	Benzonitrile	THF	−10	60
			25	60
5	Ethyl acetate	Et$_2$O	35	45
6	Ethyl benzoate	THF	0	60
7	Methyl vinyl ketone	THF	−10	60
8	Crotonaldehyde	Et$_2$O	0	60
9	2-Butanone	Et$_2$O	0	60
10	Cyclopentanone	THF	−10	60
11	Cyclohexanol	THF	0	60
12	Diisopropyl ketone	Et$_2$O	−10	30
13	Benzaldehyde	THF	0	60
14	Benzophenone	Et$_2$O	−10	120
15	Indanone	THF	−10	60
16	Fluorenone	Et$_2$O	35	60
17	Lithium benzoate	THF	65	45

[a] Yield based on recrystallized or distilled product unless otherwise noted. See

applications. The solid benzyllithium–TED complex may also be an alternative.

Synthetic Application of (Benzyllithium)$_2$–TMEDA and Benzyllithium–TED. All the synthetic application studies were run with an aged (benzyllithium)$_2$–TMEDA toluene solution that had 0.8% of the total organolithium compounds present as the tolyllithium isomers (only meta present) or with benzyllithium–TED crystalline complex free of ring isomers.

A number of standard organolithium reactions were examined using the two benzyllithium complexes. The conditions used for the various runs listed in Tables VI and VII were selected by experience to give good yields but were not optimized. (Because of the high reactivity of organolithium reagents in general, all variables of a given reaction affect the yield. In other words there is always a set of optimum conditions that may vary markedly from one reacting compound to another even though the same organolithium compound is used.)

Reaction conditions are given in the experimental section. Yields vary considerably, but most of the normal organolithium reactions occur in satisfactory-to-good yields. When readily enolizable protons are present, lower yields might be expected and were found—as with cyclopentanone. Buhler, in a recent optimization study, found the yield of the tertiary alcohols from cyclopentanone, cyclohexanone, and benzo-

of Benzyllithium–TED

Product	% Yield[a]	Product bp or mp, °C
n-amylbenzene	80	62 (0.5 mm)
2-benzylbiphenyl	32	50
2-benzylbiphenyl	20	49
benzylphenyl ketone	59	54
1,1-dibenzylethanol	25	Crude oil
1,2,3-triphenyl-2-propanol	66	83
1,2,3-triphenyl-2-propanol	0	Condensation products
1,2,3-triphenyl-2-propanol	0	Condensation products
2-benzyl-2-pentanol	65	67 (0.3 mm)
1-benzylcyclopentanol	72	56
1-benzylcyclohexanol	55	54
3-benzyl-2,4-dimethyl-3-pentanol	60	110 (1 mm)
1,2-diphenylethanol	81	65
1,1,2-triphenylethanol	99	87
1-benzyl-1-indanol	20	141
9-benzyl-9-fluorenol	83 (crude)	126
benzylphenyl ketone	44	53

experimental for literature references for physical constants when available.

phenone to be 75, 89, and 73%, respectively (21). The corresponding yields using the tertiary diamine benzyllithium complexes were both higher and lower than those results. Lower reaction temperatures and slower addition rates and perhaps a cosolvent should increase yields in most cases. The workup of the products was normal, but several washes of the product in organic solution with aqueous ammonium chloride or dilute hydrochloric acid were required to ensure complete removal of the tertiary diamine.

The solutions of (benzyllithium)$_2$–TMEDA were much more convenient to use than the solid, relatively insoluble benzyllithium–TED complex. The reaction yields vary between these two complexes. Whether this is a function of the different solvent systems used for each, lack of optimization, or the actual tertiary diamine present was not examined, however.

These results show that the organolithium complexes are synthetically useful and are similar in their reaction products to the usual organolithium compounds.

Optimum Conditions for Preparation of Allyllithium and Crotyllithium from Olefins. Like benzyllithium, allyllithium is also difficult to prepare directly from lithium metal and allyl chloride because of the excessive coupling that occurs.

$$2CH_2=CHCH_2Cl + 2Li \rightarrow CH_2=CHCH_2CH_2CH=CH_2 + 2LiCl \quad (4)$$

Some success has been reported with Reaction 4 using low temperatures (22). We have obtained up to 61% with the procedure cited in Ref. 22 with some difficulty. A number of other routes have been found for preparing allyllithium, including the cleavage of allyl ethers with lithium metal, exchange of alkyllithium compounds with allyltin compounds, and

Table VIII. Preparation of Allyllithium– and

Run	Product	Olefin	Time min	Solvent	RLi
1	Allyllithium	propene	30	cyclohexane	sec-BuLi
2			30	cyclohexane	sec-BuLi
3			150	hexane	n-BuLi
4			90	hexane	sec-BuLi
5			30	hexane	sec-BuLi
6			30	hexane	sec-BuLi
7	Crotyllithium	1-butene	90	cyclohexane	sec-BuLi
8		trans-2-butene	90	cyclohexane	sec-BuLi

[a] Yield of allyllithium–TMEDA complex based on isolated solid.
[b] GLC analysis of dimethyl sulfate derivative formed at 20° and 55° C gave 29.0

the reaction of diallylmercury with lithium metal (22–27). The metalation of propene with n- or sec-butyllithium–TMEDA complexes is one of the most convenient routes to allyllithium as shown in Table VIII.

$$CH_2=CHCH_3 + C_4H_9Li \cdot TMEDA \rightarrow$$
$$CH_2=CHCH_2Li \cdot TMEDA + C_4H_{10} \quad (5)$$

The metalation proceeded faster for Reaction 5 than toluene metalation, but it was hard to get the excess concentration necessary for rapid reaction. An aliphatic solvent was necessary for the reaction since all other solvents produced side reactions regardless of whether an aromatic or an ether was used. Propene in 50% excess was added to n-butyllithium in hexane containing a 1:1 mole ratio of TMEDA at 0°C. A larger excess of propene may produce oligomers. This solution was warmed to 30°C in a pressure apparatus. After three hours the yellow slurry was filtered to give a 65% yield of allyllithium–TMEDA; from 10 to 20% of the product remained in the filtrate. The solid, yellow–orange complex was not pyrophoric in air, although it was instantly reactive. When higher temperatures (about 50°C) were used and the propene added at atmospheric pressure, the competing metalation of TMEDA occurred, and only a trace of allyllithium formed. The lithiated TMEDA evidently did not metalate propene at a significant rate (if at all) under these conditions.

The metalation proceeded so readily with sec-butyllithium that the reaction was run to completion at −5° to 5°C in 0.5 hour. (tert-Butyllithium was usually slower in metalation reactions and led to more side reactions than did sec-butyllithium in these systems.) About 81% of the yellow–orange product precipitated from solution. The metalation reac-

Crotyllithium–TMEDA Complexes

Ratio RLi: TMEDA	T °C	% Yield
1:2	2	81
1:2	−5–5	two liquid layers
1:1	30 (15 psi)	64
1:1	−5–0	77
4:1	−7–4	gel formation
1:1	−5–5	79
1:1	0–5	92[b,c]
1:1	0–10	55[c]

± 1%, and 29.5 ± 1% terminal lithium, respectively.
[c] Yield calculated by base titration of solution.

tion with a 2:1 ratio produced two liquid layers without any precipitate. When a 4:1 ratio of *sec*-butyllithium to TMEDA was used, a gel and a solid slurry that did not break up on heating were obtained.

The optimum procedure, therefore, was to add TMEDA to *sec*-butyllithium in an aliphatic solvent such as hexane at $-5°C$. About 50% excess of propene was then added quickly at 0 to 5°C. The solution was stirred at that temperature for 30 min and then warmed quickly to 45°C to let any *sec*-butyllithium left react. The solution was then cooled to 10°C and filtered. The slurry itself can be used without isolation.

The corresponding metalation of 1-butene or *trans*-2-butene was slightly more rapid than that of propene and gave a solution of crotyllithium–TMEDA. The crotyllithium–TMEDA compound was quite soluble in aliphatic hydrocarbons, giving dark, yellow-red solutions. (These crotyllithium solutions slowly deposited yellow crystals on standing, so they should be used soon after preparation.) The metalation proceeded faster with 1-butene than with *trans*-2-butene, which may explain the lower yield obtained with the latter. A synthetic study of the crotyllithium was not made because its reactions gave at least two products, as expected from literature reports on other crotyl anions.

Reaction with dimethyl sulfate produced only 29% *trans*-2-pentene and 71% 2-methyl-2-butene. In other words about 29% of the lithium reacted from the C-1 position of the crotyl group with only trans configuration and 71% from the internal C-3 position. Essentially the same positional ratio was obtained at 20° as at 55°C.

$$(CH_3)_2SO_4 + \left\{ \begin{array}{l} CH_3CH=CHCH_2Li \\ CH_3CHCH=CH_2 \\ | \\ Li \end{array} \right\} \xrightarrow{\text{TMEDA}} \quad (6)$$

$$\left\{ \begin{array}{ll} CH_3CH_2=CHCH_2CH_3 & (29\%) \\ (CH_3)_2CHCH=CH_2 & (71\%) \end{array} \right\}$$

Table IX. Reaction of Allyllithium–TMEDA

Compound	Reagent	T °C	Solvent	Time min
Allyllithium·TMEDA	Benzophenone	20	Et_2O	75
	Cyclohexanone	20	Et_2O	45
Crotyllithium· TMEDA	Acetone	-20	Cyclohexane	20

([a]) *See* experimental for literature physical constants and references.

The reaction with acetone gave a similar result. Table IX lists the results from several reactions of reagents with allyllithium–TMEDA and crotyllithium–TMEDA.

Lithiated Tertiary Amines and Diamines and Their Reactions. The formation of lithiated TMEDA by organolithium reagents, especially n-butyllithium, was first described in 1965 by Langer (2).

$$n\text{-}C_4H_9Li + (CH_3)N(CH_2)_2N(CH_3)_2 \rightarrow$$
$$(CH_3)_2N(CH_2)_2N(CH_3)CH_2Li + C_4H_{10} \quad (7)$$
$$\text{I}$$

Since then, further work by Langer and his co-workers has shown that this is a general reaction with the higher homologs of TMEDA (5, 8). Preparation of these lithiated tertiary diamines and a number of synthetic reactions are given in those references. As further background to the possible reactions that may occur with other tertiary amines and diamines, the work of Lepley and co-workers with n-butyllithium and aromatic tertiary amines should also be examined (29–33).

One problem in working with the lithiated tertiary diamines is that they will undergo further decomposition when overheated or on standing. For example lithiated TMEDA yields lithium dimethylamide, dimethylvinylamine, and lithium acetylide (1, 2, 5, 34). Lithiated TMEDA (I) in hexane decomposes at a rate of 0.52% of the initially contained material per day at 35°C. Therefore solutions of the lithiated tertiary diamines and amines should be stored at 10°C or lower for long periods of storage.

The preparation of monolithiated N,N,N',N'-tetramethylethylenediamine (I) was studied under several sets of conditions to optimize its preparation. The lithiation of other tertiary monoamines and diamines was also examined.

TMEDA was added in 1:1 mole ratios to hydrocarbon solutions of the alkyllithium compound. The solutions were kept at a given temperature until gas evolution ceased. The dark, orange-brown solutions were analyzed using the Gilman procedure when complete reaction of the

and Crotyllithium–TMEDA

Product	Product [a]	
	mp or bp, °C	% Yield
1,1-diphenyl-3-buten-1-ol	125 (0.6 mm)	91
1-allylcyclohexanol	64 (7.5 mm)	52
2,3-dimethyl-4-penten-2-ol, and trans-2-methyl-4-hexen-2-ol	65–71 (120 mm)	11

starting alkyllithium was observed (*17*). This analysis for the carbon-bound lithium gave reproducible, consistent results for the lithiated tertiary diamines.

However inconsistencies were found in this procedure with solutions of lithiated tertiary monoamines so that only total base titrations were used for their analysis. In most cases a small amount of the decomposition product was filtered from the product solution. The solid on hydrolysis produced dimethylamine and acetylene (*34*). Product solutions up to 30 wt % in hexane were readily produced without solubility problems. No lithiation of TMEDA on the methylene carbons was observed. This was determined by preparation of the dimethylsulfate derivative of the lithiated TMEDA and examining for N,N,N',N'-tetramethyl-1,2-propanediamine in the presence of the product, N-ethyl-N,N',N'-trimethyl-1,2-ethanediamine.

Although *n*-butyllithium lithiated the TMEDA efficiently at a convenient rate, *sec*-butyllithium worked even more smoothly. Surprisingly *tert*-butyllithium was quite poor, although some of the problem was caused by the low-boiling solvent, pentane. For instance *tert*-butyllithium in refluxing pentane took 14 hours to completely react with TMEDA.

A summary of the lithiation of tertiary amines and diamines is given in Table X. The actual total base minus the base stemming from carbon-bound lithium present divided by the amount of carbon-bound lithium present was used to obtain the ratio of tertiary nitrogen to carbon-bound lithium reported in the table. Agreement between Runs 1 through 11 was surprisingly good. Yields for the same runs were also quite good—90 to 100%—with both *n*- or *sec*-butyllithium. *sec*-Butyllithium was preferred because of the milder conditions involved. The conditions become especially important when lithiation of potentially more sensitive (yet more unreactive) tertiary amines or diamines is attempted. Runs 9 through 15 show some interesting but unexploited metalation results. The lithiation rate for the reactions in Table X was observed by gas evolution, and the solutions were analyzed when the gas evolution ceased.

Reaction occurred in only a few hours with *sec*-butyllithium and trimethylamine, but the yields seemed to be lower than those obtained with TMEDA.

$$(CH_3)_3N + sec\text{-}C_4H_9Li \rightarrow (CH_3)_2NCH_2Li + n\text{-}C_4H_{10} \quad (8)$$
$$\text{II}$$

This may be because II has less thermal stability than does the lithiated TMEDA (I). The lithiation of trimethylamine has also been reported under slightly different conditions by Peterson (*35*). Analysis of these tertiary monoamines in Runs 14 to 19 were inconsistent using the Gilman double titration with benzyl chloride, and only total-base titration was used for assay of the solution. It was assumed that all decomposition products, including LiH and lithium dimethylamide, were insoluble. The lithiation of the tertiary monoamines was performed in all cases with a mole ratio of alkyllithium to tertiary monoamine of 1:2 to keep the RLi:tertiary nitrogen ratio the same as that used with the tertiary diamines. In those cases noted in Table X, the total base in solution was measured and divided by three to estimate yield. (Analysis for the unreacted starting alkyllithium was not affected in these solutions.)

In Runs 10 and 12 the precipitate was very heavy and the total soluble base was quite low. In those two instances the product was apparently insoluble or the lithiated tertiary amine structure too unstable and decomposition resulted.

Each lithiated tertiary amine or diamine will have optimum conditions for preparation to ensure complete reaction with minimum decomposition. These optimum conditions may be determined by observing the rate of gas evolution as a function of tertiary amine or diamine ratio and temperature together with analysis of the final solution or slurry.

Only a few reactions were tried with the lithiated tertiary amines and diamines; these are reported in Table XI. The yields for the coupling of organic halides with lithiated TMEDA were reasonable for a non-optimized reaction. Changes in a few conditions were made to see whether some obvious improvements could be made. In general the yields ranged from 30 to almost 50%. Alkyl bromides gave the best yields in this coupling reaction. The reaction of lithiated TMEDA (I) with a reactive ketone and an aldehyde again gave low yields. Less-reactive reagent molecules should give higher yields. The lithiated trimethylamine (II) produced a good yield of the amine alcohol with benzaldehyde, although low yields were obtained in coupling with 1-iodopropane and 1-bromooctane.

Another possible problem is in the work-up of these compounds since the products are very soluble in water and organics. Only further study can determine whether the products from the reactions of these lithiated tertiary amine and diamine compounds will give high-enough product yields to permit these novel organolithium compounds to become dependable reagents for the chemist.

Experimental

All reagents were reagent-grade chemicals. Hydrocarbon solvents were dried over sodium wire. TMEDA was dried over sodium wire and then distilled under reduced pressure. TED was tried under vacuum to the point of sublimation. Liquid tertiary amines and diamines were

Table X. Preparation of Lithiated

Run	Amine	T °C	Time hr	RLi[a]
1	TMEDA	55	1.0	sec-BuLi
2[d]		49	1.0	sec-BuLi
3		47	1.25	sec-BuLi
4[e]		55	1.0	sec-BuLi
5		60	2.5	n-BuLi
6		64	3.5	n-BuLi
7[f]		36	14.0	tert-BuLi
8[d]		45	0.75	cyclopentyl Li
9	$N,N,N'N'$-Teramethyl-1,3-butanediamine	55	0.75	sec-BuLi
10	N,N'-Dimethyl-1,4-piperazine	65	1.0	sec-BuLi
11	Triethylenediamine (TED)	65	1.0	sec-BuLi
12	N,N,N',N',-Tetramethyl-methylenediamine	63	1.5	sec-BuLi
13	N-Methylpyrrolidine[g]	55	0.5	sec-BuLi
14	N-Methylpiperidine[g]	55	1.5	sec-BuLi
15	Triethylamine[g]	65	1.0	sec-BuLi
16	Trimethylamine[g]	80[h]	5.25	n-BuLi
17		70[h]	5.5	n-BuLi
18		60[h]	3.0	sec-BuLi
19		60[h]	6.5	sec-BuLi

[a] Hexane is solvent unless otherwise stated.
[b] Analysis done using total base and Gilman benzyl chloride coupling [17].
[c] Analysis of gas from hydrolysis of yellow ppt. showed dimethylamine and acetylene present.
[d] Solvent is cyclohexane.
[e] RLi:tertiary diamine ratio is 1:2.

all distilled before use and protected from air. All lithium compounds were from Foote Mineral Co. Elemental analyses were performed by Micro-Analysis, Inc., of Wilmington, Del. All reactions were run under argon, and air-sensitive materials were also handled under argon. GLC analyses were performed on a Varian-Aerograph 90P3 Chromatograph with a thermal conductivity detector. The column used for trimethylsilyl derivatives was 5-ft, SE-30/Chromasorb W.

Tertiary Amines and Diamines

Wt % Product Conc. in Sol.	Product Sol. Anal. tert-N/RLi[b]	% Yield	Comments
19.0	2.06	94	Slight ppt., no RLi left by GLC[c]
18.5	2.06	95	Heated slightly beyond point when gas evolution ceased.
17.9	2.19	94	2% RLi left
—	—	—	5% RLi left after 60 min at 55°C; considerable ppt
24.0	2.26	90	5% RLi left
27.3	1.95	100	Heated slightly beyond point when gas evolution ceased
22.0	1.86	96	Trace RLi left, lot of ppt
30.0	1.96	100	3% RLi left, no ppt., only haze
21.4	1.86	95	No RLi left, some ppt. also formed
—	—	—	No RLi left, heavy ppt.; double titration showed no product
12.3	2.20	68	No RLi left, heavy ppt.; analysis on slurry.
—	—	—	No RLi left, heavy ppt.; double titration showed no product
7.6	3.1 (theory 3.0)	57	No RLi left; heavy ppt
14.3	—	90[i]	Only trace RLi left, mod. ppt.; double titration showed low yield.
14.4	—	98[i]	Trace RLi left, mod. ppt.; low low yield by titration
8.9	—	67[i]	No RLi left
—	—	—	5% RLi left
6.6	—	60[i]	No RLi left
—	—	—	8% RLi left

[f] Solvent is pentane.

[g] RLi:tertiary amine ratio is 1:2.

[h] Pressure reactor used.

[i] Gilman double titration does not give good results with this amine; yield based on total base in solution.

n-Butyllithium in Benzene (or Toluene). A 12-liter flask equipped with stirrer, pressure-equalizing dropping funnel, reflux condenser, and thermometer was flushed thoroughly with argon. (Nitrogen must not be used because it will react with lithium metal powder.) The flask was charged with 573 grams of 1% Na-Li powder and 6,650 ml of benzene. The reaction was first initiated with 100 ml of 1-chlorobutane while stirring. The temperature was maintained at 35°–37°C and the remainder

Table XI. Reactions of Lithiated

Run	Lithiated Compound[a]	Reagent	Addition T, °C	Total Time, min
1	TMEDA	1-iodobutane	−5	30
2		1-bromobutane	0	30
3			10	75
4		1-chlorobutane	68	90
5		1-bromobutane	0	30
6		2-bromobutane	−10	30
7		iodobenzene	10	75
8		1-bromooctane	20	30
9			−20	30
10			−10	30
11		bromocyclohexane	−5	30
12		cyclopentanone	5	50
13		valeraldehyde	−10	30
14	Trimethylamine	1-iodopropane	−10	30
15		1-bromooctane	−20	30
16		valeraldehyde	−10	30
17		benzaldehyde	−10	60

[a] Reagent added as a molar equiv. based on the lithiated TMEDA or trimethylamine present; solvent was hexane in all cases.

of the 3,310 grams of 1-chlorobutane was added over 3.5-hrs. The slurry was stirred 2 hr more before filtration. The yellow filtrate analyzed to be 26.99% n-butyllithium. The Gilman correction was 0.28% (17). The yield was 83% with no wash on the filter cake. (Actual yield estimated to be about 90%.)

Lithiation of Benzene by n-Butyllithium–TMEDA. A 500-ml flask equipped with stirrer, reflux condenser, and thermometer was purged with argon and then charged with the required amount of 24.4% n-butyllithium in benzene (120 to 200 ml). The stirred solution was heated to the desired temperature and the heat removed. Purified TMEDA was then added in the proper molar ratio (30 to 50 ml) using a large syringe over 0.5 to 1.0 min while cooling with a heptane-dry ice bath to maintain proper temperature. (Do not use rapid addition for preparations larger

TMEDA and Trimethylamine

RLi Used for Lithiation of tert-Amine	Product	% Yield	°C, bp[b]
sec-BuLi	N-n-pentyl-N,N',N'-trimethyl-1,2-ethanediamine	40	182
		49	182
		33	61 (4.5 mm)
		18	
n-BuLi		44	
sec-BuLi	N-(2-methylbutyl)-N,N',N'-trimethyl-1,2-ethanediamine	33	52 (2.9 mm)
	N-benzyl-N,N',N'-trimethyl-1,2-ethanediamine	30	69 (0.3 mm)
	N-n-nonyl-N,N',N'-trimethyl-1,2-ethanediamine	40	85 (0.5 mm)
		37	
n-BuLi		30	
	N-(cyclohexylmethyl)-N,N',N'-trimethyl-1,2-ethanediamine	18	69 (0.5 mm)
sec-BuLi	N-(hydroxycyclopentylmethyl)-N,N',N'-trimethyl-1,2-ethanediamine	28	81 (1.5 mm)
n-BuLi	N-(2-hydroxyhexyl)-N,N',N'-trimethyl-1,2-ethanediamine	39	84 (0.7 mm)
	N,N,N-n-butyldimethylamine	32	92
sec-BuLi	N,N,N-n-nonyltrimethylamine	12	52 (2.9 mm)
n-BuLi	condensation product	—	—
	N,N,N-(1-hydroxyl-1-phenylethyl) dimethylamine	49	72 (0.25 mm)

[b] *See* experimental for literature references and physical constants when available.

than 250 ml—it is too vigorous.) The evolution of gas was observed as it passed out of the system through an oil bubbler. Periodically a 2-ml sample of the solution was withdrawn and injected into a serum bottle containing 2 ml of trimethysilyl chloride and 10 ml of diethyl ether. This derivative solution was hydrolyzed, washed, and dried over anhydrous sodium sulfate. The ether layer was analyzed by GLC, and the amounts of n-butyltrimethylsilane and phenyltrimethylsilane were determined.

$$\% \text{ Unreacted } n\text{-butyllithium} = \frac{(\text{Area})_{\text{TMS butyl}}}{\frac{Mw_{\text{TMS butyl}}}{Mw_{\text{TMS phenyl}}}(\text{Area})_{\text{TMS phenyl}} + (\text{Area})_{\text{TMS butyl}}}$$

When all of the gas had evolved, the heat was removed, the solution weighed, and a sample allowed to react with trimethylsilyl chloride and analyzed by GLC.

$$\% \text{ Yield} = \frac{(\text{Area})_{\text{TMS butyl}}}{(\text{moles BuLi})(130)(\text{area})_{\text{benzene}}}$$

[(Initial wt benzene present) − (moles BuLi)78]

Recommended Preparation of Phenyllithium–TMEDA in Benzene. A 5-liter flask fitted with a thermometer, stirrer, reflux condenser, and pressure-equalizing dropping funnel was flushed with argon and charged with 878 grams (13.65 moles) of n-butyllithium (27.1%) in benzene and 2,560 ml of benzene. (It is not necessary to use n-butyllithium at this concentration because the 1:1 complex is not soluble in benzene beyond about 19% phenyllithium at room temperature. The extra benzene need not be added when an *in situ* reaction is planned.) The slow addition of 1,960-ml (13.0 moles) of TMEDA was begun. The temperature was allowed to rise to 50°C with about one-third of the TMEDA added. The gas evolution was vigorous at this point, and the rest of the TMEDA was added over 1.5 hrs as butane evolution gradually slowed. (Good cooling in the condenser was necessary to prevent the carryover of benzene with the butane.) The solution was then heated cautiously to 79°C then cooled quickly under argon. Yield was 100%—6,100 grams of an 18.89% phenyllithium solution. Phenyllithium:TMEDA mole ratio was 1.04 by Gilman double titration (*17*).

Thermal Stability of Organolithium–TMEDA Complexes. About 400 ml of the organolithium–TMEDA complex was placed in a bottle and then analyzed using the Gilman double titration (*17*). The bottle was placed in a circulating constant-temperature oil bath at the desired temperature (usually 35°C) and left for 30 days. The solution was reanalyzed.

% Decomposition of active material/day

$$= \frac{(\text{Net contained})_{\text{initial}} - (\text{Net contained})_{\text{final}}}{(\text{Net contained})_{\text{initial}} (30 \text{ days})}$$

Preparation of (Benzyllithium)$_2$–TMEDA in Toluene. A 12-liter flask equipped with pressure-equalizing dropping funnel, stirrer, reflux condenser, and thermometer was purged with argon and charged with 7,096 grams of 22.8% n-butyllithium in toluene. The system was sealed from the air by an oil bubbler under slight positive argon pressure. About 30% of 1,470 grams of TMEDA (1% excess) was added over 40 min to bring the reaction to 56°C with vigorous gas evolution. The complete addition took 95 min. The flask was heated to 70°C and then cooled to room temperature. The clear, deep red-orange solution weighed 7,100 grams; it was analyzed as 36.0 wt % benzyllithium by Gilman double titration with 101% yield. Analysis by GLC using trimethylsilyl chloride derivative and mesitylene as an internal standard gave 35.5% concentration and a yield of 100%.

Ring-Isomer Content in Benzyllithium–TMEDA Complexes. Only solutions of benzyllithium–TMEDA complexes with little or no butyllithium were analyzed. A 2-ml aliquot of the solution in a syringe was injected into a small bottle containing 10 ml of ether and 3 ml of dimethyl sulfate under argon. The reaction was vigorous. After 20 min the solution was washed with water, dried over anhydrous Na_2SO_4, and analyzed by GLC using a 10-ft, 10% Triton-S305/Chromasorb W at 90°C. (Only partial resolution of meta and para isomers was obtained, although it could have been observed whether a relatively small amount of para isomer was present.) The percent of ring isomer was calculated from the total amount of ring isomer plus the ethylbenzene present.

Preparation of Benzyllithium–TED Solid Complex. A 500-ml flask equipped with a stirrer, thermometer, and reflux condenser after purging with argon was charged with 184 grams 15.2% n-butyllithium in hexane, and 47.9 grams TED (2% deficiency) was added. The solution was then heated after adding 48.2 grams of toluene (12% excess). The solution was refluxed 3.25 hrs. It was then cooled to 10°C, stirred for 1 hr, and filtered under argon. The yellow crystals were dried for 6 hrs (1 mm) at room temperature. The yield was 89% or 80.6 grams of solid complex.

Preparation of Allyllithium–TMEDA Solid Complex. A 1-liter flask equipped with a gas inlet tube, stirrer, and thermometer was purged with nitrogen and charged with 437 grams of 10.0% sec-butyllithium in hexane. The solution was cooled to $-5°C$, and 77.7 grams of TMEDA were added with cooling to maintain $-5°C$. About 43.0 grams (50% excess) of propene were dissolved in the solution in 30 min. The solution was then added to a 1-liter heavy-wall pressure bottle, sealed, and heated to 45°C with mechanical shaking (a Parr low-pressure reactor). The solution was then cooled to 10°C and filtered. The yellow-orange precipitate was rinsed with pentane and blown dry with argon. The yield was 90 grams of the solid complex desired or 80%. (Pressure is optional in final warm-up.)

Preparation of Crotyllithium–TMEDA. A 500-ml flask equipped with a stirrer, thermometer, condenser, and gas-inlet tube was purged with argon and charged with 191 grams of 12.2% sec-butyllithium in cyclohexane. The solution was cooled to $-5°C$, and 41.4 grams (2% deficiency) of TMEDA was added while cooling to maintain $-5°C$. To this solution 30.6 grams (50% excess) of 1-butene was added at 0°C in 20 min. The solution was allowed to warm to 25°C over 30 min, heated to 40°C quickly, and then cooled to room temperature. The dark, red-orange solution titrated as $3.39N$ base or $1.13M$ with a volume of 297 ml, corresponding to a yield of 92%. GLC analysis of the trimethylsilyl derivative of this solution gave the composition of the solution as $29.5 \pm 1\%$ 1-lithio-2-butene (terminal) and $70.5 \pm 1\%$ 2-lithio-3-butene (internal) at 55°C. Analysis of the solution at 20°C gave 29.0% terminal.

Preparation of Lithiated N,N,N',N'-Tetramethylethylenediamine. A 500-ml flask equipped as above was charged with 157 grams of 11.4% sec-butyllithium in cyclohexane. The solution was cooled to 0°C and 32.5 grams of TMEDA was added while stirring with further cooling to maintain 0°C. The solution was heated to 49°C for 1 hr (when gas evolution completely stopped), then cooled to room temperature. The dark, brown-red solution was almost clear and weighed 175 grams. GLC analysis of the trimethylsilyl derivative showed no *sec*-butyllithium remaining. Analysis of the solution by Gilman double titration gave a net product concentration of 18.5% with the mole ratio of total base to carbon-bound lithium of 2.06. The yield based on this titration was 95%.

Preparation of Lithiated Trimethylamine. After purging a 500-ml heavy-wall bottle with argon, it was charged with 175 grams (0.325 mole) of 11.9% *sec*-butyllithium in cyclohexane. This solution was cooled under argon to −10°C, and 38.4 grams (0.65 mole) of trimethylamine was dissolved in the solution with cooling to maintain −10°C. The bottle was sealed off, placed in a Parr laboratory pressure reactor, and heated to 60°C for 3 hrs with shaking. The clear, dark, brown-yellow solution was cooled to ambient temperature and analyzed. The solution weighed 190 grams, and contained no *sec*-butyllithium by GLC analysis of the trimethylsilyl chloride derivative. Total base analysis of the filtered solution was 19.2% indicating a product concentration of at least 6.6% for a 60% yield.

Preparation of Other Lithiated Tertiary Amines and Diamines. The procedure followed for these compounds was the same as that used for the lithiated TMEDA described above.

Synthetic Reactions of Organolithium-Tertiary Diamine Complexs. When (benzyllithium)$_2$–TMEDA, phenyllithium–TMEDA, crotyllithium–TMEDA and lithiated TMEDA or trimethylamine solutions were used, from 0.4 to 0.5 mole of the organolithium complex in the hydrocarbon solvent was added to a 500-ml flask under argon and cooled to the appropriate temperature. In the cases of benzyllithium–TED and allyllithium–TMEDA the solid complexes were added as a solvent. A molar equivalent of the particular reactant was added *via* a pressure-equalizing dropping funnel over a given interval with stirring at a constant temperature. (About 10% excess of the organolithium compound was used.) At a selected time the solution was hydrolyzed with excess water. The organic layer was separated from the aqueous layer, washed two or three times with 10% aqueous ammonium chloride, and dried over anhydrous Na_2SO_4. The solvent was removed by distillation at reduced pressures. The crude product was purified by recrystallization or distillation. In all cases the infrared spectra were compared with knowns or were examined to ascertain their agreement with the assigned structures. The physical constants were also compared with literature values when possible. Analyses were made in cases where doubt existed concerning the structure because starting materials having similar structures may have been contaminants; the results were in agreement.

PHENYLLITHIUM–TMEDA IN BENZENE. *Benzophenone.* Added benzophenone dissolved in benzene at 5°C over 1 hr; let warm 1 hr; hydrolyzed, used extra benzene during work-up; recrystallized from 1:1 methanol-ethanol; mp 161°C (lit. mp 162.5°C); 95%, triphenylcarbinol (*36*).

Cyclohexanone. Added cyclohexanone dissolved in THF at 0°C in 1 hr; let warm 1 hr; hydrolyzed; recrystallized from heptane; mp 60°C (lit. mp 61°C); 59%, 1-phenylcyclohexanol (*37*).

(BENZYLLITHIUM)$_2$–TMEDA IN TOLUENE. *Trimethylsilyl chloride.* Added trimethylsilyl chloride dissolved in ether at −5°C over 1 hr; let warm 0.5 hr; hydrolyzed; distilled product; bp 59°–62°C (5 mm) (lit. bp 184°–5°C); 81%, benzyltrimethylsilane (38).

1-Bromobutane. Addition at −20°C over 1 hr; let warm 0.5 hr; hydrolyzed; distilled product; bp 61°–64° (0.5 mm) (lit. bp 81°C (202 mm)); 84%, n-amylbenzene (36).

Benzonitrile. Added benzonitrile at −5°C over 1 hr; let warm 1.5 hrs; used 20% HCl for hydrolysis; recrystallized from ethanol; mp 53°–54°C (lit. mp 60°C); 16% crude, benzylphenyl ketone (36).

Ethyl acetate. Added 0.5-mole equivalents ethyl acetate at 0°C over 1 hr; let warm 0.5 hr; hydrolyzed; distilled; bp 122°–125°C (0.4 mm) [lit. bp 122°C (0.4 mm)]; 43%, dibenzylethanol (39).

Ethyl benzoate. Added 0.5-mole equivalents of ethyl benzoate at 5°C over 1 hr; let warm 0.5 hr; hydrolyzed, recrystallized from heptane; mp 55°–79°C (crude) (lit. mp 86°C); 22%, 1,2,3-triphenyl-2-propanol (40).

Cyclopentanone. Added cyclopentanone at −40°C over 6 min; let warm to 0°C; hydrolyzed; distilled, recrystallized from heptane; mp 55°C (lit. mp 59°C); 35%, 1-benzylcyclopentanol (37).

Cyclohexanone. Added cyclohexanone at −5°C over 1 hr; let warm for 1 hr; hydrolyzed; recrystallized from heptane; mp 55°C (lit. mp 61°C); 59%, 1-benzylcyclohexanol (37).

2-Butanone. Added 2-butanone at −15°C over 1 hr; let warm for 0.5 hr; hydrolyzed; distilled; bp 92°–94°C (0.5 mm) [lit. bp 110°–112°C (4 mm)]; 53%, 2-benzyl-2-butanol (41).

Diisopropyl ketone. Added diisopropyl ketone at −5°C over 1 hr; let warm 0.5 hr; hydrolyzed; distilled; bp 96°C (0.1 mm) [lit. bp 99°C (1 mm)]; 82%, 3-benzyl-2,4-dimethyl-3-pentanol (42).

Benzaldehyde. Added benzaldehyde at −35°C over 20 min; hydrolyzed; distilled; bp 62°C (0.5 mm) [lit. bp 81°C (10 mm)]; 80% n-amylbenzene (36).

Benzophenone. Added benzophenone dissolved in toluene at −15°C over 18 min; let warm over 50 min; hydrolyzed; recrystallized from ethanol; mp 86°C (lit. mp 89°C); 92%, 1,1,2-triphenylethanol (36).

BENZYLLITHIUM–TED. *1-Bromobutane.* Added 1-bromobutane to complex dissolved in ether at reflux over 1 hr; refluxed overnight; hydrolyzed; distilled; bp 62°C (0.5 mm) (lit. bp 81°C); 80%, n-amylbenzene (36).

Bromobenzene. Added mole equivalents bromobenzene to complex dissolved in diethyl ether at reflux over 65 min; stand overnight; hydrolyzed; recrystallized from methanol; mp 50°C (lit. mp 56°C); 32%, 2-benzylbiphenyl (44).

Iodobenzene. Same procedure as for bromobenzene; mp 49°C (lit. mp 56°C); 20%, 2-benzylbiphenyl (43).

Benzonitrile. Added benzonitrile to complex dissolved in THF at −10°C over 1 hr, stirred at −10°C for 0.5 hr; let warm for 1 hr; hydrolyzed normally, but added 20% HCl to crude solid product in isolation to complete hydrolysis of imine; recrystallization from ethanol; mp 53°C (lit. mp 60°C); 59%, benzylphenyl ketone (36).

Ethyl acetate. Added 0.5-mole equivalent ethyl acetate to the complex dissolved in diethyl ether at 35°C over 45 min; let warm overnight; hydrolyzed; oil residue isolated after removal of volatiles; infrared indicated that the tertiary alcohol was present but impure; 25% crude; 1,1-dibenzylethanol (*39*).

Ethyl benzoate. Added 0.5-mole equivalent ethyl benzoate to complex dissolved in THF at 0°C over 0.5 hr; let warm over 1.5 hr; hydrolyzed; recrystallized from ethanol; mp 83°C (lit. mp 86°C); 66%, 1,2,3-triphenyl-2-propanol (*39*).

Methyl vinyl ketone. Added methylvinyl ketone to complex dissolved in THF at −10°C over 50 min; let warm over 1 hr; hydrolyzed; distilled; pot temperature maintained below 75°C, yet apparent decomposition or rearrangement occurred; pot residue had a bp > 78°C (0.2 mm); infrared indicated considerable hydroxyl and carbonyl groups present in pot residue.

Crotonaldehyde. Added crotonaldehyde to complex dissolved in diethyl ether at 0°C over 20 min; let warm over 1 hr; hydrolyzed; no product, only polymer isolated.

2-Butanone. Added 2-butanone to complex dissolved in diethyl ether at 0°C over 40 min; let warm over 1 hr; hydrolyzed; distilled; bp 67°C (0.3 mm) [lit. bp 110°–112°C (14 mm)]; 65%, 2-benzyl-2-butanol (*41*).

Cyclopentanone. Added cyclopentanone to complex dissolved in THF at −10°C over 1 hr; let warm over 1 hr; hydrolyzed; recrystallized from heptane; mp 56°C (lit. mp 59°C); 1-benzylcyclopentanol (*37*).

Cyclohexanone. Added cyclohexanone to complex dissolved in diethyl ether at −10°C over 1 hr; let warm overnight; hydrolyzed; recrystallized from heptane; mp 55°C (lit. mp 61°C); 45%, 1-benzylcyclohexanol (*37*).

Diisopropyl ketone. Added diisopropyl ketone to complex dissolved in diethyl ether at −10°C in 5 min; let warm to 10°C over 18 min; hydrolyzed; distilled; bp 110°C (1 mm) [lit. bp 99°C (1 mm)]; 60%, 3-benzyl-2,4-dimethyl-3-pentanol (*42*).

Benzaldehyde. Added benzaldehyde to complex dissolved in THF at 0°C over 1 hr; let warm over 2 hrs; hydrolyzed; recrystallized from ethanol; mp 65°C (lit. mp 67°C); 81%, 1,2-diphenylethanol (*36*).

Benzophenone. Added benzophenone dissolved in diethyl ether to complex dissolved in diethyl ether at −10°C over 0.5 hr; stirred 0.5 hr more at −10°C, then let warm overnight; hydrolyzed; recrystallized from methanol; mp 87°C (lit. mp 89°C); 99%, 1,1,2-triphenylethanol (*36*).

Indanone. Added indanone dissolved in THF to complex dissolved in THF at −10°C for 45 min; let warm over 1.5 hrs; hydrolyzed; distilled and recrystallized from ethanol; mp 141°C (lit. mp 155°C); 20%, 1-benzyl-1-indanol (*44*).

Fluorenone. Added fluorenone dissolved in diethyl ether to complex dissolved in diethyl ether at 35°C over 25 min; hydrolyzed; recrystallized from cyclohexane; mp 133°–136°C (lit. mp 143°C); 83%, 9-benzyl-9-fluorenol (*45*).

Lithium benzoate. Added lithium benzoate to complex dissolved in THF at 35°C, heated to reflux for 25 min, cooled, stirred 1 hr before hydrolysis; hydrolyzed; recrystallized from ethanol; mp 56°C (lit mp 60°C); 44%, benzylphenyl ketone (*36*).

ALLYLLITHIUM–TMEDA. *Benzophenone.* Added benzophenone dissolved in diethyl ether to slurry of the complex in diethyl ether at 20°C over 0.5 hr; stirred 1 hr; hydrolyzed; distilled bp 128°C (0.6 mm) [lit. bp 135°C (0.5 mm)]; 91%, 1,1-diphenyl-3-buten-1-ol (46).

Cyclohexanone. Added cyclohexanone to slurry of the complex in diethyl ether at 20°C over 20 min; stirred 20 min; hydrolyzed; distilled; bp 64°–68°C (8 mm [lit. bp 81°C (15 mm)]; 52%, 1-allylcyclohexanol (47).

CROTYLLITHIUM–TMEDA IN HEXANE. Added acetone to complex in solution at −20°C over 0.5 hr; let warm over 40 min; hydrolyzed; distilled; bp 65°–71°C (120 mm) [lit. bp 133°–135°C (760 mm); 142°C. (760 mm)]; 11%, 2,3-dimethyl-4-penten-2-ol and *trans*-2-methyl-4-hexen-2-ol; infrared indicated no cis olefin bonds (48).

LITHIATED $\dot{N},N,N',N'$-TETRAMETHYLETHYLENEDIAMINE (I) IN HEXANE. These reactions were acidified after hydrolysis and the organic layer discarded. The aqueous layer was made basic with NaOH solution and extracted with ether. The ether layer was distilled after drying over anhydrous Na_2SO_4.

1-Bromobutane. Added one-mole equivalent TMEDA to 12% *sec*-butyllithium in hexane and heated at 55°C for 1 hr, cooled to 0°C and added 0.95-mole equivalent of 1-bromobutane over 0.5 hr; let warm to 20°C; hydrolyzed; distilled; bp 182°C (760 mm) [lit. bp 183°C (760 mm)]; 49%, N-n-pentyl-N,N',N'-trimethylethyl-1,2-ethanediamine (49). Analytically calculates for $C_{10}H_{24}N_2$: C, 69.75; H, 13.95; N, 16.29. Found: C, 69.75; H, 13.89; N, 16.49.

2-Bromobutane. Added 1-mole equivalent TMEDA to 11.4% *sec*-butyllithium in hexane at 0°C, heated to 50°C for 1 hr and then cooled to 0°C; added 1-mole equivalent 2-bromobutane over 20 min; let warm over 0.5 hr; worked up; hydrolyzed; distilled; bp 52°C (2.9 mm); 33%, N-(2-methylbutyl)-N,N',N'-trimethyl-1,2-ethanediamine.

Iodobenzene. Added 1-mole equivalent TMEDA to 12% *sec*-butyllithium in hexane at 0°C, heated at 55°C for 1.5 hrs; cooled to 0°C and added 1-mole equivalent iodobenzene over 15 min; stirred 1 hr at 20°C; hydrolyzed; distilled, bp 69°C (0.3 mm) [lit. bp 116°–119°C (10 mm)]; 30%, N-benzyl-N,N',N'-trimethyl-1,2-ethanediamine (49). Analytically calculated for $C_{12}H_{20}N_2$: C, 75.00; H, 10.42; N, 14.58. Found: C, 75.01; H, 10.32; N, 14.41.

1-Bromooctane. Added 1-mole equivalent TMEDA to 13% *sec*-butyllithium in hexane at 0°C, heated at 50°C for 1.25 hrs, cooled to 20°C; added 1-mole equivalent 1-bromooctane over 0.5 hr, stirred 0.5 hr more; hydrolyzed; distilled; bp 85°C (0.5 mm) [lit. by 135°–139°C (11 mm)]; 40%, N-n-nonyl-N,N',N'-trimethyl-1,2-ethanediamine (50).

Bromocyclohexane. Added 1-mole equivalent TMEDA to 15.5% *n*-butyllithium in hexane at 0°C, heated to reflux 3.75 hrs and cooled to 0°C; added 1-mole equivalent bromocyclohexane over 20 min at 0°C; warmed slowly over 0.5 hr; hydrolyzed; distilled; bp 69°C (0.5 mm); 18%, N-(cyclohexylmethyl)-N,N',N'-trimethyl-1,2-ethanediamine.

Cyclopentanone. Added 1-mole equivalent TMEDA to 11.9% *sec*-butyllithium at 0°C, heated to 50°C for 1.25 hrs, cooled to 0°C and added 1-mole equivalent cyclopentanone over 20 min; stirred 20 min at 20°C; hydrolyzed; distilled; bp 81°C (1.5 mm); 28%, N-(1-hydroxycyclopentylmethyl)-N,N',N'-trimethyl-1,2-ethanediamine.

Valeraldehyde. Added 1-mole equivalent TMEDA to 15.5% *n*-butyllithium in hexane at 0°C, heated at reflux 3.75 hrs, cooled to −10°C, and added 1-mole equivalent valeraldehyde over 20 min; let warm over 0.5 hr; hydrolyzed; distilled; bp 84°C (0.7 mm); 39%, *N*-(2-hydroxylhexyl)-*N,N',N'*-trimethyl-1,2-ethanediamine.

LITHIATED TRIMETHYLAMINE (II) IN HEXANE. Lithiation of trimethylamine was done in a pressure reactor during the heating step for actual lithiation of the tertiary amine.

1-Iodopropane. Added 2 mole equivalents trimethylamine to 15.1% *n*-butyllithium in hexane at 0°C; solution heated at 70°C for 5 hrs and cooled to −10°C; 0.85-mole equivalent 1-iodopropane added over 20 min; warmed over 0.5 hr; hydrolyzed; distilled; bp 92°C (760 mm) [lit. bp 95°C (760 mm)]; 32%, *N,N,N-n*-butyldimethylamine (*51*).

1-Bromooctane. Added 2-mole equivalents trimethylamine to 13.2% *sec*-butyllithium in hexane, heated to 60°C for 3 hrs, cooled to −10°C, and added 1-mole equivalent 1-bromooctane over 20 min; let warm to 20°C; hydrolyzed; distilled; bp 52°C (2.9 mm) [lit. bp 209°C (741 mm)]; 33%, *N,N,N*-dimethylnonylamine (*52*).

Benzaldehyde. Added 2 mole equivalents trimethylamine to 11.9% *sec*-butyllithium in cyclohexane at −10°C, heated to 50°C for 6 hrs, cooled to −10°C; added 0.85-mole equivalent benzaldehyde over 0.5 hr; let warm over 0.5 hr; hydrolyzed; distilled; bp 72°C (0.25 mm) [lit. bp 170°C (760 mm)]; 49%, (1-hydroxy-1-phenylethyl)dimethylamine (*53*).

Acknowledgment

I thank E. D. Kuehn and K. B. Lynskey who assisted in the laboratory work.

Literature Cited

1. Eberhardt, G. G., Butte, Jr., W. A., *J. Org. Chem.* (1964) **29**, 2928.
2. Langer, Jr., A. W., *Trans. N.Y. Acad. Sci.* (1965) **27**, 741.
3. Screttas, C. G., Eastham, J. F., *J. Amer. Chem. Soc.* (1965) **87**, 3276.
4. Rausch, M. D., Ciappenelli, D. J., *J. Organometal. Chem.* (1967) **10**, 127.
5. Langer, Jr., A. W., U.S. Patent **3,536,679** (1970).
6. Crawford, R. J., Erman, W. F., Broaddus, C. D., *J. Amer. Chem. Soc.* (1972) **94**, 4298.
7. Agami, C., *Bull. Soc. Chem. Fr.* (1971) 1619.
8. Mallan, J. M., Bebb, R. L., *Chem. Rev.* (1969) **69**, 693.
9. Langer, Jr., A. W., *Polym. Prepr., Am. Chem. Soc., ACS Div. Polym. Chem.* (1966) 137.
10. Eberhardt, G. G., Davis, W. R., *J. Polym. Sci., Part A* (1965) **3**, 3753.
11. Stucky, G. D., *Amer. Chem. Soc., Div. Polym. Chem., Preprint*, **13** (2), 644 (New York, Aug. 1972).
12. Eberhardt, G. B., Butte, Jr., W. A., U.S. Patent **3,321,479** (1967).
13. Gilman, H., Pacevitz, H. A., Baine, O., *J. Amer. Chem. Soc.* (1940) **62**, 1514.
14. Gilman, H., Schwebke, G. L., *J. Org. Chem.* (1962) **27**, 4259.
15. Seyferth, D., Weiner, M. A., *J. Org. Chem.* (1959) **24**, 4797.
16. Chalk, A. J., Hoogeboom, T. J., *J. Organometal. Chem.* (1968) **11**, 615.
17. ASTM Standard Method No. E-233 (*n*-Butyllithium Analysis).
18. Broaddus, C. D., *J. Org. Chem.* (1970) **35**, 10.

19. West, R., Jones, P. C., *J. Amer. Chem. Soc.* (1968) **90**, 2656.
20. Gau, G., *Bull. Soc. Chim. Fr.* (1972) 1942.
21. Buhler, J. D., *J. Org. Chem.* (1973) **38**, 904.
22. Magruder, W. J., *Dissert. Abstr.* (1966) **27**, 759B.
23. Eisch, J. J., Jacobs, A. M., *J. Org. Chem.* (1963) **28**, 2145.
24. Seyferth, D., Weiner, W. A., *J. Org. Chem.* (1959) **26**, 4797.
25. Lampher, E. J., *J. Amer. Chem. Soc.* (1957) **79**, 5578.
26. Seyferth, D., Juta, T. J., *J. Organometal. Chem.* (1967) **8**, P13.
27. Johnson, C. S., Weiner, M. A., Waugh, J. S., Seyferth, D., *J. Amer. Chem. Soc.* (1961) **83**, 1306.
28. Langer, Jr., A. W., U.S. Patent **3,541,149** (1970).
29. Lepley, A. R., *et al., J. Org. Chem.* (1966) **31**, 2047.
30. *Ibid*, 2051.
31. *Ibid*, 2055.
32. *Ibid*, 2061.
33. *Ibid*, 2064.
34. Smith, W. N., unpublished data.
35. Peterson, D. L., *J. Amer. Chem. Soc.* (1971) **93**, 4027.
36. "Handbook of Chemistry and Physics," 53rd ed., The Chemical Rubber Co., Cleveland, O. 1972.
37. Stach, D., Winter, W., *Arzneimettel-Forsch.* (1962) **12**, 194; *Chem. Abstr.* (1962), **57**, 16453.
38. Brook, A. G., *et al., J. Amer. Chem. Soc.* (1960) **82**, 5102.
39. Gamboa, J. M., Ossorio, R. P., Rapun, R., *An. Real Soc. Espan. Fis. Quim.* (1961) **57B, 607C**; *Chem. Abstr.* (1962), **57**, 700.
40. Rampart, P., Amagat, P., *Ann. Chim.* (1927) **8**, 263.
41. Warnick, P., Saunders, Jr., W. H., *J. Amer. Chem. Soc.* (1962) **84**, 4095.
42. Frank, C. E., Foster, W. E., *Ind. Eng. Chem.* (1954) **46**, 1019.
43. Chel'tsova, M. A., *et al., Izv. Akad. Nauk SSSR, Ser. Khim.* (1965) **1**, 124; *Chem. Abstr.* (1965) **62**, 11707.
44. Nizamuddin, S., Ghosal, M., Chudhury, D. N., *J. Indian Chem. Soc.* (1965) **43**, 569.
45. Cadogan, J. I. G., Hey, D. H., Sandersen, W. A., *P. Chem. Soc.* (1960) 3203.
46. Vozza, J. F., *J. Org. Chem.* (1959) **24**, 720.
47. Huet, J., *Bull. Soc. Chim. Fr.* (1964) 2677.
48. Kochi, J., *J. Org. Chem.* (1963) **28**, 1969.
49. Grail, G. F., *et al., J. Amer. Chem. Soc.* (1952) **74**, 1313.
50. Shepherd, R. A., Wilkinson, R. G., *J. Med. Pharm. Chem.* (1962) **5**, 823.
51. Braun, H. C., Berneis, H. L., *J. Amer. Chem. Soc.* (1953) **75**, 10.
52. King, H., Work, T. S., *J. Chem. Soc.* (1942) 401.
53. Klosa, J:. *J. Prakt, Chem.* (1963) **21**, 1.

RECEIVED March 12, 1973.

3

Stereochemical Properties of N-Chelated Alkali Metal Complexes

GALEN STUCKY

University of Illinois, Urbana 61801

> *This review examines the structure and bonding of Group I organometallic complexes. The results demonstrate that ground-state geometries of neutral delocalized molecules are not good models for the reduced species that form upon reduction with Group I metals, and that for aromatic species a significant loss of aromatic character occurs. Solid-state structural results are consistent with observations made via ESR in solution for the naphthalene radical anion. In general the symmetries of the carbanion molecules are consistent with the symmetries of the highest-occupied molecular orbitals obtained from semiempirical molecular orbital theory. A model is presented for predicting the position of the metal atom with respect to the delocalized carbanion in contact-ion pairs.*

Alkali metals and their organic complexes form an important class of uniquely reactive species that have found increasing use in organic and inorganic syntheses. For example organolithium reagents, in addition to undergoing the remarkable reactions listed in Table I, display the chemistry typical of organomagnesium systems, with the added benefit that organolithium reagents are generally more reactive.

The chemical reactivities of the alkali metal organometallic compounds (RM) vary widely depending on metal M, basicity of the solvent systems used, and steric and electronic properties of the organic group R. In many reactions an important factor is the stabilization resulting from formation of a delocalized carbanion system as in the polymerization of dienes or aromatic substituted ethylenes, and in Reactions 3, 4, 5, and 10 in Table I. It is primarily with these delocalized carbanion systems that this review is concerned although saturated organolithium compounds are discussed briefly.

There have been numerous attempts to elucidate both the nature of the ion-pair interactions and the delocalized carbanion geometries of chelated alkali metal organic systems in solution (18). There is now apparent agreement that this equilibrium:

$$\text{RMB}_n \underset{\text{Contact}}{\overset{S}{\rightleftarrows}} \text{R}^- \mid \text{MB}_n\text{S}^+ \atop \text{Solvent-separated}$$

(R = organic group, M = alkali metal,
B = coordinated base, and S = solvent)

Table I.[a] **Reactions of Alkali Metals and Their Organic Complexes**

Reaction	Reference
1) $\text{Li} + \text{N}_2 \xrightarrow{25°C} \text{Li}_3\text{N}$	1
2) $[\text{TMEDLi}]\text{-}n\text{-Bu} + \text{H}_2 \xrightarrow{1 \text{ atm}} [\text{TMEDLi}]\text{H} + n\text{-BuH}$	2, 3
3) $(\text{C}_6\text{H}_5)_3\text{CNa} + \text{N}_2\text{O} \rightarrow (\text{C}_6\text{H}_5)_3\text{CN}_2\text{ONa} \xrightarrow{\text{EtOH}}$ $(\text{C}_6\text{H}_5)_3\text{COH} + \text{N}_2 + \text{EtONa}$	4
4) $[\text{TMEDLi}]\text{-}n\text{-Bu} + \text{CH}_3\text{C}_6\text{H}_5 \rightarrow$ $[\text{TMEDLi}]\text{CH}_2\text{C}_6\text{H}_5 + n\text{-BuH}$	5
5) $[\text{TMEDLi}]\text{CH}_2\text{C}_6\text{H}_5 + n(\text{CH}_2{=}\text{CH}_2) \rightarrow$ $[\text{TMEDLi}](\text{CH}_2\text{—CH}_2)_n\text{CH}_2\text{C}_6\text{H}_5$	6–8
6) $2[\text{TMEDLi}]\text{-}n\text{-Bu} + (\text{C}_5\text{H}_5)_2\text{Fe} \rightarrow$ $2([\text{TMEDLi}]\text{C}_5\text{H}_5)_2\text{Fe} + 2\text{-}n\text{-BuH}$	9, 10
7) $\text{RLi} + \text{C}_6\text{H}_5\text{F} \rightarrow [(\text{benzyne})] + \text{LiCl} + \text{RH}$	11, 12
8) $\text{RLi} + \text{CH}_2\text{Cl}_2 \rightarrow [\text{CHCl:}] + \text{LiCl} + \text{RH}$ carbenoid	13
9) $\text{R}'\text{Li} + \text{CH}_3\text{R}_3\text{P}^+\text{Cl}^- \rightarrow [\text{CH}_2{=}\text{PR}_3] + \text{R}'\text{H} + \text{LiCl}$ (ylide)	14–15
10) $2[\text{TMEDLi}]\text{-}n\text{-Bu} + \text{C}_6\text{H}_5\text{CH}_2\text{—CH}_2\text{C}_6\text{H}_5 \rightarrow$ $(\text{TMEDLi})_2[\text{C}_6\text{H}_5\text{CH}{=}\text{CHC}_6\text{H}_5]^{2-} + 2\ n\text{-BuH}$	16
11) $\text{K} + \text{CH}_2{=}\text{CH—CH}_2\text{—CH}_2\text{—CH}{=}\text{CH}_2$ $\xrightarrow[\text{H}_2\text{O}]{\text{THF}} \text{CH}_3\text{—CH}{=}\text{CH—CH}{=}\text{CH—CH}_3$	17

[a] TMED = tetramethylethylenediamine; n-Bu = C_4H_9; Et = C_2H_5.

describes much of the solution chemistry of alkali metal organometallic reactions. With more polar solvents the reaction shifts to the right, and the formation of solvent-separated ion pairs may predominate.

As catalysts, solvent-separated organometallic complexes behave much like the heavier alkali metal complexes in that branched polymerization products form. Contact-ion pairs on the other hand favor stereoregular polymerization. Three important conclusions from earlier studies of the catalytic behavior of N-chelated organolithium reagents are (*19, 20, 21, 22*): (1) the most effective catalysts are those formed with tertiary amine groups coordinated to a lithium atom; (2) the existence of the carbanion as a discrete species is highly unlikely, and ion pairs or partially covalently bonded species are suggested in N-chelated organolithium reagents; and (3) catalytic activity of the RLi–amine reagent increases with increasing amine concentration up to two moles of amine per mole of lithium. Greater amine concentrations do not affect the catalyst's activity.

Because of the exceptional reactivity and chemical importance of the amine-chelated organolithium reagents, their stereochemical properties are of special significance. Our interest centered on six points:

a) What is the relative position and orientation of the metal–amine group with respect to a given delocalized carbanion in a contact-ion pair?

b) What effect does the coordinated base have on the stereochemistry of these systems?

c) Is it possible to develop a physical model of the metal–carbanion interaction that will allow prediction of the stereochemistry of any given contact pair?

d) What effect does the occupancy of carbanion antibonding orbitals have upon the geometry of the carbanion?

e) In those cases where it is possible to form representative element metal–π-group complexes comparable with organometallic transition-metal complexes, what structural differences do d orbitals cause?

f) Are the solid-state molecular structures reasonable models to use in interpreting NMR, ESR, and electronic spectral results obtained for solutions?

The specific purpose of this review is to study these and related questions.

Synthesis and Isolation of Complexes

The reaction of toluene and biphenyllithium in tetrahydrofuran gives about a 1% yield of benzyllithium (*23*). By contrast, yields of 80% or greater are easily obtained *via* the metalation reaction

$$C_6H_5CH_3 + n\text{-BuLi(benzene)} + TMED \rightarrow C_6H_5CH_2LiTMED$$

The same reaction gives excellent yields with fluorene, triphenylmethane, indene, and other benzylic precursors. Also, reducing reactions,

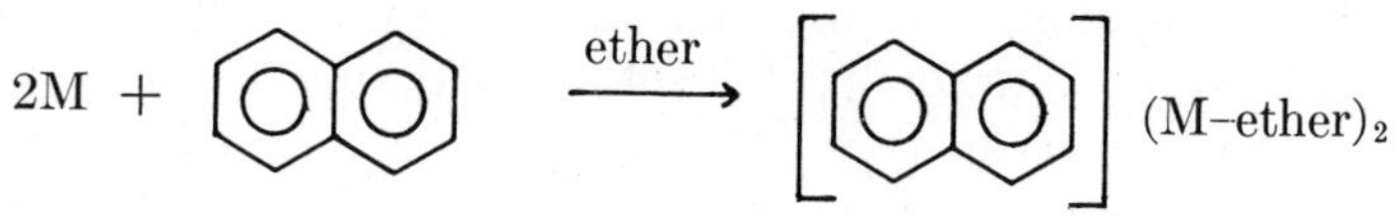

2M + [naphthalene] $\xrightarrow{\text{ether}}$ [naphthalene]²⁻ (M–ether)₂

frequently give poor yields of the desired complex and a mixture of products but reactions like

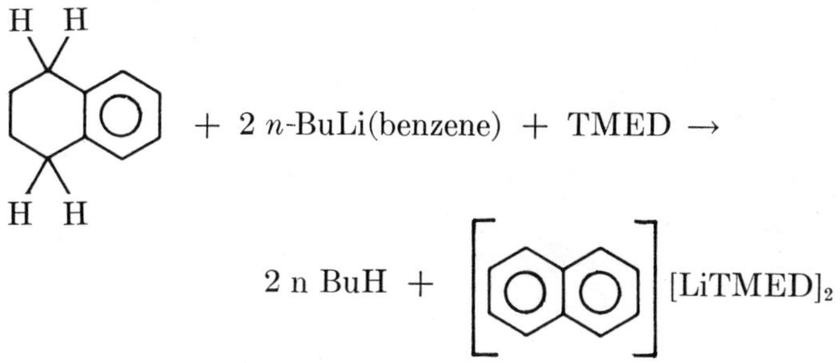

[1,4,5,8-tetrahydronaphthalene] + 2 n-BuLi(benzene) + TMED →

2 n BuH + [naphthalene]²⁻ [LiTMED]₂

are generally clean and give good yields of the dianion for a large number of systems (24, 25). The above method of synthesis enables preparation of anions of unsaturated hydrocarbons that are not themselves isolable. For example, the dianion of pentalene can be prepared (26) by

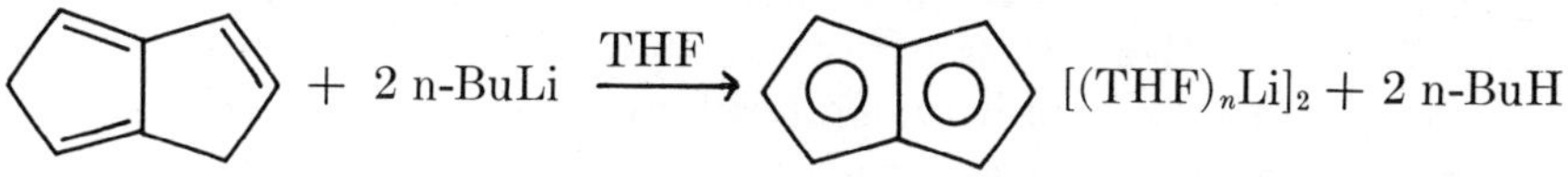

[dihydropentalene] + 2 n-BuLi $\xrightarrow{\text{THF}}$ [pentalene]²⁻ [(THF)$_n$Li]₂ + 2 n-BuH

Subsequent reaction of this dianion with nickelocene gives bis[pentalenyl-nickel(II)] (27).

A potentially useful synthesis for unsaturated organic compounds from saturated ones is:

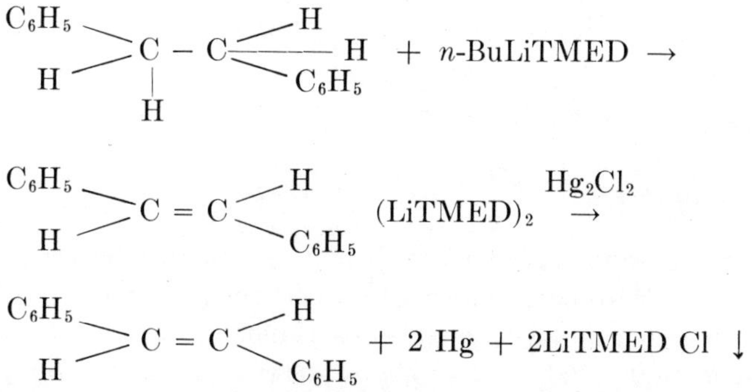

$$\begin{array}{c} C_6H_5 \\ \diagdown \\ C-C \\ H | \\ H \end{array} \begin{array}{c} H \\ \diagup \\ H \\ \diagdown \\ C_6H_5 \end{array} + n\text{-BuLiTMED} \rightarrow$$

$$\begin{array}{c} C_6H_5 \\ \diagdown \\ C=C \\ H \end{array} \begin{array}{c} H \\ \diagup \\ \diagdown \\ C_6H_5 \end{array} \text{(LiTMED)}_2 \xrightarrow{Hg_2Cl_2}$$

$$\begin{array}{c} C_6H_5 \\ \diagdown \\ C=C \\ H \end{array} \begin{array}{c} H \\ \diagup \\ \diagdown \\ C_6H_5 \end{array} + 2\text{ Hg} + 2\text{LiTMED Cl} \downarrow$$

Removing hydrogen atoms from adjacent carbon atoms, as shown above for stilbene, is generally difficult synthetically. Preliminary work (28) has shown that dianions of acenaphthalene can be prepared by treating acenaphthene with n-BuLi in THF. Unfortunately the technique has limited applicability—for example, similar treatment of 9,10-dihydrophenanthrene does not give dianion formation (28). We have routinely used a modified version of this reaction with the base TMED for years (24, 25)—as above in the synthesis of naphthalene dianion. The dianion of phenanthrene with TMED can also be readily prepared by this reaction:

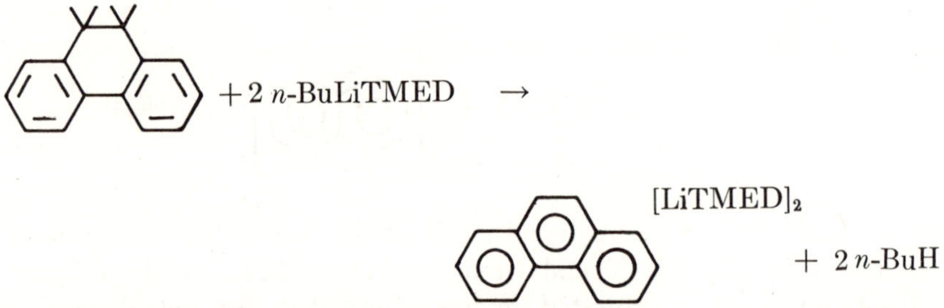

As this synthesis and that of the stilbene dianion show, there appear to be no particular problems associated with the removal of protons from adjacent carbon atoms with this reagent. The results suggest that the unusual ability of n-butyl or tert-butyl LiTMED can be used to abstract protons from adjacent carbon atoms in preparing cyclobutadiene dianion via this reaction:

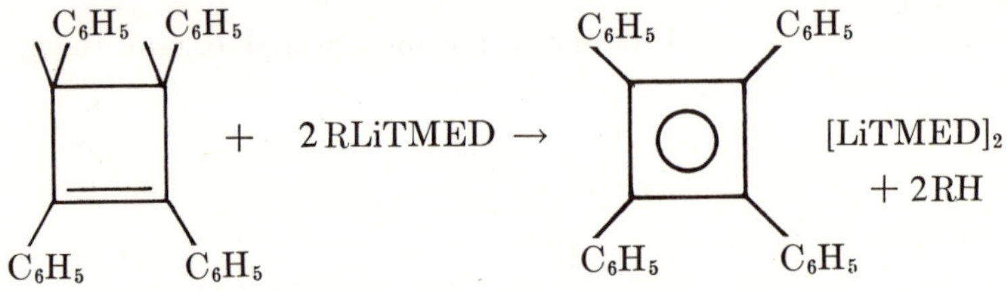

Numerous other possibilities exist.

Structural Properties

Structural features of the π carbanion organometallic complexes are discussed here in three sections: (1) π-carbanion geometry, (2) metal-base coordination, and (3) metal–π-carbanion geometry. The primary tool used in these studies was single-crystal x-ray crystallography. Selected results from NMR and ESR studies obtained in our laboratories and elsewhere are given where appropriate.

Single-crystal x-ray studies of the contact-ion pair complexes shown in Table II have been completed. The data were measured with a Picker automated diffractometer, and the structures were solved by direct methods. Hydrogen atom positions were included in all the structures but usually not refined. Refinements of the structures were made using a full-matrix, least-squares technique with neutral-atom scattering factors.

If the wave function for the highest occupied molecular orbital (HOMO) of the complex is expressed in the form $\psi_{HOMO} = \phi_{(carbanion)} + \phi_{metal\text{-}ligand}$, $\phi_{(carbanion)}$ would not necessarily have the same symmetry as the HOMO of the isolated carbanion in the absence of the metal atom perturbations. However in all the systems we have examined, this is

Table II. Results of Single-Crystal X-ray Studies of Contact-Ion Pair Complexes

Compound		Space group	Z	Rw
$C_6H_5CH_2Li \cdot N(CH_2)_3N$	toluene	$P2_12_12_1$	4	0.083
$(C_6H_5)_3CLi \cdot TMEDA$ [a]	triphenylmethane	$P2_1/c$	4	0.068
$C_{13}H_9Li \cdot [N(CH_2)_3CH]_2$	fluorene	$P2_1/c$	4	0.073
$C_9H_7Li \cdot TMEDA$	indene	$P1$	2	0.062
$C_{13}H_9K \cdot TMEDA$	fluorene	$P2_1/c$	4	0.028
$C_{12}H_8[Li \cdot TMEDA]_2$	acenaphthylene	$Fdd2$	4	0.059
$C_9H_8[Li \cdot TMEDA]_2$	naphthalene	$P2_1/c$	2	0.057
$C_{13}H_{10}[Li \cdot TMEDA]_2$	anthracene	$P2_1/c$	4	0.068
$C_{26}H_{16}[Li \cdot TMEDA]_2$	bifluorene	$C2/c$	4	0.045
$[C_9H_8Al(CH_3)_2Na(THF)_2]_2$ [b]	naphthalene	$P2_1/c$	2	0.080
$[C_{13}H_{10}Al(CH_3)_2Na(THF)_2]_2$	anthracene	$P2_1/c$	2	0.080
$C_5H_5MgBr \cdot TEEDA$ [c]	cyclopentadiene	$Pna2_1$	8	0.083
$[C_4H_5Li \cdot TMEDA]_2$	[1.1.0]bicyclobutane	$C2/m$	2	0.108
$[C_6H_5CH{=}C_6H_5CH] \cdot [LiTMEDA]_2$	trans-stilbene	$C2/m$	2	0.061

[a] TMEDA = tetramethylethylenediamine.
[b] THF = tetrahydrofuran.
[c] TEEDA = tetraethylethylenediamine.

the case. Structural results for the contact-ion pair are therefore useful in determining the symmetry properties of the HOMO of the carbanion.

As the first example consider the geometry of the neutral naphthalene molecule (29) compared with that of naphthalene in [Li(TMED)]-naphthalene (Figure 1) (30). The changes in two of the bond lengths are very significant with a decrease of 0.072 A in the 2-3 bonds (1.415 to 1.343 A) and an increase of 0.051 A in the 1-2 bonds. The symmetry of the HOMO in the free dianion, as given by simple Hückel calculations, is also shown in figure 1.

Table III. Coefficients of HOMO (INDO)

Orbital		C_{2h} [naphthalene]$^{2-}$	C_{2h} [naphthalene]$^{2-}$ [Li(H$_2$O)$_2$]
C_1	s	0.03	0.01
	P_x	−0.10	−0.09
	P_y	−0.03	−0.01
	P_z	0.41	0.39
C_2	s	−0.03	−0.03
	P_x	−0.01	0.01
	P_y	−0.04	−0.02
	P_z	−0.24	−0.22
C_3	s	−0.03	−0.03
	P_x	−0.01	−0.01
	P_y	−0.04	−0.02
	P_z	−0.24	−0.22
C_4	s	0.03	0.01
	P_x	0.10	0.09
	P_y	−0.03	−0.02
	P_z	0.41	0.39
C_5	s	−0.05	−0.10
	P_x	−0.04	−0.04
	P_y	0.00	0.00
	P_z	0.01	0.02
C_6	s	−0.05	−0.10
	P_x	0.04	0.04
	P_y	0.00	0.00
	P_z	−0.01	−0.02
C_7	s	0.03	0.01
	P_x	0.10	0.09
	P_y	0.03	0.02
	P_z	−0.42	−0.39
C_8	s	−0.03	−0.03
	P_x	−0.01	−0.01
	P_y	0.04	0.02
	P_z	0.24	0.22
C_9	s	−0.03	−0.03
	P_x	0.01	0.01
	P_y	0.04	0.02
	P_z	0.24	0.22
C_{10}	s	0.03	0.01
	P_x	−0.10	−0.09
	P_y	0.03	0.02
	P_z	−0.42	−0.39

3. STUCKY Stereochemical Properties of the Naphthalene Dianion

C_{2h} 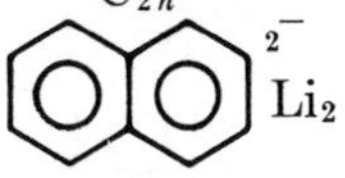 $\begin{array}{c}2^-\\ \text{Li}_2\end{array}$	D_{2h} HUMO 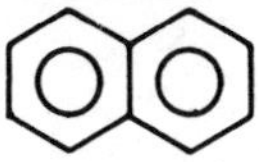
0.00	0
−0.08	0
−0.01	0
0.36	0.42
−0.02	0
0.01	0
−0.02	0
−0.21	−0.28
−0.02	0
−0.01	0
−0.02	0
−0.21	0.28
0.00	0
0.08	0
−0.01	0
0.35	0.42
−0.11	0
−0.04	0
0.00	0
0.01	0
−0.11	0
0.04	0
0.00	0
−0.01	0
0.00	0
0.08	0
0.01	0
−0.35	−0.42
−0.02	0
−0.01	0
0.02	0
0.21	0.28
−0.02	0
0.01	0
0.02	0
0.21	0.28
0.00	0
−0.08	0
0.01	0
−0.35	−0.42

Table III.

Orbital		C_{2h} 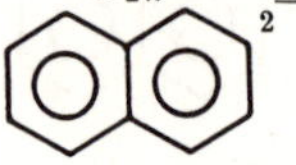 naphthalene^{2-}	C_{2h}  naphthalene^{2-} [Li(H$_2$O)$_2$]
H$_1$	s	0.00	−0.02
H$_2$	s	0.06	0.05
H$_3$	s	0.06	0.05
H$_4$	s	0.00	−0.02
H$_5$	s	0.00	−0.02
H$_6$	s	0.06	0.05
H$_7$	s	0.06	0.05
H$_8$	s	0.00	0.02
Li$_1$	s		0.10
	P$_x$		0.03
	P$_y$		−0.20
	P$_z$		−0.01
B$_1$	s		0.04
	P$_x$		−0.02
	P$_y$		0.01
	P$_z$		0.02
B$_2$	s		−0.04
	P$_x$		−0.02
	P$_y$		0.01
	P$_z$		0.03
B$_3$	s		−0.04
	P$_x$		0.02
	P$_y$		−0.01
	P$_z$		−0.03
B$_4$	s		0.04
	P$_x$		0.02
	P$_y$		−0.01
	P$_z$		0.03

Note particularly the bonding contribution to the 2-3 bond and the antibonding contribution to the 1-2 bond. The differences in bond lengths suggest that the isolated carbanion symmetry is retained in the complex. This bonding feature of the [Li(TMED)]$_2$naphthalene complex was further confirmed by INDO molecular orbital calculations (31, 32); the results, in more detail, are given in Table III for both the free carbanion and naphthalene fragment in [LiB$_2$]$_2$naphthalene. Because of computer storage limitations, NH$_3$ or H$_2$O was used to simulate the coordinated amine ligand. The symmetry of the HOMO of the free carbanion is B$_{2u}$(D$_{2h}$) or B$_u$(C$_{2h}$). The dianion in the contact-ion pair is distinctly

Continued

C_{2h}	D_{2h} HUMO
−0.03	0
0.05	0
0.05	0
−0.03	0
−0.03	0
0.05	0
0.05	0
0.03	0
0.24	
0.01	
−0.24	
0.10	

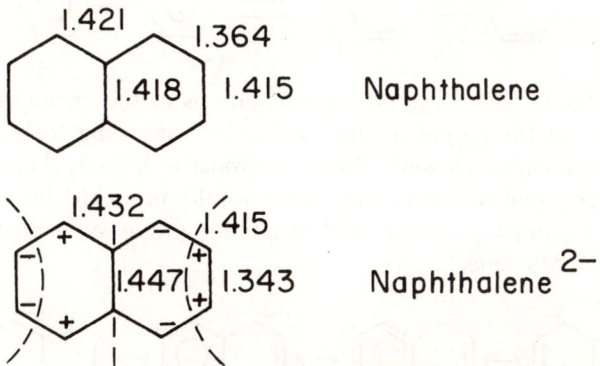

Proceedings of the Royal Society

Figure 1. Comparison of the geometries of the naphthalene molecule (29) geometry with that of the naphthalene fragment in (LiTMED)$_2$naphthalene

nonplanar, and the correct representation is B_u in the point group C_{2h}. The presence or absence of coordinated base does not affect the ordering of the carbanion energy levels although it significantly affects the amount of lithium s character used in the HOMO. The net charge for the naphthalene group in the monolithio compounds is $-0.23e^-$; in the dilithio complex the net charge for the naphthalene group is $-0.31e^-$. This suggests that bond orders are not greatly different in the two reduced naphthalene fragments.

Walsh and Pearson (33, 34) have suggested that information about the point-group symmetry of the excited state of a neutral molecule can often be ascertained from the anion's geometry. The first excited state of a molecule containing n electrons should belong to the same point group as the ground state of a similar molecule having $n + 1$ or $n + 2$ electrons. The extra one or two electrons are assumed to be in that molecular orbital that becomes occupied in the excited state. For example, adding electrons to a planar aromatic ring should cause some atoms to bend out of the plane and the two-electron reductions of benzene would give the dianion of cyclohexadiene:

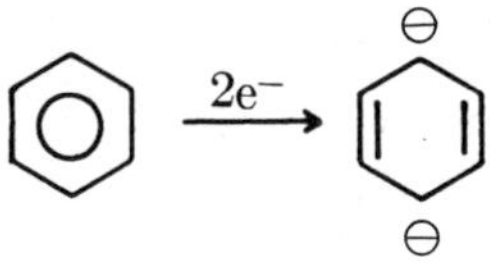

Similarly, for biphenyl:

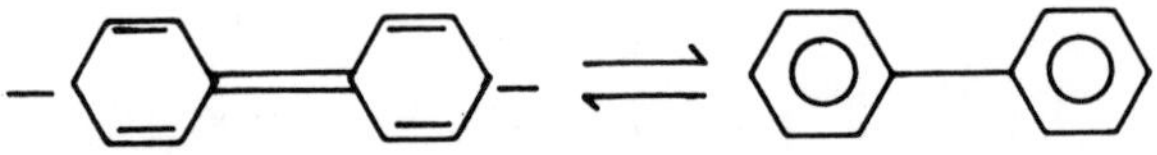

These should then be good representations of the geometries of the excited states of the parent neutral molecules, according to Walsh's and Pearson's arguments. However in most aromatic hydrocarbon anions, a single-valence bond-resonance structure would probably be enough to describe the geometry and predict the point group symmetry—for example, for naphthalene,

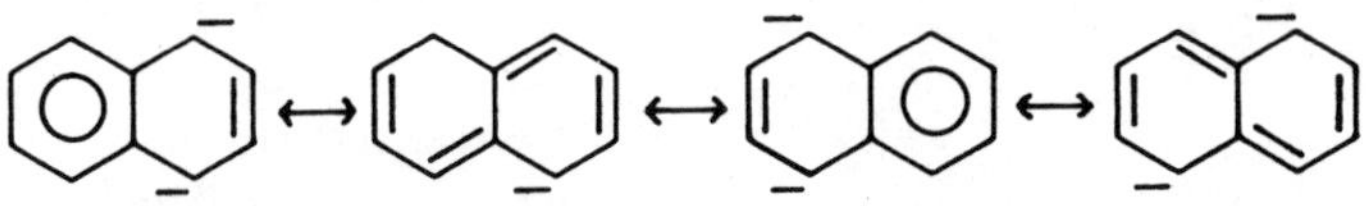

The predicted geometry for the first excited state of naphthalene has been calculated by Fujimura, Yamaguchi, and Nakajima (35); see Table IV. The calculations do not take into account the possible nonplanarity of the excited state molecule nor are electron–electron repulsions expected to be the same for the first excited state and the dianion. However similarities in the trends of the bond lengths found for naphthalene (TMEDLi)$_2$naphthalene and calculated for the B_{2u}(excited state of neutral naphthalenes are encouraging (Table IV).

The structure observed for the naphthalene group in [Li(TMED)]$_2$-naphthalene can best be described in terms of a delocalized molecular orbital (MO) picture

Table IV. Naphthalene Bond Lengths

	1–2	1–9	2–3	9–10
[a] $^1B_{1u}$	1.399	1.421	1.430	1.457
[a] $^1B_{2u}$	1.427	1.417	1.382	1.427
[b] Dianion	1.444(8)	1.432(9)	1.343(8)	1.447(10)
[b] Naphthalene (B_{2u})	1.364(5)	1.421(5)	1.415(5)	1.418(5)

[a] Theoretical values from Ref. 35.
[b] Observed values from Ref. 35.

with predominant contributions from the two valence-bond structures. An alternative resonance form of naphthalene more closely describes the geometry of the naphthalene fragment in $[Na(C_4H_8O_2]_2[Al(CH_3)_2$-$C_{10}H_8]_2$ (36, 37). This reagent is readily prepared by this reaction:

$$2Al(CH_3)_3 + 2Na + THF + \text{naphthalene} \rightarrow Na(THF)_2[Al(CH_3)_2\text{naphthalene}]$$

The geometry of the complex is shown in Figure 2, and the relevant dimensions of the naphthalene fragment are given in Figure 3. These bond distances and the puckered nature of the six-membered ring (on the right, above) suggest a predominant valence bond structure of

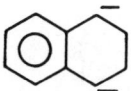

Before leaving the geometry of naphthalene anion, possible implications of the above to the structural properties of solvated alkali metal contact-ion pairs in solution are considered. Details of naphthalene anion geometry in solution are unavailable, and inferences from spectroscopic studies must be relied on. Attempts have been made to fit the hyperfine coupling constants of alkali metal–organic-radical anion systems by varying the structural parameters of the metal complex (38, 39). We observe that the molecular geometry found for the dianion should be closely related to that of the monoanion since the same molecular orbital is used with two- and one-electron occupation, respectively. In this framework the results obtained by Pedersen and Griffin for the hyperfine coupling constants of the protons of the naphthalene radical monoanion are given in Table V.

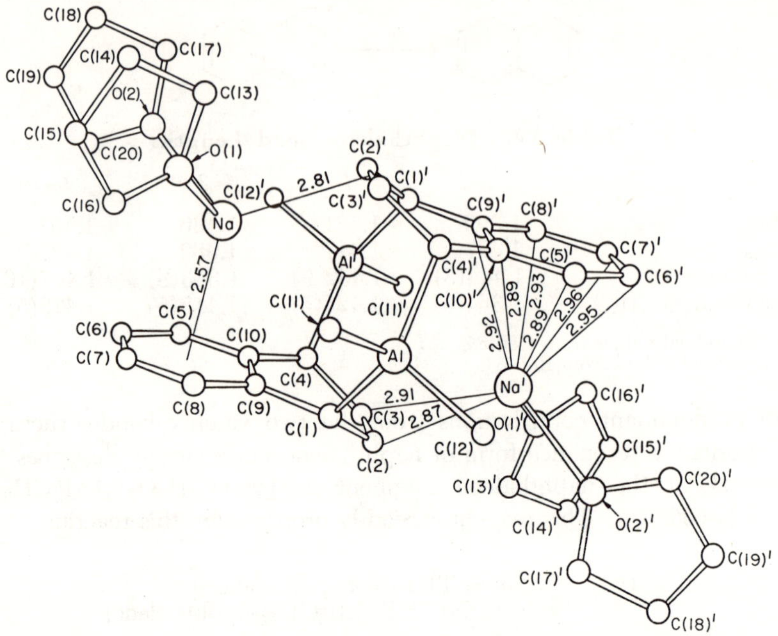

Journal of the American Chemical Society

Figure 2. Molecular geometry of $[Na(C_4H_8O)_2]_2[(CH_3)_2AlC_{10}H_8]_2$ (37)

Table V. INDO Hyperfine Coupling Constants *(30)* for the Napthalene Radical Anion

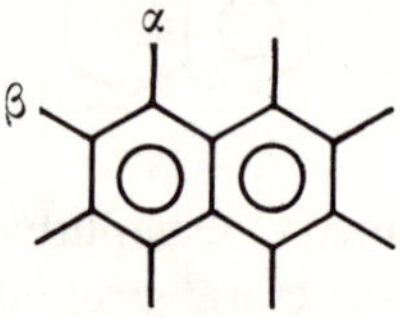

	Calculated				
	Free anion		*Complex*		*Obsd*[a]
	[b]	[c]	[c]	[d]	
Li			0.19	0.22	0.1–0.4
α	−5.3	−5.8	−5.5	−5.3	4.9
β	−0.88	1.60	1.72	−0.86	1.83

[a] Refs. *40, 41.*

[b] Assuming planar D_{2h} symmetry, with bond lengths as determined in the crystal structure of $[LiTMEDA]_2$ naphthalene and C—H distances of 1.08 A.

[c] Naphthalene geometry as observed in the crystal structure of $[LiTMEDA]_2$ naphthalene and C—H distances of 1.08 A.

[d] Ref. *38.*

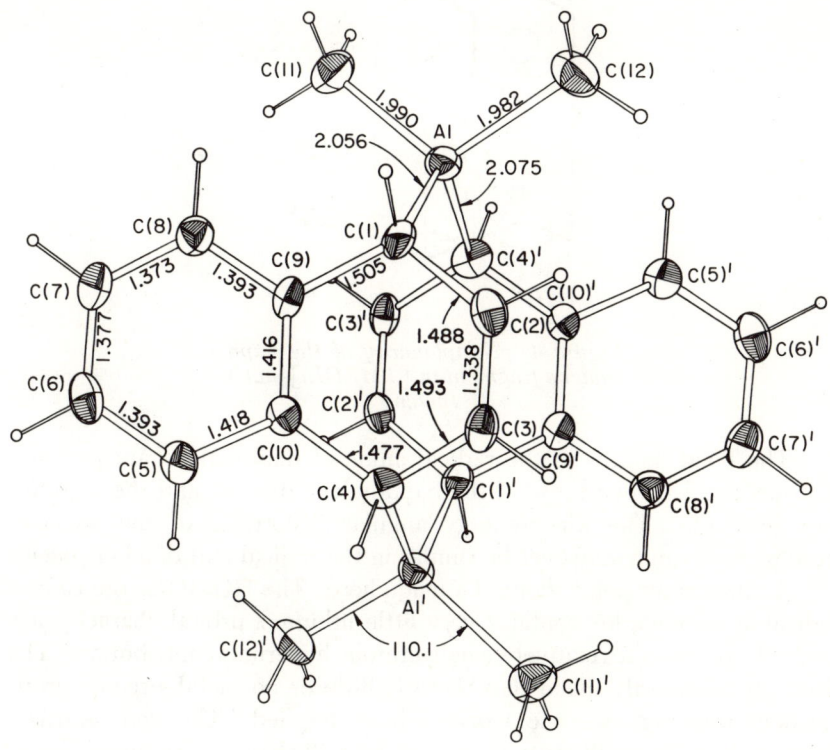

Journal of the American Chemical Society

Figure 3. Geometry of the naphthalene fragment in $[Na(C_4H_8O_2)_2]_2[(CH_3)_2\text{-}AlC_{10}H_8]_2$ (37)

The observed values of the hyperfine coupling constants (*40, 41*) are in the last column of the table. Pedersen and Griffin used the geometry of the neutral naphthalene molecule to calculate proton hyperfine coupling constants. The first column lists the proton hyperfine coupling constants calculated using a planar geometry for the isolated naphthalene anion, but the bond lengths are taken from the crystal structure of the dianion. The results of Pedersen and Griffin are not significantly changed. In fact, the naphthalene molecule is definitely nonplanar in naphthalene [LiTMED]$_2$, as shown by Figure 4. When these angular distortions observed for the dianon are included, very large changes (282%) are observed in the β coupling constant from -0.88 to 1.60 *vs.* an observed value of 1.83). Including the lithium atom in the position found in the crystal structure further increases algebraically the magnitude of the hyperfine coupling constants of the α and β protons in the direction suggested by the data.

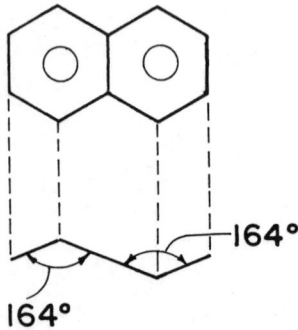

Figure 4. Nonplanarity of the naphthalene fragment in $[TMEDLi]_2$naphthalene

The bond lengths in the radical anion of naphthalene are probably intermediate between those of the naphthalene dianion and the naphthalene molecule. The lower-energy angular distortions of the molecule from planarity may however be similar in the radical and dianion species.

An important point should be made here. The HOMO of the contact radical anion complex contains very little lithium s orbital character and correctly predicts a relatively small lithium hyperfine contribution. This does not necessarily mean that there is little or no metal–organic group covalent bonding as some workers have implied. The comparatively large lithium $2p$ coefficients in the HOMO (Table III) suggests in fact that overlap of the lithium $2P$ orbital with the π-carbanion-orbital may indeed be significant.

Several approximations are involved in these calculations. We assumed, for example, that the position of a lithium–base group with respect to the naphthalene molecule is the same in the radical anion as one of the lithium base groups in the dianion, and that the base group exchanges rapidly on the ESR time scale from one ring to the other. The ring hyperfine coupling constants are then taken to be the average of those calculated for the two static structures. Also, no attempt was made to take into account any effects of solvent-separated ⇌ contact equilibria. The distortions in the naphthalene radical anion contact-ion pair may not be the same as those of the solvent-separated species. However, the experimental proton hyperfine coupling constants of the naphthalene radical anion are relatively independent of the counter ion or solvent used. The point remains that it is almost certainly incorrect to ignore angular distortions resulting in nonplanarity of the parent aromatic system when interpreting ESR data, as has been done by all workers in the field until now. Moreover, the nonplanar molecules are good models for delocalized π-carbanions in reactions with electrophilic reagents, and even in solvent-separated systems solvent dipolar interactions may cause significant distortion from planarity.

The structure of the anthracene molecular fragment in [LiTMED]$_2$-anthracene is given in Figure 5 (*42*). The dashed lines indicate the modal points in the HOMO. The signs of the changes in bond distances again are all correctly predicted by the symmetry of the HOMO. However, the magnitudes of the changes are somewhat less than in [LiTMED]$_2$naphthalene, as expected since the charge is delocalized over 18 nuclear centers in naphthalene and 24 in anthracene. Assuming equal delocalization over each enter, the bonding and antibonding contributions to a given bond in anthracene should be about 75% of those in naphthalene. The change in length of the 2,3 bond going from naphthalene to naphthalene^{2-} is 0.072 A, while the corresponding change in anthracene is 0.049 A (or 68%) of that in naphthalene. The only other similar bond in the two systems is the 1,2 bond, for which there is a 62% factor. The analogy is crude since the electron populations vary from one atom pair to another.

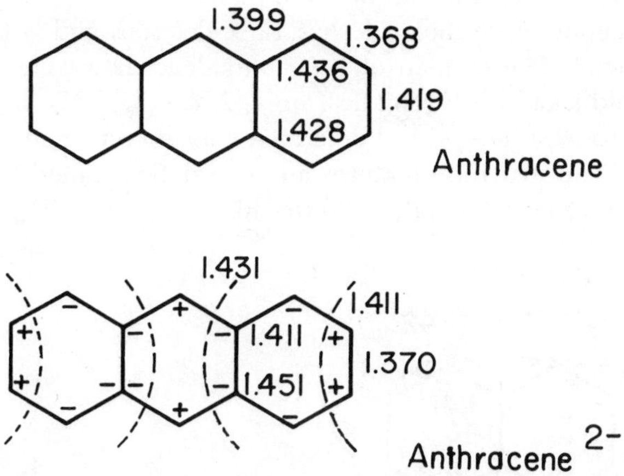

Anthracene

Anthracene^{2-}

Proceedings of the Royal Society

Figure 5. Comparison of the geometries of the anthracene (29) molecule and the anthracene group in [TMEDLi]$_2$anthracene

Like the naphthalene dianion, the anthracene dianion is distinctly nonplanar (Figure 6). However, there is some indication from the results obtained for this complex that the angular distortions observed in both the naphthalene and anthracene dianions may be caused by interactions of the unsaturated carbanion with the LiTMED group. There are three preferred sites in anthracene to which the lithium atom is most likely to

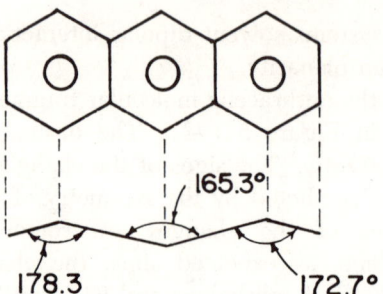

Figure 6. Nonplanarity of the anthracene fragment in [TMEDLi]$_2$anthracene

coordinate—over the central six-membered rings, and the two positions slightly to the outside of the centers of the outer two rings. In the crystal structure, the two lithium atoms are located over the central six-membered ring and over one of the outside six-membered rings. These two ring systems display the most distortion from planarity.

The acenaphthalene molecule has 12 π electrons and is thus a nonaromatic system. The dianion of acenaphthalene, however, contains 14 π electrons and should be aromatic ($4n + 2, n = 3$). We have isolated [LiTMED]$_2$acenaphthalene and determined its molecular structure (Figure 7) (43). The structural features and correlations noted for naphthalene and anthracene are equally valid for this molecule.

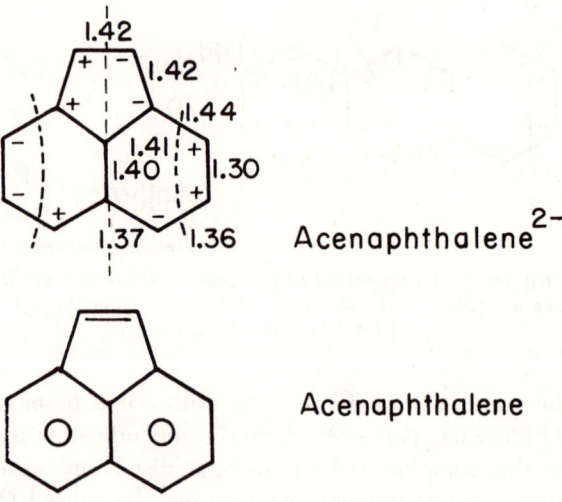

Figure 7. Molecular structure of acenaphthalene in [LiTMED]$_2$acenaphthalene

Many photochemical studies of cis-trans isomerization of stilbene have been made. Herkstroeter and Hammond (*44*) have measured rates of energy transfer to stilbenes and α-methylstilbenes from triplet sensitizers that have lower triplet energies than those of the olefins. For *cis*-stilbene, the energy-transfer rates do not fall off as fast with decreasing sensitizer energy as predicted; this and other evidence suggests that energy transfer to stilbene does not produce an excited state with the same geometry (*45–53*) as the ground-state olefin. A perpendicular twisted geometry has been suggested for this lower-energy excited state and also for the stable geometrical form of the radical anion and dianion of ethylene (*45–53*). The supporting data, however, are not strong, and a planar configuration is also possible. Simple Hückel MO considerations indicate that the highest unoccupied MO of ethylene should be antibonding with respect to the C=C bond. This implies that excitation to this MO should result in a lengthening of this bond and a lower-energy barrier to the intraconversion between cis and trans isomers *via* rotation about the C=C bond of the anion. The molecular structure of the anion in (TMEDLi)$_2$stilbene shows that all the atoms and the phenyl groups of the dianion are coplanar. There appears to be, however, some crystallographic disorder in the system, and other details are uncertain. The lithium atom may be an important factor in determining a planar sterochemistry, and related systems are being studied.

One such molecule is bisfluorenylidene (Structure I, below), an example of a sterically hindered olefin that on the basis of its chemistry, has been suggested to have partial diradical character. The unsubstituted bisfluorenylidene was first reported (*54, 55*) to exist in a folded configuration (Structure II). However, recent studies have shown that the biplanar molecule possesses the geometry shown in Structure III with a long C=C distance of 1.39 A (*56*).

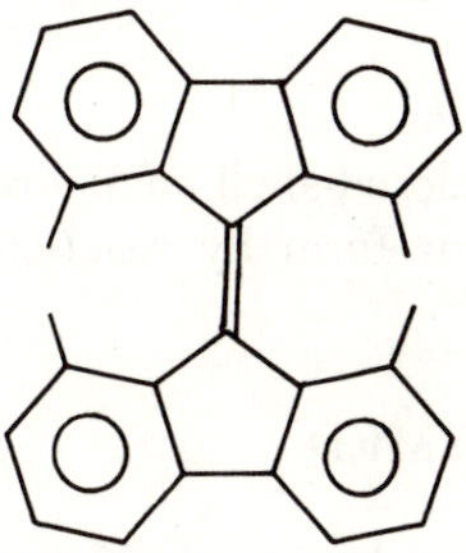

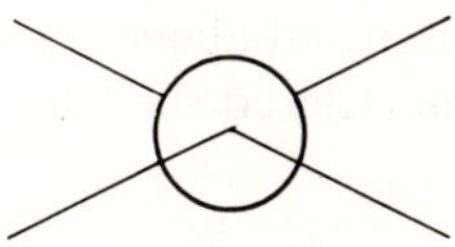

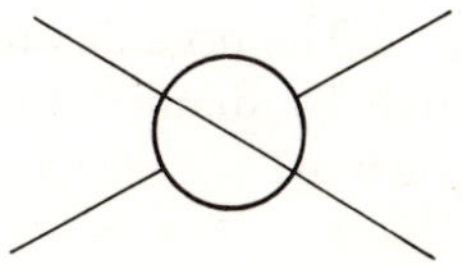

The molecular structure of the trans-substituted complex 1,1′-bisisopropoxycarbonyl-9,9′-bisfluorenylidene (*57*) is consistent with the chemical interpretation of diradical character in that the C=C bond length is 1.39 A and that there is a dihedral angle of 52° between the two "unfolded" fluorenyl planes (Structure III). The differences in bond lengths in the neutral (see, 57) and dianion species (Figure 8) follow the symmetry predicted by CNDO II calculations (43). The C=C bond length is increased to 1.493 A, or 0.1 A greater than that in the neutral compound. The alternating bond lengths around the periphery of the fluorenyl rings coincide with the bonding–antibonding character of the HOMO of the dianion. The dihedral angle between the mean fluorenyl planes of the dianion is not very different from that of the parent molecules—that is, 52° in 1,1'-bisisopropoxycarbonyl-9,9'-bisfluorenylidene, 48° in the dianion, and 42° in bisfluorenylidene.

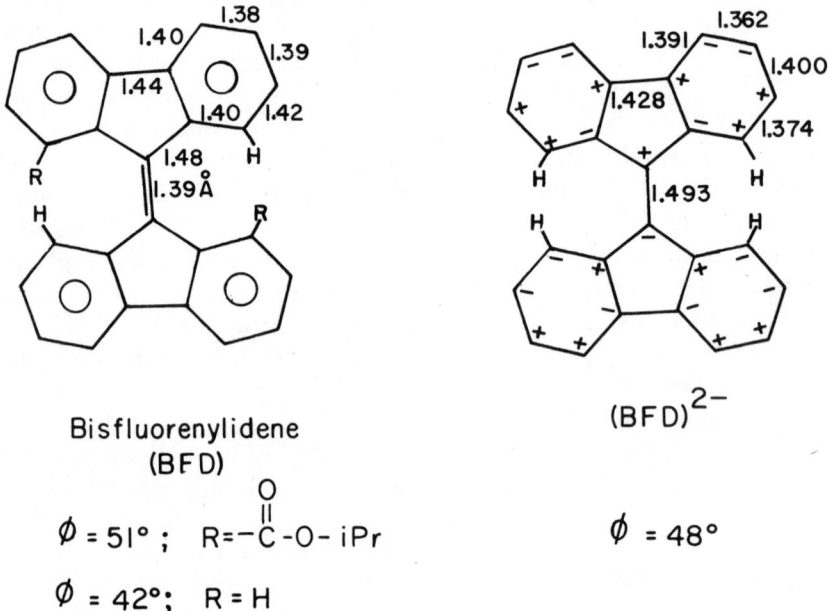

Figure 8. Molecular structure of the bisfluorenylidene fragment in [TMEDLi]$_2$-bisfluorenylidene

The organic anions discussed above achieve a closed-shell configuration by dianion formation. Closed-shell monoanions form by reactions such as:

$$R_3CH + n\text{-BuLiTMED} \rightarrow R_3C^-Li^+TMED$$

The simplest example of this type of system is the benzyl derivative, $C_6H_5CH_2LiB_2$, where B is a coordinated tertiary amine. Two structural possibilities for the $C_6H_5CH_2$ moiety are a sigma system in which the lithium atom replaces one of the toluene protons and a delocalized π anion. The geometry of the benzylic anion (58) proves that the latter is a more correct representation:

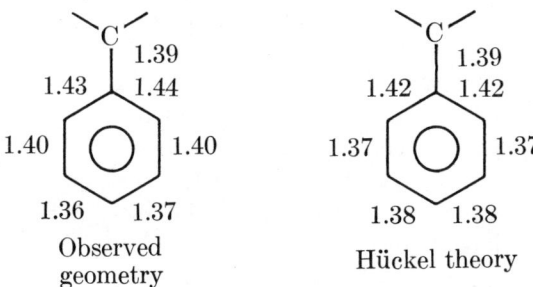

Observed geometry

Hückel theory

In fact, there is a good correlation with the bond orders predicted by Hückel theory for the anion. The triphenylmethyl carbanion shows the same delocalized character with some deviations resulting from the presence of the lithium atom (59). Each phenyl ring of the triphenylmethyl carbanion is planar (within experimental error) although only one benzyl fragment is planar. The C(phenyl)—C(methyl)—C(phenyl) angles are 117.0(6)°, 122.8(7)°, and 118.3(6)°, respectively, indicating the sp^2 character of the methyl carbon atom. Maximum π stabilization should occur when the phenyl groups are coplanar with the methyl carbon atom. However, it is sterically impossible for the triphenylmethyl group to attain a planar configuration, and two models for the twisted geometries have been proposed (60). One is the symmetrical propeller geometry in which each phenyl ring is twisted by some angle θ from the mean plane of the four central carbon atoms. The other model is the nonsymmetrical form in which one of the three rings is twisted in a direction opposite to the other two. A recent study by Hoffmann (61) and co-workers, based on extended Hückel calculations, predicts an equilibrium twist angle of 25° ± 2° for the triphenylmethyl carbanion. The triphenylmethyl carbanion in [LiTMED](C_6H_5)$_3$C has the propeller geometry with average twist angles for the three phenyl rings of 31.7°. This agrees with Hoffmann's theoretical value, particularly considering the fact that a C(methyl)—C(phenyl) bond length of 1.50 A was used in the calculations. However, the experimental twist angles for the three phenyl rings are not equal; they vary according to their interaction with the lithium cation with values of 19.7°, 30.6°, and 44.8°. The degree of π-electron delocalization between the benzylic carbon atom and the phenyl groups should depend on the twist angle, with most delocalization occurring when the phenyl ring is coplanar with the mean plane of the four central carbon atoms of the carbanion. Increasing π delocalization is, in fact, suggested by C(methyl)—C(phenyl) bond distances of 1.448(9), 1.462(13), and 1.488(10)A for twist angles of 19.7, 30.6, and 44.8°, respectively. The C(methyl)—C(phenyl) bond distance predicted by Hückel theory is 1.43 A for a planar D_{3h} system. This distance should increase with increasing twist angle as found and also as noted for the olefenic C=C bond distance in the structure of the substituted bisfluorenylidene. An increase in the C(methyl)—C(phenyl) bond order should be accompanied by a decrease in the bond order of ring C—C bonds involving C(phenyl). The average distance for bonds of this type in benzyllithium (58) is 1.44(1) A, while the corresponding distances in $(C_6H_5)_3C^-$ are 1.418(9), 1.409(10), and 1.399(9) A for twist angles of 19.7°, 30.6°, and 44.8°, respectively.

Summary: Carbanion Geometries

The results cited show that ground-state geometries of the neutral delocalized molecules are not good models for the corresponding anion. With aromatic species—naphthalene, anthracene, etc.—a significant loss of aromatic character has been observed. ESR results for the napththalene monoanion suggest that distortions from planarity may be a significant structural feature for aromatic anion systems even in solution. The interaction of the carbanion with its counter ion undoubtedly plays an important role in the extent of this distortion. Nevertheless, energy barriers to such distortions are greatly reduced from those in the parent aromatic hydrocarbon, and models for reactions of electrophiles with their systems should take this effect into consideration.

The symmetries of the carbanion molecules that we have studied are consistent with the symmetries of the HOMO's predicted by Pople's INDO theory (*31*). The net effect on bond lengths depends on the extent of delocalization in the carbanion and the atomic coefficients of the HOMO of the carbanion. Even in relatively large molecular systems sizable and chemically significant geometrical changes may occur when the cation or anion forms, when the HOMO is primarily localized on 10 or fewer atoms. An example of this is the metal-porphyrin systems. Changes in the metal–electron distribution result in geometric changes that promote the allosteric behavior of hemoglobin proteins (*62, 63*). As this review notes, oxidation or reduction of the porphyrin ligand could also produce important geometric changes.

The chlorophyll cation radical is a most likely candidate for the active light-gathering site of photosynthesis, and a π cation radical has been proposed as the active species for the iron porphyrin system in catalase and horse-radish peroxidase (*64–66*). Two ground states observed for metal porphyrin cation systems are $^2A_{2u}$ and $^2A_{1u}$ (67, 68). The former is characterized by spin density on the meso-carbon and nitrogen atom while the $^2A_{1u}$ state should have a large density (about 60%) at the carbon atoms adjacent to the nitrogen atoms. For both these electronic states, the density associated with the HOMO's is localized over a small number of atomic centers, and, particularly in the $^2A_{2u}$ state, significant changes might occur in the geometric configuration about the metal atom when the porphyrin cation is formed.

Metal–Base Coordination and Metal–π-Carbanion Geometry

The propensity of atoms to achieve a rare-gas configuration has resulted in useful generalizations that can be often applied to transition and representative element systems. The application of these rules to organoaluminum chemistry is particularly interesting in, for example, the series [$(CH_3)_3Al$]$_2$, $(CH_3)_2AlC_5H_5$, [$(CH_3)_2AlCl$]$_2$, and [$(C_6H_5)_2AlC \equiv CC_6H_5$]$_2$.

The trimethylaluminum dimer attains a closed-shell configuration only in a molecular orbital sense with electron-deficient Al—C(bridge) bonds of bond order 1/2 (69, 70). The gas-phase structure of $(CH_3)_2$-AlC_5H_5, as recently determined by Drew and Haaland (71), is shown in Figure 9. The C_5H_5 group is not pentahapto but trihapto, and Haaland has probably correctly interpreted this to indicate that only three electrons of the C_5H_5 group are needed for the Al atom to achieve an inert-gas configuration. The Al atom in [$(C_6H_5)_2Al \equiv CC_6H_5$]$_2$ (72) and [$(CH_3)_2AlCl_2$]$_2$ (73) uses all of its low-energy orbitals and achieves a closed-shell configuration in yet another way. The structure for [$(C_6H_5)_2$-$AlC \equiv CC_6H_5$]$_2$ indicates that the π electrons of the acetylenic fragment are used to form a four-center, eight-electron system. In [$(CH_3)AlCl_2$]$_2$ (73), the observed geometrical features are consistent with normal Lewis

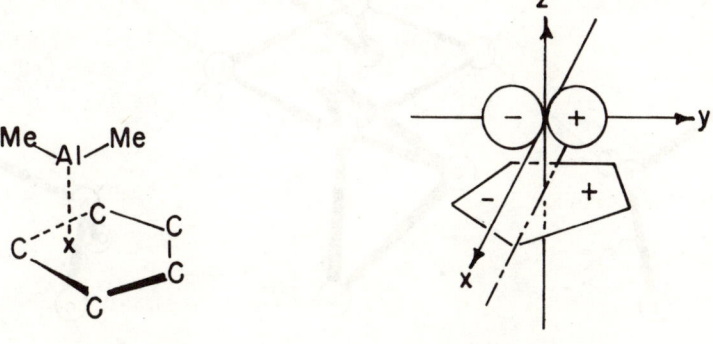

Chemical Communications

Figure 9. Molecular structure of $(CH_3)_2AlC_5H_5$ (71)

pair Al—Cl bonds and each chlorine atom acting as a three-electron donor. Similar examples can be quoted for beryllium compounds—for example, $[Be(CH_3)_2]_n$ (74), $[BeCl_2]_n$ (75), $CH_3BeC_5H_5$ (76), and $[CH_3Be(C \equiv CCH_3)N(CH_3)_3]_2$ (77). An octet of electrons about the beryllium atom in $CH_3BeC_5H_5$ can be achieved when one electron is contributed by the methyl group and five by the cyclopentadienyl radical —that is, the Cp group is pentahapto rather than trihapto, as in $CpAlMe_2$. This geometry actually has been observed. $[Be(N(CH_3)_2)_2]_3$ (78) is an example of yet another way in which a metal atom can achieve co-ordinative saturation. Here, a filled p orbital of the nitrogen atom in a dimethylamino group is used to form a dative π bond with the empty p orbital of a sp^2 hybridized beryllium atom so that both the nitrogen and beryllium atoms are coplanar with their other coordinated atoms and with each other.

It would not be too surprising if similar bonding considerations were to apply to lithium since it is a neighbor of beryllium and a second-row element. Certainly the structures of $(LiCH_3)_4$ and $[LiC_6H_{11}]_6$ (79) are consistent with the bonding descriptions for Be and Al as is the stereo-chemistry of the solvated lithium complex [TMEDLi–bicyclobutyl]$_2$ (80) (Figure 10). This configuration is almost exactly analogous to the four-center, electron-deficient geometries of the Group IIa and IIIa

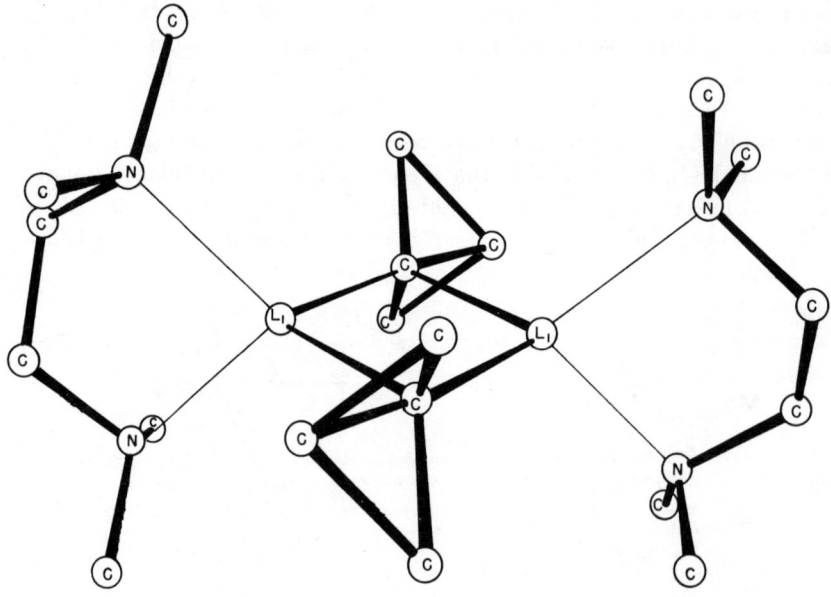

Chemical Communications

Figure 10. Molecular structure of TMEDLi[1.1.0]bicyclobutyl (80)

metals. The molecule thus provides an example of a dimeric electron-deficient organolithium compound with a saturated tertiary carbon atom in a bridging position.

In compounds containing unsaturated organic groups, several bonding modes and resulting sterochemical configurations are possible. For example, benzyllithium–(tertiary amine) could exist as an electron-deficient bridged system; a bridged system, but not electron-deficient, in which the benzyl π electrons are used to form normal two-center, two-electron bonds; a monomeric σ complex in which the lithium atom occupies one of the coordination sites of the metal hydrogen atoms in the parent toluene molecule; and a π-carbanion with the lithium cation coordinated to the most electron-rich portion of the carbanion. The remainder of this review deals with specific details of metal-base and metal–π-carbanion interactions in unsaturated organometallic complexes.

Metal–Base Interactions

Almost without exception, the preferred coordination number of lithium is 4 (*81*). As noted elsewhere in this paper, the catalytic behavior of RLi(base)$_n$ is apparently optimized with $n = 2$. We find in fact that it is difficult not to isolate compounds with lithium/base ratios of 1:2. Complexes such as RLi(base)$_{3\text{ or }4}$ are frequently less stable thermally than are complexes of RLi(base)$_2$.

General features of the lithium atom's coordination are probably best shown in compound A, $C_6H_5CH_2Li(N(CH_2)_6N)$ (*59*) and compound B, $C_{13}H_9Li(NC_7H_{13})_2$ (*83*). In both compounds two monodentate tertiary amine ligands are coordinated to each lithium atom. Details of the geometry about the lithium atom in compound A are presented in Figure 11. The N—Li—N angles are 118.6° in compound A and 123.7° in compound B. Furthermore, when a vector, V, is drawn from the lithium atom to the point of closest approach to the mean plane of the carbanion, the N—Li—V angles are also almost 120°—for example, 119.5° and 121.5° in compound A. In short, geometrically, the lithium atom can be considered to be sp^2 hybridized and three-coordinate, with two nitrogen atoms occupying two coordination sites and the delocalized carbanion in the third position.

The structural features of the bidentate chelate $(CH_3)_2N(CH_2)_2$-$N(CH_3)_2$(TMED) require a considerably smaller N—Li—N angle. The expected direction of the nitrogen lone pair electrons of chelated TMED can be calculated when the lone-pair distribution is assumed to be coincident with the mean vector obtained by vectorially averaging the two N—CH$_3$ vectors and the N—CH$_2$ vector of the ligand. A survey of N—N distances in ethylenediamine complexes suggests that the N—N

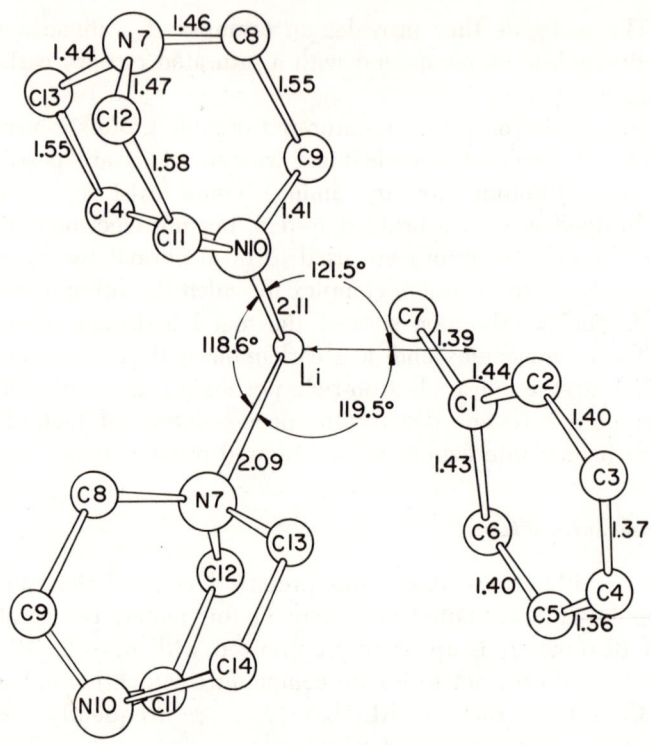

Journal of the American Chemical Society

Figure 11. Molecular geometry of $C_6H_5CH_2Li(N(CH_2)_6N)$ (58)

distance is relatively constant—about 2.9 ± 0.1 A. This implies that the lone-pair orbitals of the two nitrogen atoms make an angle of 85° ± 3° with each other. As Figure 12 shows, effectiveness of TMED as a chelating group for larger atoms such as potassium must be considerably less than for lithium since the nitrogen atom lone pairs cannot be directed effectively toward the potassium atom. This agrees with our observation that the TMED group in $C_{13}H_9KTMED$ (84) is not chelated but acts as a bridge for two potassium atoms.

With respect to the metal–base interaction, one function of the base apparently is as a variable electron source that can compensate for electron loss to the anion, thus maintaining a relatively constant charge on the metal. Table VI gives the relative carbanion stabilities as inferred from the pK_a of the most acidic proton of a number of hydrocarbons (85). The metal–nitrogen atom distance (Table VII) closely follows the observed or expected trends in anion stability. Thus, the ordering of pK_a values for the lithium compounds, $C_6H_5CH_3 < (C_6H_5)_3CH < C_{13}H_{10}$

(fluorenyl), is paralleled by decreasing lithium nitrogen distances. The same trend exists for Mg complexes with a pK_a order of $2CH_4 < CH_4$, $R_2NH < C_2H_6$, HBr. Finally, the changes in the aluminum complexes are particularly striking with $3CH_4 < 2CH_4$, HI < 3HCl and a corresponding overall decrease in Al—N bond lengths of 0.14 A.

In short, as the ability of an R· group (R· = halogen atom or organic radical) to accept an electron from the representative metal atom increases, the metal–base distance decreases. The metal–base distance could also depend on the strength of the base. Dicoordinated $N(CH_2)_6N$ (Dabco) should be a weaker base toward lithium than monocoordinated $N(CH_2)_6CH$ (quinuclidine), and indeed the Li—N distance is greater for Dabco than for quinuclidine. However, quinuclidine should be a stronger base toward Mg than triethylamine, but the Mg-Quin distance is 0.09 A greater than the Mg-triethylamine distance. Thus, the differences in base strengths in Table VII are not great enough to differentiate strongly between metal–base interactions, and the anion effect is the predominating factor.

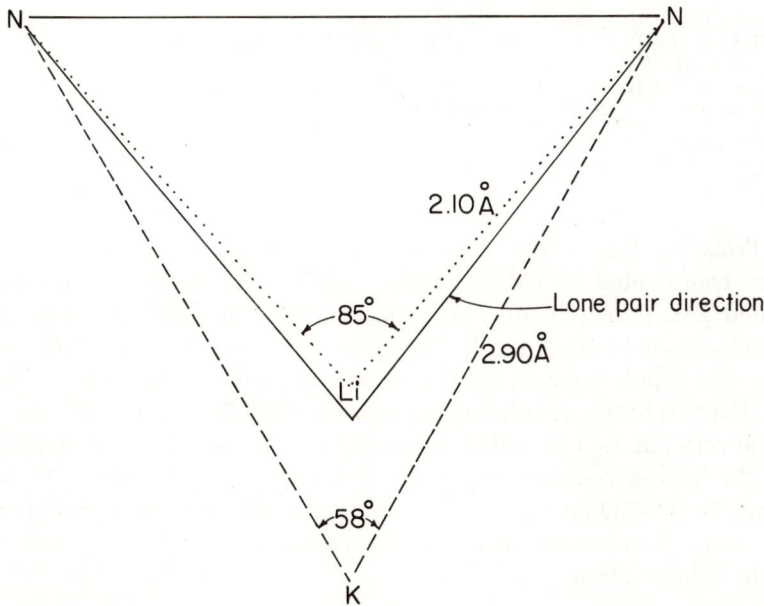

Figure 12. Hypothetical N–K–N angle for a potassium atom chelated to TMED compared with that found for TMEDLi complexes. The direction of the lone pair electrons is the mean vector of the two $N-CH_3$ vectors and the $N-CH_2$ vector for each nitrogen atom.

Table VI. Relative Carbanion Stabilities (85)

Group	pK_a of R—H
Vinyl	37
Phenyl	37
Allyl	36.5
Benzyl	35
Triphenylmethyl	31
Fluorenyl	27
Indenyl	23
Cyclopentadienyl	18

Table VII. Metal–Base Distances

Compound	Li—N(A)	Compound	Mg—N(A)
BzLiDabco[a]	2.10	$(CH_3)_2$Mg Quin[d]	2.24
$(Ph)_3$CLiTMED[b]	2.08	CH_3MgTRMED[e]	2.19
Fliquin[c]	2.03	C_2H_5MgBrTea[f]	2.15

Compound	Al—N
$(CH_3)_3Al \cdot N(CH_3)_3$[g]	2.10
$(CH_3)_3Al \cdot$Quin[h]	2.06
$(CH_3)_2AlI \cdot N(CH_3)_3$[h]	2.01
$Cl_3AlN(CH_3)_3$[i]	1.96

[a] Bz = $C_6H_5CH_2^-$; Dabco = $N(CH_2)_6N$ (58).
[b] Ph_3C = $(C_6H_5)_3C^-$; TMED = $(CH_3)_2N(CH_2)_2N(CH_3)$ (59).
[c] F = $C_{13}H_6^-$ (83).
[d] Quin = $N(CH_2)_6N$ (86).
[e] Trmed = $[CH_3N(CH_2)_2N(CH_3)_2]^-$ (87).
[f] Tea = $(C_2H_5)_3N$. Ref. (136).
[g] Ref. (88).
[h] Ref. (89).
[i] Ref. (90).

Probably the best estimate the charge distribution in a lithium cation coordinated to TMED is given via an INDO calculation of the isolated TMEDLi⁺ (Figure 13) fragment (91). The hydrogen atoms are predicted to have small residual negative charges of less than 0.01 electron. The dipole moment for this moiety is predicted to be 5.5 Debye. The INDO charge distribution in neutral TMED (Figure 13) should not be very different from that given above, with the important exception that the hydrogen atom charges are all between −0.04 and −0.06 electrons. The TMED proton chemical shifts generally agree with this charge difference. A relatively small positive charge of +0.36 e^- is predicted for the lithium atom.

Metal–π-Carbanion Interaction

Considerable theoretical and experimental effort has been expended to predict the position of the metal atom relative to the delocalized carbanion system in π-carbanion complexes. One early effort was made by

Dixon (*92*) who investigated the ^{7}Li and proton NMR of fluorenyllithium in many solvents. More recent and detailed NMR studies were done by McKeever (*93*), Okamoto and Yuki (*94*), Grutzner, Lawlar, and Jackman (*95*), and Sandell (*96*).

Grutzner's studies led to these conclusions:

1) In addition to the solvent-separated ⇌ contact ion-pair equilibrium there is a secondary process in contact ion-pair systems (such as fluorenyl) that may result from higher aggregate formation, change of degree of solvation, movement of the cation within the ion pair, or change in the vibrational structure of the ion pair.

2) $(C_6H_5)_3C^-$ exists in the symmetrical propeller form or is rapidly interconverting between such structures that will have in the same symmetrical form. The inclination angle is about 30°.

Kranzer and Sandel interpreted their NMR studies of benzyliclithium systems in diethyl ether in terms of the allylic structure:

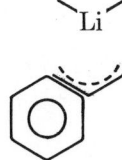

Extensive studies were also done on model systems where the metal hyperfine coupling constants were calculated as a function of both the height of the atom above the unsaturated anion plane and its horizontal position. The most recent work in this area was done by Canters, Corvaja, and de Boer (*97*), Takeshita and and Hirota (*98*), Pedersen and Griffin (*99*), and Goldberg and Bolton (*100*). Attempts also have been made to use electronic spectra to determine the role of the metal atom in these molecular complexes (*99, 100*).

We concern ourselves first with the question of the expected height of the metal atom above the delocalized carbanion plane. Figure 14 shows the structure of $C_5H_5MgBr \cdot [(C_2H_5)_2N(CH_2)_2N(C_2H_5)_2]$ (*103*), which contains a magnesium atom coordinated in a pentahapto configuration to a π-cyclopentadienyl ring. So long as the bonding forces are reasonably similar, it should be possible to find a set of empirical metal radii that would enable us to predict the metal–carbon distance relative to that in the magnesium compound for any other metal pentahapto π-cyclopentadienyl compound. Much success is attained by using the atomic radii obtained from the elemental metals (Table VIII). It is unreasonable to expect one set of radii to fit all oxidation states, and the values quoted are restricted to the 0, I, or II oxidation states except for Ti(IV).

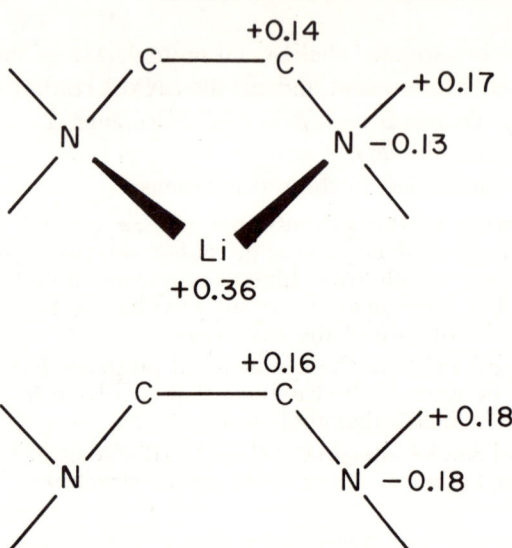

Figure 13. INDO charge distributions for TMEDLi⁺ and TMED

Table VIII. Comparison of Metal–Cyclopentadienyl Distances (103)

Metal	Formal Oxidation State	$M-\pi C^a$	$(Mg-C)-(M-C)$	Δr^b
Ti	4,2	(2.41)^c, 2.43	0.14, 0.12	0.13
V	1,2	(2.28)^d, 2.30^e	0.27^d	0.26
Nb	1	2.45^f	0.10	0.14
Cr	2,2	2.25^g, 2.22^e	0.30^g	0.33
Mo	2	2.35^h	0.20	0.21
W	1	2.36^i	0.19	0.21
Mn	1	2.17^j	0.38	0.34
Re	2	2.28^k	0.27	0.23
Fe	2	2.04^l	0.51	0.34
Ru	1,2	2.26^m, 2.21^n	0.34^n	0.26
Co	0,2	(2.07)^o, 2.13^e	0.48^o	0.35
Rh	1	2.18^p	0.37	0.26
Ni	2,2	2.20^q, 2.20^e	0.35^i	0.36
Pd	2	2.26^r	0.29	0.23
Cu	1	2.24^s	0.31	0.32
Be	2	1.50^t	0.71	0.48
Mg	2	2.55^u		

^{a} Values in parentheses are calculated from the relation $(r_{m\text{-ring}})^2 + (1.21)^2 = (r_{m\text{-}\pi_c})^2$, where 1.21 A = distance from π_c to the center of the ring when $r_{c\text{-}c} = 1.43$ A.
^{b} $\Delta r = \gamma mg - r_m$; where r_i = metallic radii with ligancy = 12 (105).
Refs. ^{c} 106; ^{d} 107; ^{e} 108; ^{f} 109, 110; ^{g} 111; ^{h} 112, 113; ^{i} 114; ^{j} 115; ^{k} 116; ^{l} 117; ^{m} 118; ^{n} 119; ^{o} 120; ^{p} 121; ^{q} 122; ^{r} 123; ^{s} 124; ^{t} 125; ^{u} 103; ^{v} 126.

Only for two systems is there a large deviation in the observed metal–carbon distance relative to magnesium from the distance predicted. In both cases the observed metal–ring distance is much shorter than expected. The first exceptions are the iron and cobalt families and the second are the beryllium compounds $(C_5H_5)_2Be$ *(104)* and $C_5H_5BeCH_3$ *(76)*. The deviation for ferrocene and cobaltocene is not particularly surprising. Ferrocene is a remarkably stable organometallic compound, and relatively strong metal ligand covalent bonding is commonly given as the reason for this unusual stability. $(C_5H_5)_2Be$ has an interesting structure *(104)*. Gas-phase electron diffraction results and a large dipole moment (2.46 Debye) are consistent with a structure where the beryllium atom is 1.50 Å from one C_5H_5 ring and 2.00 Å from the second. The 1.50 Å value is peculiar not only to $(C_5H_5)_2Be$ but is also the ring–metal distance in $C_5H_5BeCH_3$. This distance is about 0.25 Å less than that predicted from an extension of the metallic radii and the C_5H_5Mg distance. This suggests that there is a relatively strong covalent contribution to the bonding in unsaturated beryllium compounds.

No structural data are available on Group Ia cyclopentadienyl compounds. Nevertheless, there is some indication that similar considerations

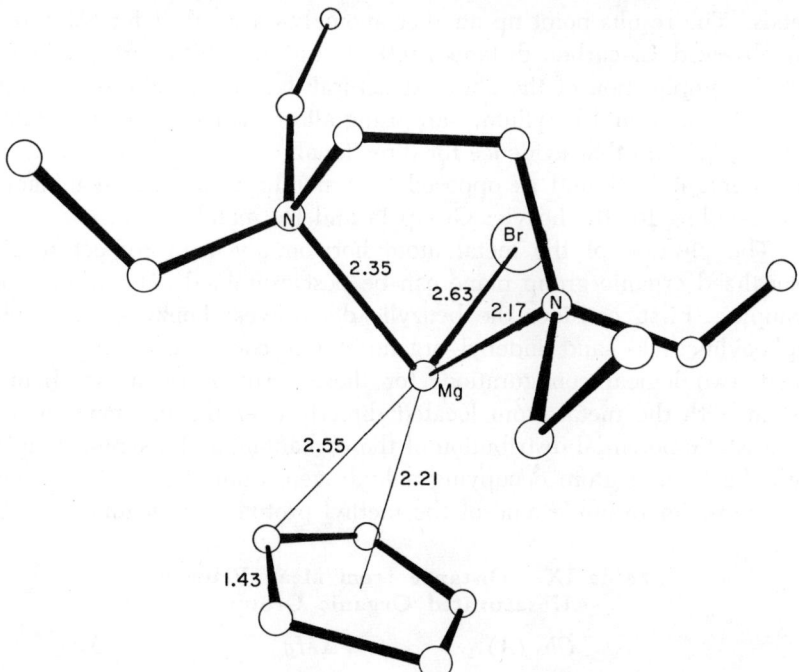

Journal of Organometallic Chemistry

Figure 14. Molecular structure of $C_5H_5MgBr[(C_2H_5)_2N(CH_2)_2N(C_2H_5)_2]$
(103)

may apply to other carbanion systems. Figure 15 gives the chromium–aromatic distances in the structure of $(CO)_3CrC_6H_6$ and compares the sodium atomic radius with that of Cr. The difference in the Cr–aromatic distance in $(CO)_3CrC_6H_6$ (2.25 A) and Na–aromatic distance in $[Na(THF)_2][(CH_3)_2Al)naphthalene]$ (2.90 A) agrees with the differ-

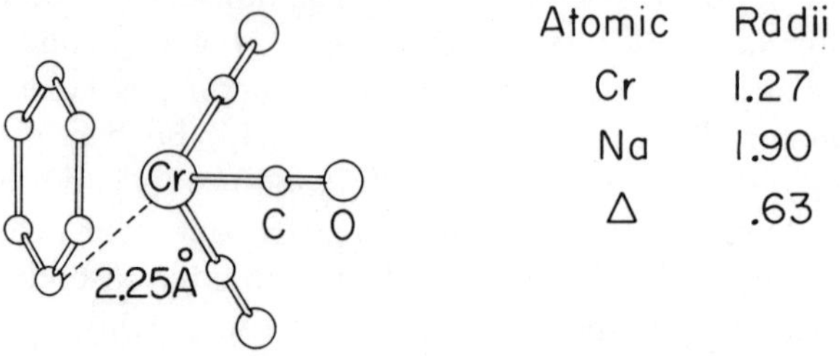

	Atomic Radii
Cr	1.27
Na	1.90
Δ	.63

Journal of the American Chemical Society

Figure 15. Molecular structure of $(CO)_3CrC_6H_6$ and comparison of atomic radii of Na and Cr (127)

ence in the metallic radii of Na and Cr. Table IX gives the average disolvated metal to π-carbanion distances we have observed for Group Ia metals. The results point up another anomaly—a good fit for Na and K but observed Li–carbon distances 0.19 A less than those are predicted.

The implication of the above structural data is that the second-row metals—lithium and beryllium—are using all of their low-energy orbitals ($2s$ and $2p$) and that evidence for directional, covalent bonding of these metals might be found as opposed to a nondirectional, predominantly ionic bonding for the heavier Group Ia and IIa metals.

The position of the metal atom horizontally with respect to the delocalized organic group plane can be best examined *via* a number of examples. First, consider the benzylic derivatives: benzyl–, fluorenyl–, triphenylmethyl–, and indenyl–organolithium compounds. As already noted, two logical configurations for these systems are a π-carbanion system with the metal atom located directly over the minimum in the electrostatic potential distribution of the carbanion, and a sigma complex with the lithium atom occupying a hydrogen atom site in the organic precursor—for example, one of the methyl proton sites in toluene. The

Table IX. Distance from Mean Plane of Unsaturated Organic Group

	Obs (A)	ΔMg	Δr
Mg	2.21	—	—
K	2.95	−0.74	−0.75
Na	2.55	−0.34	−0.30
Li	2.00	0.24	0.05

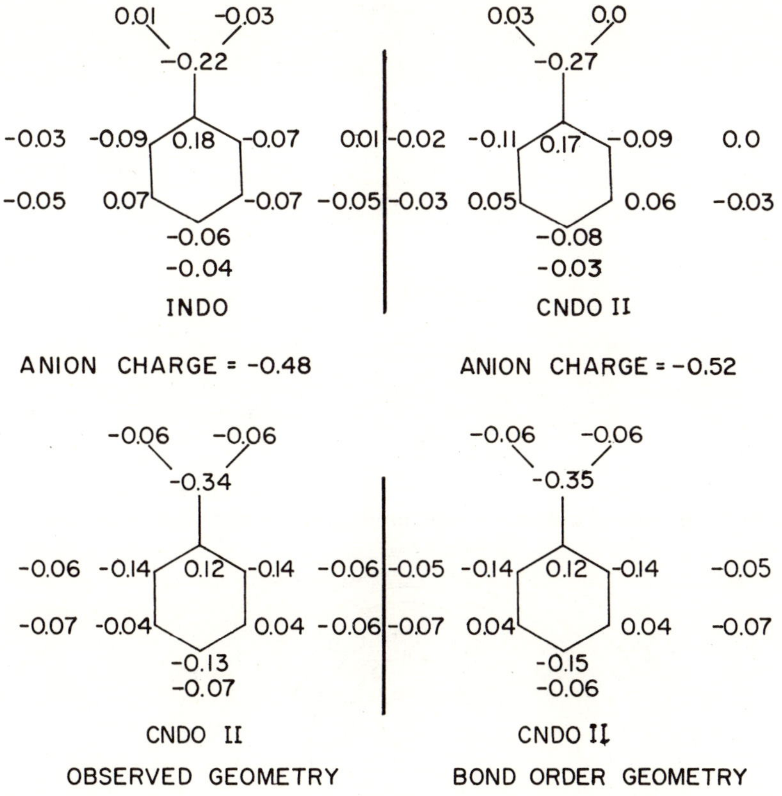

```
       0.01   -0.03                    0.03    0.0
          \   /                          \    /
          -0.22                          -0.27
           |                              |
-0.03 -0.09/0.18\-0.07   0.01|-0.02 -0.11/0.17\-0.09    0.0
-0.05  0.07\    /-0.07  -0.05|-0.03  0.05\    /0.06    -0.03
           -0.06                          -0.08
           -0.04                          -0.03
           INDO                          CNDO II

     ANION CHARGE = -0.48            ANION CHARGE = -0.52


       -0.06  -0.06                   -0.06   -0.06
          \   /                          \    /
          -0.34                          -0.35
           |                              |
-0.06 -0.14/0.12\-0.14  -0.06|-0.05 -0.14/0.12\-0.14   -0.05
-0.07 -0.04\    /0.04   -0.06|-0.07  0.04\    /0.04    -0.07
           -0.13                          -0.15
           -0.07                          -0.06
          CNDO II                        CNDO II

     OBSERVED GEOMETRY              BOND ORDER GEOMETRY
```

Journal of the American Chemical Society

Figure 16. Net atomic charge distribution for the benzyl carbanion. The top two figures are for the metal complex, $(NH_3)_2LiCH_2C_6H_5$, and indicate a net charge on the organic group of about -0.5 electron. The bottom two figures are for the isolated benzyl carbanion (83).

sigma model can be immediately eliminated since it is inconsistent with both the position of the metal atom and the geometries of the organic groups.

CNDO/INDO estimates of the net atomic charge distribution for the benzyl and fluorenyl carbanions are given in Figures 16 and 17 (25). The electrostatic potential energy distribution at 2.0 Å above the mean plane for these distributions of point charges are shown in Figures 18 and 19. The electrostatic model predicts that the lithium atom would be located over the potential energy minima in the two carbanions—that is, on a normal to the fluorenyl plane that intersects the plane just inside the 9 position, and a normal to the mean benzylic plane that intersects the plane about 0.4 Å for $C(7)$ on the $C(1)—C(7)$ bond. In fact, the observed position of the lithium atom is about 1.5 Å from the pre-

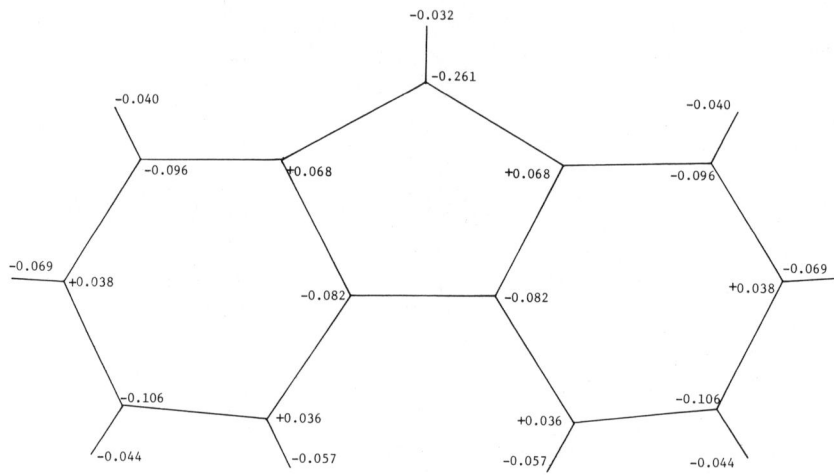

Figure 17. CNDO II charge distribution for isolated fluorenyl carbanion (83)

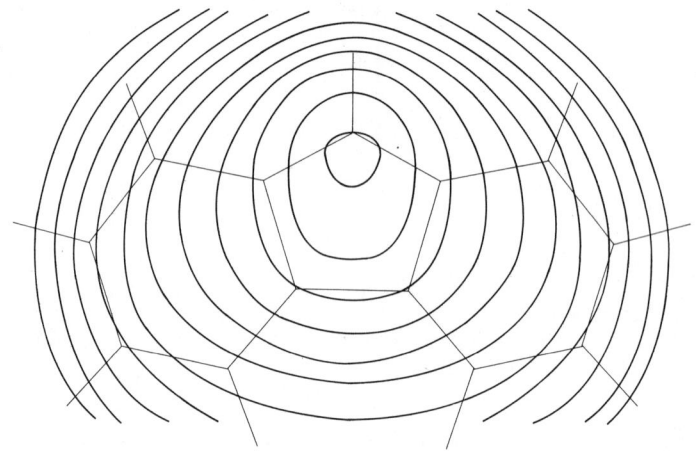

Figure 18. Potential energy surface at 2.0 A above the plane of the fluorenyl carbanion calculated from the CNDO II atomic charge distribution of Figure 17. Contour lines are drawn at levels of 0.02 eV (83).

dicted fluorenyl position and 0.75 A from the predicted benzylic position (Figures 20 and 21).

A purely electrostatic model of TMEDLi-triphenylmethyllithium suggests that the lithium atom would be located on the threefold axis of the $C(C_6H_5)_3$ group. A projection of the lithium atom position onto the mean C—C(phenyl)$_3$ plane is shown in Figure 22. In short, all the benzylic carbanion systems have the allylic geometry indicated here:

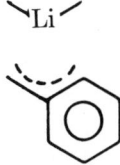

The electrostatic model predicts correctly that the closest approach of the metal will be to the most highly charged carbon atom in the carbanion and that the geometry of the organic group will approximate that of the π-carbanion. However, it does not correctly predict the allylic geometry of the complexes. In this regard a comparison of the structure of a benzylic transition metal complex (Figure 23) (*128*) with that of

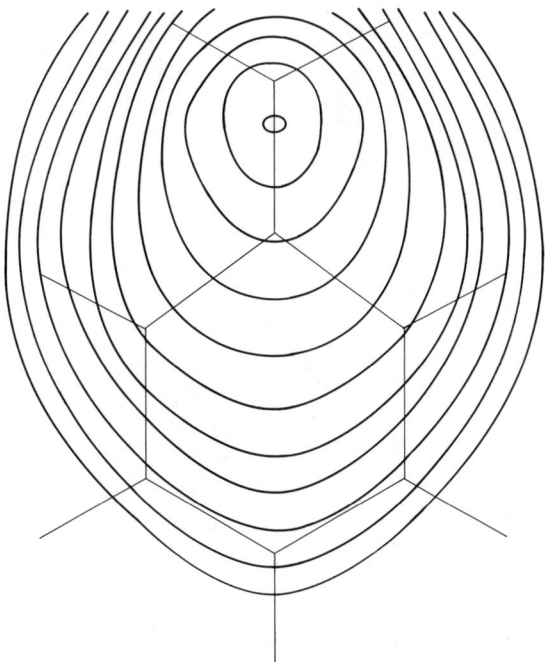

Journal of the American Chemical Society

Figure 19. Potential energy surface at 2.0 A above the mean plane of the benzyl group of the isolated benzyl carbanion and calculated from the CNDO II atomic charge distribution. Contour lines are drawn at levels of 0.02 eV (83).

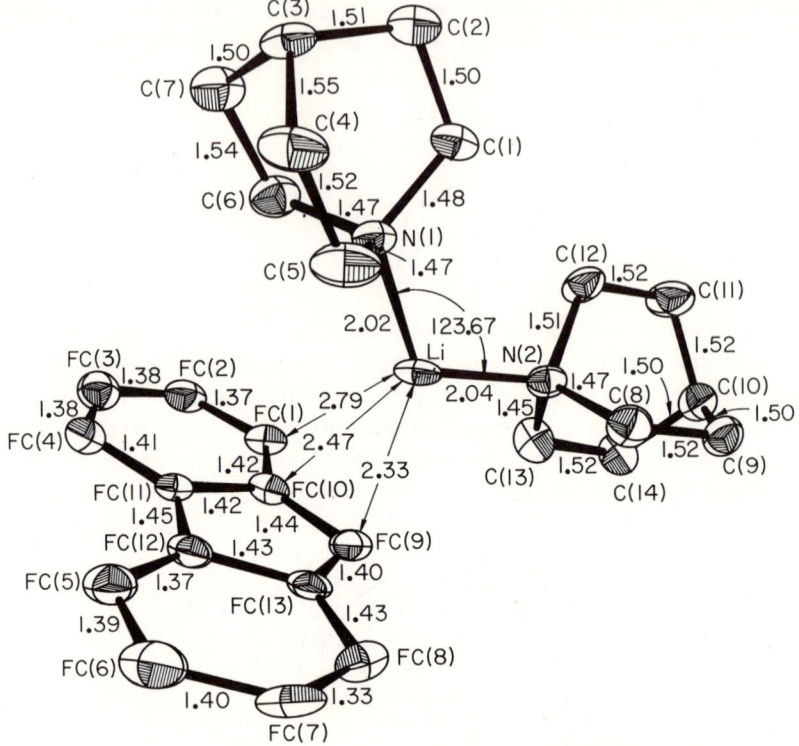

Figure 20. Molecular structure of $[CH(CH_2)_3NLi]C_{13}H_9$ (83)

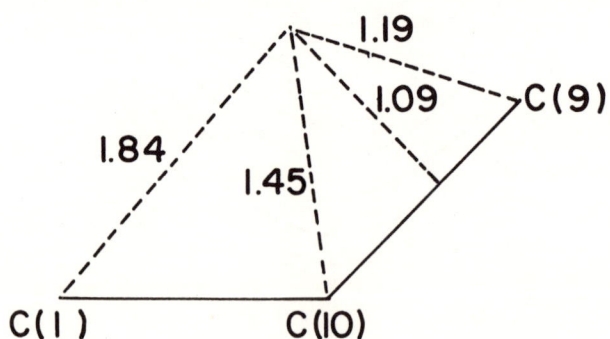

Figure 21. Projection of lithium atom position onto the mean plane of the fluorenyl carbanion. Distances in angstroms (83)

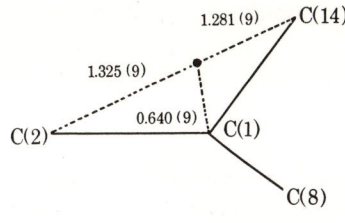

Figure 22. Projection of the lithium atom position onto the mean plane of the central four carbon atoms of the triphenylmethylcarbanion (59)

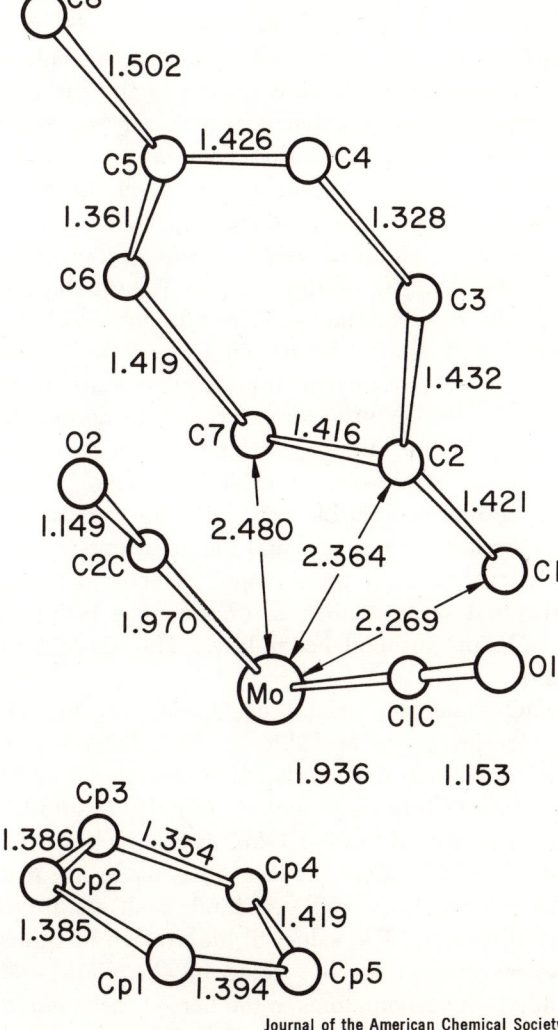

Figure 23. Molecular structure of p-methyl-π-benzyl-π-cyclopentadienyldicarbonylmolybdenum (128). Compare with Figure 24

benzyllithium (Figure 24) shows that the positioning of the two coordinated metal fragments is very similar.

The observed geometrical features can be rationalized this way. In a simple valence-bond picture the lithium atom is pictured as being sp^2 hybridized, with two sp^2 orbitals to bond with the two nitrogen atoms (the observed N—Li—N angle is about 120° for monodentate amine bases), and the remaining sp^2 orbital is coordinated to the aromatic ring. The unhybridized p orbital parallel to the ring plane can then form a bond with the π cloud of the carbanion. An additional structural requirement dictated by this model is that the N—Li—N group be oriented so that the lithium p orbital parallel to the carbanion plane can overlap favorably with the carbon p orbitals normal to the carbanion plane.

To investigate this interaction it is necessary to examine the symmetry of the molecular orbitals that contain substantial contributions from these atomic orbitals. The bond lengths in the carbanions in the structures examined are close to those expected from simple Hückel bond orders for the isolated carbanion. As noted, the carbanion π-orbital contribution to the HOMO of the complex is probably very similar to the HOMO of the free carbanions. The symmetries of the HOMO's of the free benzyl and fluorenyl carbanions and the orientation of the N—Li—N groups with respect to the carbanions are shown in Figures 25 and 26. The N—Li—N group is positioned to permit the appropriate symmetry overlap of the lithium p orbital, which is parallel to the carbanion plane and the appropriate p_z orbitals of carbon atoms in the plane.

Are these results compatible with (1) the calculated overlap integrals between the lithium p orbital and the unsaturated carbon p_z orbitals and (2) the contribution of these atomic orbitals to the HOMO of the complex? The best we can do is to examine the results of INDO and CNDO II calculations for model complexes. The results for benzyllithium are covered first.

The INDO molecular orbital coefficients for the HOMO of the benzyl carbanion are given in Table X. As indicated, the symmetry of this orbital is appropriate for overlap between the C_2 and C_7 p_z orbitals and a combination of lithium p_x and p_y orbitals. In addition, the contribution of the C_1 p_z orbital to the HOMO is relatively small. The HOMO of the complex $Li(NH_3)_2C_7H_7$ (Table X) is made up of the HOMO of the carbanion and the Li p_x and p_y orbitals with a substantially smaller proportion of lithium p_z. The values of the overlap integrals, $Li(p_x,p_y)$—$C(p_z)$ and, for comparison, the value of the $C(p_z)$—$C(p_z)$ overlap integral for adjacent carbon atoms in the benzyl carbanion are also given. Therefore, both symmetry and INDO calculations are consistent with a substantial degree of three-center carbanion–metal bonding of the type described.

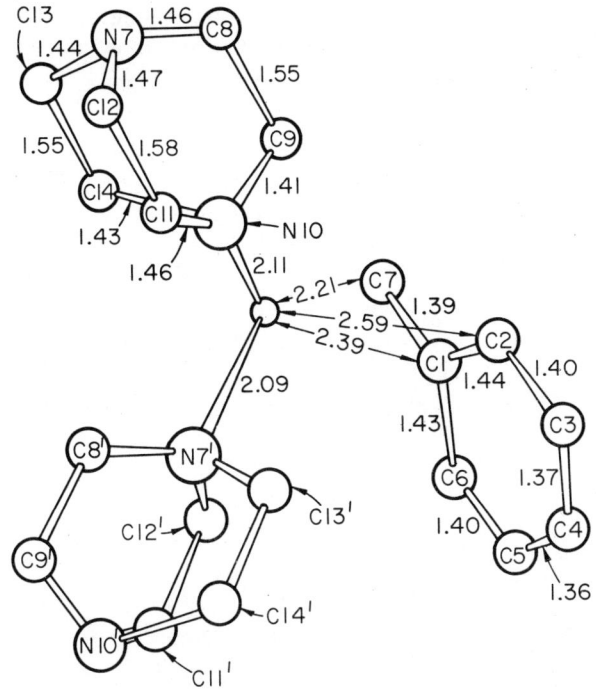

Figure 24. Molecular structure of $C_6H_5CH_2LiN(CH_2)_3N$ (58)

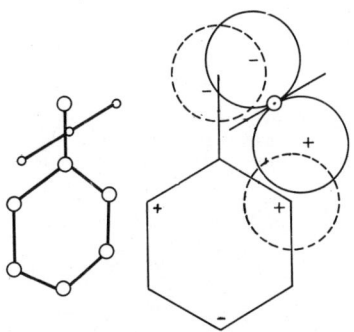

Figure 25. Symmetry of the highest occupied molecular orbital (carbon P_z atomic orbitals) of the isolated benzyl carbanion and orientation of the LiN_2 group with respect to the carbanion plane. The figure to the left is the projection as viewed along the bisector of the N–Li–N angle, while that to the right is the projection onto the mean plane of the benzyl carbanion (83).

Table X. INDO Molecular Orbital Results for the Highest Occupied Molecular Orbitals in the Benzyl Carbanion and in the Complex $(NH_3)_2LiC_7H$ (83)

Benzyl Carbanion

$$\Psi^* = 0.04p_z{}^1 - 0.41p_z{}^2 + 0.00p_z{}^3 + 0.41p_z{}^4 + 0.00p_z{}^5 - 0.41p_z{}^6 + 0.69p_z{}^7$$

Benzyl Li(NH$_3$)$_2$

$$\Psi^* = -0.17p_y{}^{Li} - 0.24p_y{}^{Li} - 0.08s^{Li} - 0.07p_z{}^1 - 0.36p_z{}^2 + 0.10p_z{}^3 + 0.39p_z{}^4 + 0.00p_z{}^5 - 0.40p_z{}^6 + 0.65p_z{}^7$$

Overlap Integrals

C—Li	S_{ab}	C—Li	S_{ab}
$p_z{}^1 - p_x$	0.11	$s^1 - p_z$	0.35
$p_z{}^1 - p_y$	0.09	$s^2 - p_z$	0.29
$p_z{}^2 - p_y$	0.02	$p_z{}^1 - s$	0.12
$p_z{}^2 - p_y$	0.14	$p_z{}^2 - s$	0.11
$p_z{}^7 - p_x$	0.13	$p_z{}^7 - s$	0.13
$p_z{}^7 - p_y$	0.07	C—C	
$s^7 - p_z$	0.39	$p_z{}^1 - p_z{}^2$	0.23
$s^7 - s$	0.32	$s^1 - s^2$	0.39
$s^1 - s$	0.28	$s^2 - p_y{}^3$	0.41
$s^2 - s$	0.25		

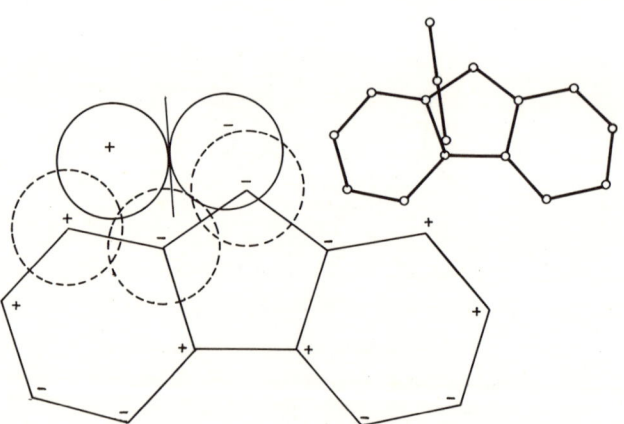

Journal of the American Chemical Society

Figure 26. As in Figure 25, but for fluorenyllithium. The projection as viewed along the bisector of the LiN$_2$ group is to the upper right. The lower-left projection is along the normal to the mean plane of the fluorenylcarbanion (83).

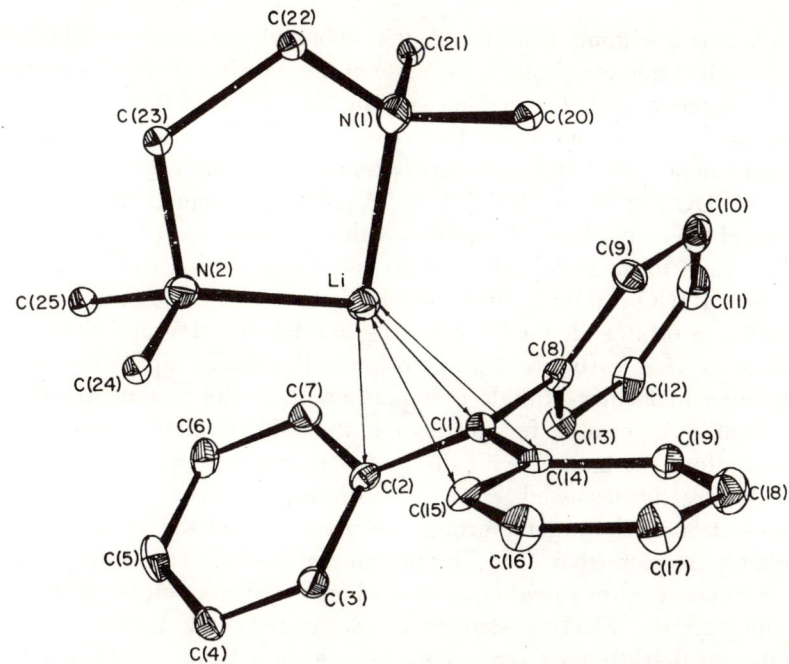

Journal of the American Chemical Society

Figure 27. Molecular structure of TMEDLi C(C_6H_5)$_3$ (59)

Table XI. CNDO Molecular Orbital Results for the Highest Occupied Molecular Orbital in the Fluorenyl Carbanion and in the Complex Li(NH$_3$)$_2$C$_{13}$H$_9$ (83)

Fluorenyl Anion

$$\Psi = -0.30p_z^1 - 0.09p_z^2 + 0.32p_z^3 + 0.13p_z^4 + 0.13p_z^5 + 0.32p_z^6 - 0.09p_z^7 - 0.30p_z^8 + 0.61p_z^9 + 0.06p_z^{10} - 0.31p_z^{11} - 0.31p_z^{12} + 0.06p_z^{13}$$

Fluorenyl Li(NH$_3$)$_2$

$$\Psi = 0.05s^{Li} + 0.17p_x^{Li} + 0.01p_y^{Li} - 0.07p_z^{Li} - 0.28p_z^1 - 0.06p_z^2 + 0.28p_z^3 + 0.13p_z^4 + 0.15p_z^5 + 0.37p_z^6 - 0.04p_z^7 - 0.33p_z^8 + 0.58p_z^9 + 0.05p_z^{10} - 0.26p_z^{11} - 0.32p_z^{12} - 0.06p_z^{13}$$

Overlap Integrals

C—Li	S_{ab}	C—Li	S_{ab}
p_z^1 — p	0.11	s^1 — s	0.22
p_z^1 — p	0.07	s^{10} — p_z	0.34
p_z^9 — p	0.12	s^1 — p_z	0.24
p_z^9 — p	0.06	p^{10} — s	0.12
p_z^{10} — p	0.02	p^1 — s	0.09
p_z^{10} — p	0.14	p^9 — s	0.13
s^9 — p	0.38		
s^9 — s	0.29		
s^{10} — s	0.27		

A corresponding analysis of the fluorenyl carbanion (Table XI) leads to the same conclusions. Because of the twisting of the phenyl rings in the triphenylmethyl carbanion and the interaction of the lithium atom with two phenyl rings instead of one (Figure 27), a similar localized similar valence bond picture of this system is not as easily described. However, CNDO results for $(NH_3)_2LiHC(C_6H_5)_2$ also suggest that the interaction of the empty lithium p orbital with delocalized π carbanion orbitals is important to the stabilization of di- and triaryl methyl complexes.

The INDO results further indicate that a σ interaction involving the sp^2 lithium orbital directed toward the unsaturated ring and the combinations of ring carbon s and p_z orbitals that make up lower energy carbanion molecular orbitals is important. It is this interaction that is apparently responsible for positioning the lithium atom closest to the carbon atom (or atoms) with the largest net atomic charge.

As already discussed, even if the disolvated lithium atom gives up one electron to the organic group, the net charge at the lithium atom is probably no more than 0.4. The lithium sp^2 orbital in a neutral lithium B_2 fragment can thus readily form a σ bond with electron transfer to the organic group. Electron density is transferred back to the empty p orbital on the lithium atom *via* the three-center bond described above.

Another way to study this model is to compare the expected electron density on the unsaturated system in the contact ion pair with that in the solvent-separated ion pair. Fluorenyllithium, one of the more ionic systems, is considered here, and the charge distributions inferred from NMR results are examined.

The proton chemical shifts for the fluorenyl carbanion (92, 129) are given in Table XII. Both $FLi(DME)_n$ in DME and $FLi(THF)_n$ in THF exist in solution as solvent-separated ion pairs (130, 131). This is consistent with the observation that the fluorenyl proton chemical shifts are nearly identical in these solvents. The 1:1 dimethoxyethane (DME) adduct in benzene probably has the same structural configuration as the 2:1 quinuclidine adduct (Figure 20) although a rapid equilibrium will

Table XII. Proton Chemical Shifts for Fuorenyl Metal Systems *(83)*

Compound	Solvent	$\mu_1{}^a$	μ_2	μ_3	μ_4	μ_9
$FLi(DME)_n$	DME	7.21	6.72	6.32	7.78	5.87
$FLi(THF)_n$	THF	7.20	6.72	6.30	7.78	5.80
$FLi \cdot 3THF$	C_6D_6	7.65	7.25	6.91	8.24	6.00
FLiDME	C_6D_6	7.80	7.36	7.04	8.35	6.27
$FNa(THF)_n$	THF	7.376	6.899	6.546	8.004	6.035
$FK(THF)_n$	THF	7.272	6.808	6.542	7.865	5.893
$FRb(THF)_n$	THF	7.247	6.806	6.441	7.815	5.884

a Chemical shifts downfield from TMS.
b F = fluorenyl

exist in solution with the lithium atom's going from the FC(1) to the FC(8) side of FC(9); cf. solution NMR studies noted earlier (96). The proton shifts for FLiDME in C_6D_6 and for FLi(THF)$_n$ in THF or FLi(DME)$_n$ then represent the two extremes of contact and solvent-separated species.

The approximation $\rho_{NMR} = \delta_{\text{benzene}}/k$ can be used where δ_{benzene} is the proton chemical shift relative to benzene, $k = 10.7$, and ρ_{NMR} = electron density at the carbon atom bonded to the proton (133).

The results, along with charges for the isolated and contact ion pair fluorenyl carbanion as calculated by CNDO methods, are given in Table XIII. The net charge predicted for carbon atoms C_1 through C_9 by the NMR method is -0.72 and -0.22 in the solvent-separated and contact ion pairs, respectively. The corresponding CNDO values are -0.52 and -0.32. A large reduction in charge upon formation of the contact ion pair is predicted in both cases. It is unlikely that this is simply the result of charge migration to C_{10}, C_{11}, or the hydrogen atoms since the CNDO results indicate a decrease in charge density at all of these positions.

The only reasonable assumption is that charge density has been transferred from the anion to the lithium atom in the contact ion pair. The three-center mechanism provides a convenient mechanism for this exchange. In agreement with this, the charge on the lithium atom in $(NH_3)_2Li$ fluorenyl is calculated by the CNDO technique to be neutral or even slightly negative, $-0.05e^-$ compared with the value expected for the isolated disolvated cation of 0.35.

Table XIII. Fluorenyl Ring Atom Charges as Estimated by the Method of Ref. 87 and CNDO II Calculation (83)

Atom	Solvent-Separated		Disolvated Contact Ion Pair			
	ρ_{CNDO}	ρ_{NMR}[a]	ρ_{CNDO}	ρ_{NMR}	$\Delta\rho_{CNDO}$	$\Delta\rho_{NMR}$
C_1	-0.10	-0.06	-0.05	-0.00	-0.05	-0.06
C_2	0.04	-0.07	0.02	-0.01	0.02	-0.06
C_3	-0.11	-0.11	-0.06	-0.04	-0.05	-0.07
C_4	0.04	-0.22	0.02	0.03	0.02	-0.05
C_9	-0.26	-0.22	-0.18	-0.18	-0.08	-0.04
C_{10}	0.07		0.07		0.0	
C_{11}	-0.08		-0.05		-0.03	
C(1)H	-0.04		-0.01		-0.03	
C(2)H	-0.07		-0.03		-0.04	
C(3)H	-0.04		-0.02		-0.02	
C(4)H	-0.06		-0.02		-0.04	
C(9)H	-0.03		0.02		-0.05	

[a] $\rho_{nmr} = \delta_{\text{benzene}}/10.7$ corrected for ring current in neighboring ring by $\delta_1 = 12.0 (a^2/R^3)$.

Our goal in this analysis is not to convince anyone of the large or small amount of covalent bonding in these compounds but to arrive at a model that can be used to predict the structural chemistry of these systems. To what extent have we succeeded? The above discussion suggests that for benzylic systems, the lithium atom is closest to the atom that has the highest formal negative charge (sigma interaction), and the N—Li—N group is positioned on or close to a nodal surface in the HOMO of the carbanion in a way that gives the appropriate symmetry for the overlap of a p orbital normal to the N—Li—N plane and the HOMO of the carbanion. The HOMO of the isolated indenyl carbanion is predicted to have a nodal plane (see Figure 28) with the maximum charge density on C(1) and C(3). A geometry with the N—Li—N plane normal to the nodal surface in the HOMO can therefore be excluded, and instead a geometry with the lithium atom approximately over the midpoint of the five-membered ring with the N—Li—N plane about parallel to the nodal plane can be predicted. In the observed geometry (Figure 28) the N—Li—N plane is within 14° of fitting this model.

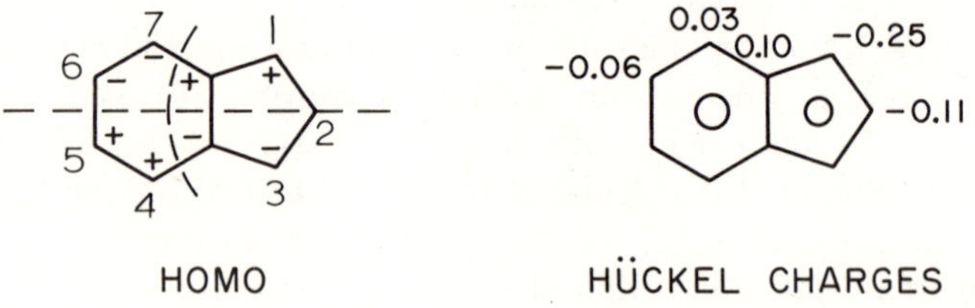

HOMO HÜCKEL CHARGES

Figure 28. Symmetry of the HOMO of the indenyl carbanion

The overlap of the sodium and potassium $3p$ and $4p$ orbitals with the carbon $2p$ orbitals should not be nearly as effective as that of the lithium $2p$. Those metals should therefore form complexes that fit the electrostatic model described in Figure 29 more closely. The potassium ion in (TMED)K fluorenyl is disolvated but not chelated by bridging TMEDA groups. The result is a polymeric system in the solid state with each fluorenyl group coordinated to two potassium atoms (Figure 30). The projection of one of the potassium atoms onto the fluorenyl plane is at a point within the periphery of the five-membered ring and also within 0.2 Å of the position predicted for the electrostatic model.

We can carry this further and ask: given a positive point charge in the position noted above for one potassium atom, where is the next minimum in the electrostatic potential—that is, where will the second potassium atom be most likely to coordinate? The lowest potential energy minimum calculated for a fluorenyl group with one interacting K atom is on the opposite side of the fluorenyl ring and vertically above the plane just to the inside of the C_{11}—C_{12} bond of the five-membered

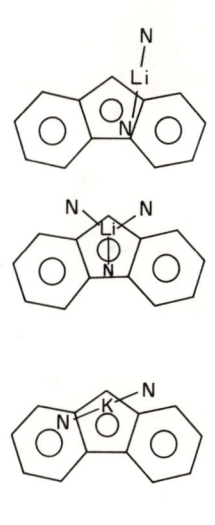

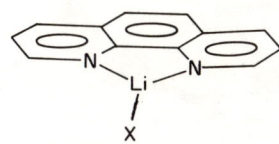

Figure 29. Factors affecting the position of an alkali metal atom with respect to the fluorenyl carbanion

ring. The observed position of the potassium atom is also within 0.2 Å of that predicted by this calculation.

The model also predicts that the position of the lithium atom should vary with solvation number. If a trisolvated lithium atom were formed, the empty metal p orbital no longer would be available, and the bonding site again would probably be that predicted by the electrostatic model (Figure 29).

The last group of complexes considered here are the closed-shell dianion organometallic complexes. One of the simplest is that of naphthalene—[(TMED)Li]$_2$C$_{10}$H$_8$. Again, an electrostatic model based on the minimum in the dianion potential energy surface has been used to predict the geometry of this complex. The INDO charge distribution is given in Figure 31, with the most negative carbon atoms being C_1, C_4, C_5, and C_8. Using the criteria developed for the monoanion systems, the lithium atom position should be adjacent to one or more of these atoms. The potential energy distribution (100) appears in Figure 32 and suggests that there is a double minimum in the electrostatic potential energy located on either side of the C(9)—C(10) bond. The observed configuration is shown in Figure 33. The lithium atom assumes a symmetrical position with respect to the ring. However, it is 2.26 Å from RC(3), 2.27 Å from RC(2), 3.32 Å from RC(1), 2.33 Å from RC(4), and 2.66 Å from

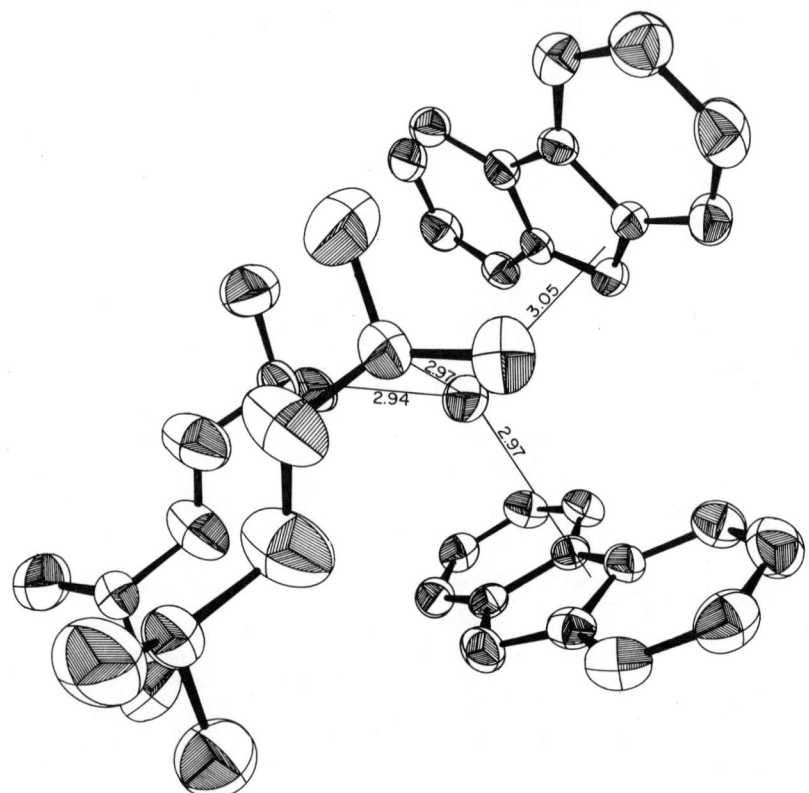

Figure 30. Molecular structure of $TMEDKC_{13}H_9$

both RC(5) and RC(5). These values show that the lithium atom is not located over the minimum in Figure 32 but instead is shifted to the outer part of the naphthalene ring and closer to the RC(2)—RC(3) bond. The chromium atom in $(CO)_3Cr$ naphthalene is also shifted to the outer edge of the naphthalene molecule (134).

The bonding of the empty p orbital of the lithium atom to the dianion molecule is not determined solely by the symmetry of the HOMO of the dianon; instead, it seems to be the result of a number of MO's of the complex. For example, at least four particular eigenfunctions involving lithium $p_{x\,y}$–carbon p_z overlap appear to be important in the naphthalene complex. These are shown at the left in Figure 34 in order of increasing stability. The top two correspond to the two HOMO's of the isolated anion and suggest that the lithium atoms should be positioned to the outside edges of the naphthalene molecule as observed. From our experience with the monoanion systems we expect that if the empty p orbital of the lithium atom were to interact with the carbon p_z orbitals, it would do so in a 1-3 or 2-4 fashion—that is, across a set of three atoms. The B_{2g}, Au, and

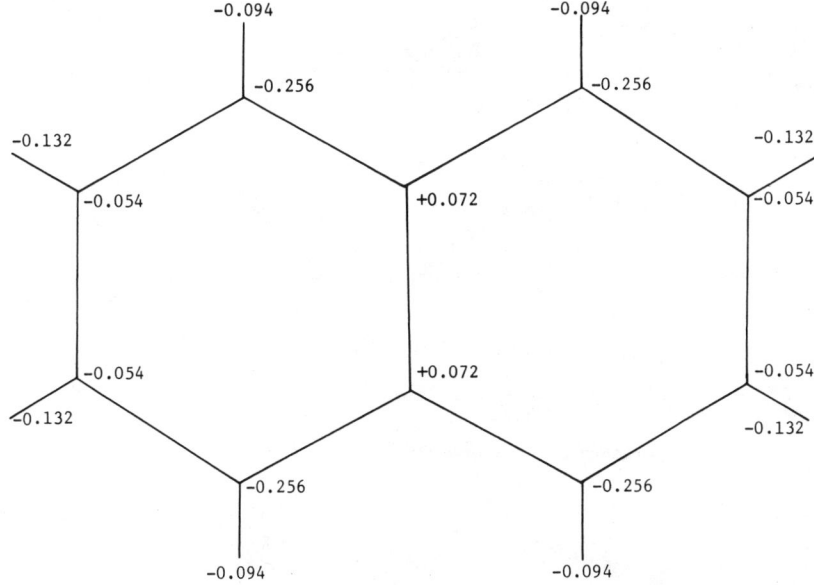

Figure 31. INDO charge distribution for the naphthalene dianion (30)

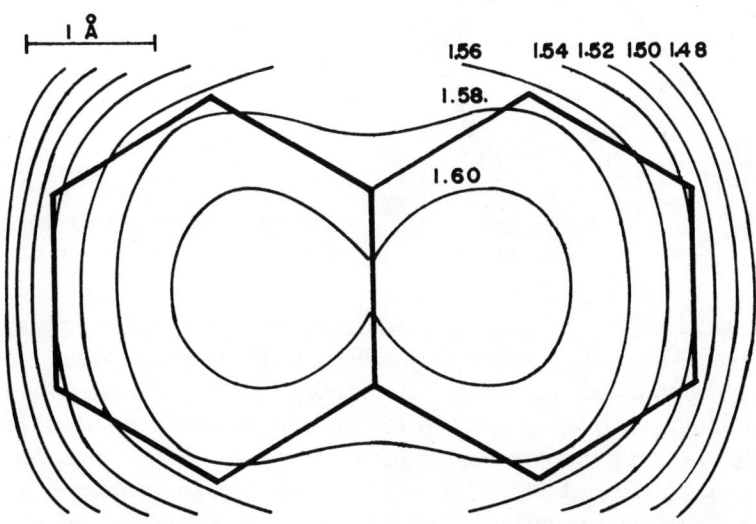

Figure 32. Potential energy distribution for the naphthalene dianion: electrostatic attraction between a positive point charge and the naphthalene anion where the positive charge is 3 Å above the nuclear plane of naphthalene. Energy in units of $\beta°_{oo'}$ (100).

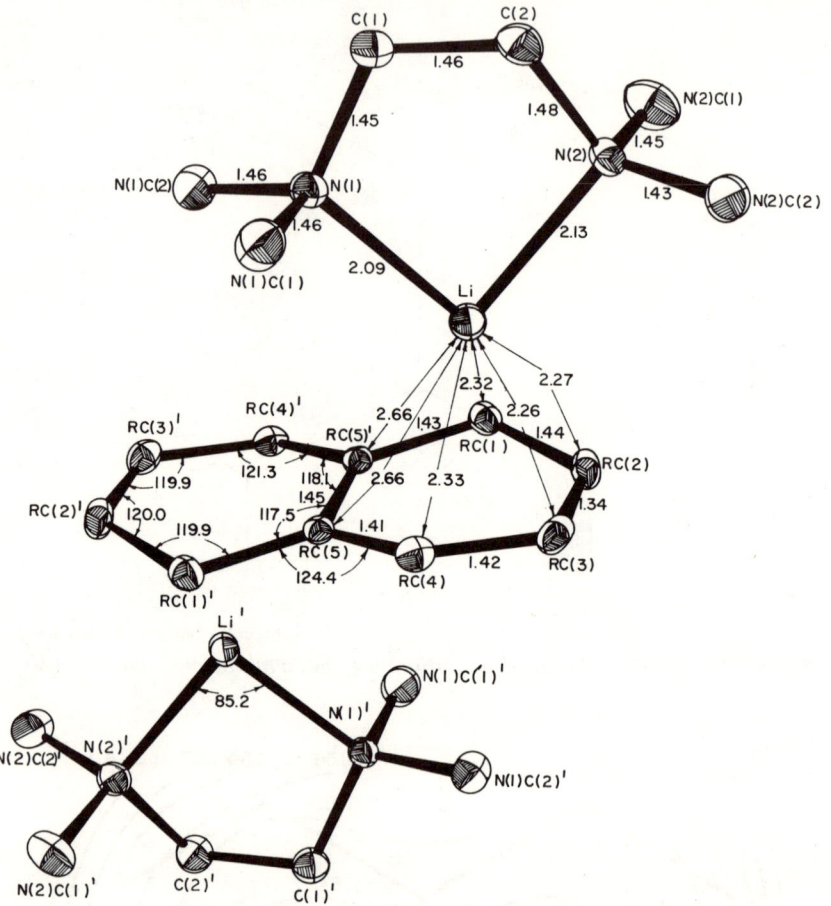

Journal of the American Chemical Society

Figure 33. *Molecular geometry of (TMEDLi)$_2$naphthalene (30)*

B_{3g} eigenfunctions in Figure 34 have the appropriate symmetry for either a 1-3 or 2-4 interaction, with opposite signs on the p_z orbitals at the 1 and 3 or 2 and 4 positions. The value of the overlap integrals calculated between the lithium p orbitals parallel to the mean carbanion plane and the C_1 or the $C_2 p_z$ orbitals are, however, about equal. (The value for the Lip_{11}—$C_1 p_z$ or the Lip_{11}—$C_2 p_z$ overlap is about 0.15, compared with a $C_1 p_z$—$C_2 p_z$ overlap value of 0.22). The Au and B_{3g} eigenfunctions, which correspond primarily to a 1,4 lithium-bridged ring, may therefore be responsible for pulling the C_1 and C_4 carbon atoms toward the lithium atom and the resulting puckering of the naphthalene dianion. The two HOMO's in Figure 34 exclude an orientation of the N—Li—N plane that is perpendicular to the long axis of the naphthalene molecule because of

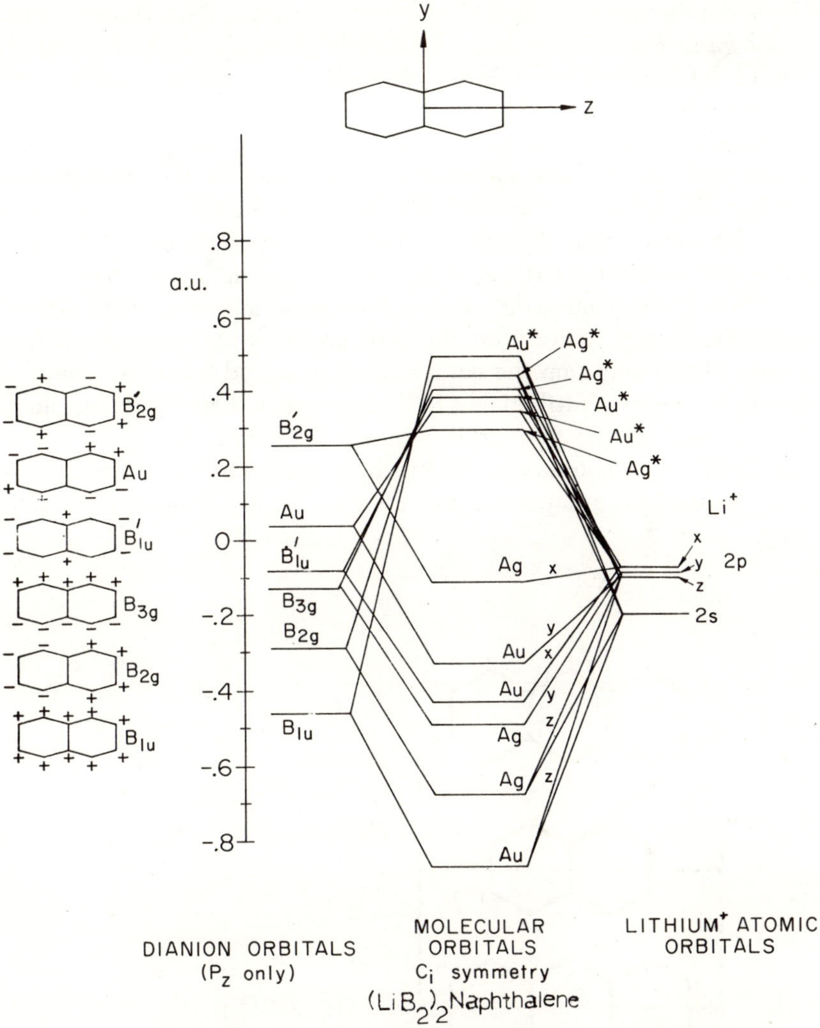

Figure 34. Partial correlation diagrams for (TMEDLi)$_2$naphthalene

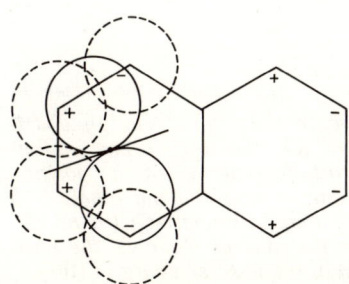

Figure 35. Orientation of N–Li–N fragment with respect to the naphthalene fragment in Li(TMEDLi)$_2$naphthalene

the nodes through C(9) and C(10) for both states. The actual orientation is shown in Figure 35 with the N—Li—N plane about 18° from the long axis of the naphthalene molecule. Nevertheless, the above analysis is still probably an oversimplification of the valence features of (TMEDLi)$_2$-naphthalene. For example, the partial correlation diagram (Figure 34) shows that considerable stabilization is obtained by the interaction of a lithium sp_z hybrid orbital with the B$_{2g}$ naphthalene molecular orbital. The latter has a node through the 9 and 10 carbon atoms and would also favor shifting the Li(TMED) fragment towards the 2 and 3 carbon atoms.

The CNDO charge distribution for the anthracene dianion places the maximum charge density on the 5-10 positions and the second largest charge distributions on the equivalent 1, 4, 6, and 9 carbon atoms. The potential energy distribution (*100*) is given in Figure 36. By analogy to naphthalene, the first Li(B$_2$) fragment should then go adjacent to the 5 and 10 carbon atoms—that is, over the central six-membered ring. The second LiB$_2$ group must then choose one of the outer six-membered rings.

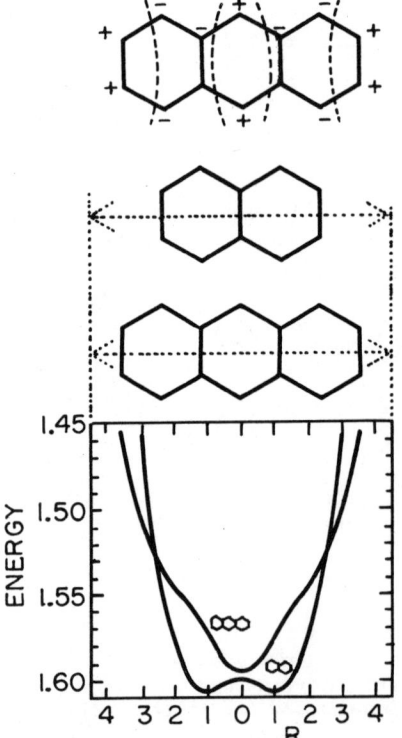

Figure 36. Potential energy distribution for the anthracene anion: electrostatic energy of a positive thracene or naphthalene anion 3 A point charge associated with an an-above the nuclear plane vs. the position of the positive charge (*100*).

Again there are four molecular orbitals (Figure 37) that can overlap with the lithium atom p orbital parallel to the mean plane of the carbanion. The symmetry of the HOMO of anthracene is similar to that of naphthalene. A covalent contribution to the bonding suggests that, as in naphthalene, the lithium atom is not exactly in the center of the middle six-membered ring but displaced toward the node in the HOMO that passes through this ring. Since the coefficients of the atomic p_z orbitals of C_{11}, C_{12}, C_{13}, and C_{14} in the HOMO of the carbanion are about half as large as the coefficients of C_2 and C_3 in naphthalene, the displacement toward these atoms might not be as large. The actual displacement in this direction is shown in Figure 38. Since the symmetry of the MO's in Figure 37 for anthracene are basically the same as those in

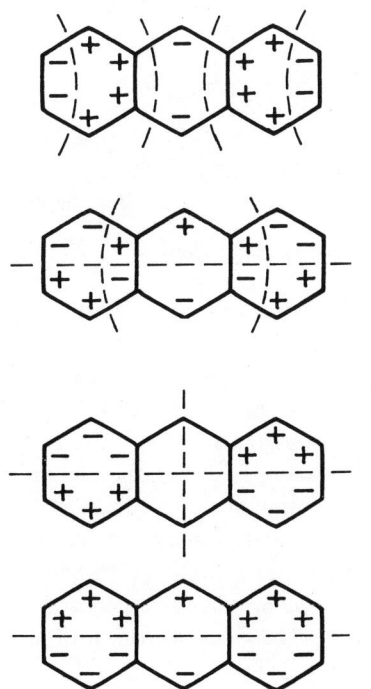

Journal of Physical Chemistry

Figure 37. Molecular orbitals for the anthracene dianion. The orbitals have the appropriate symmetry to overlap with the lithium p orbitals that are parallel to the mean plane of the dianion (100) (see Figure 34).

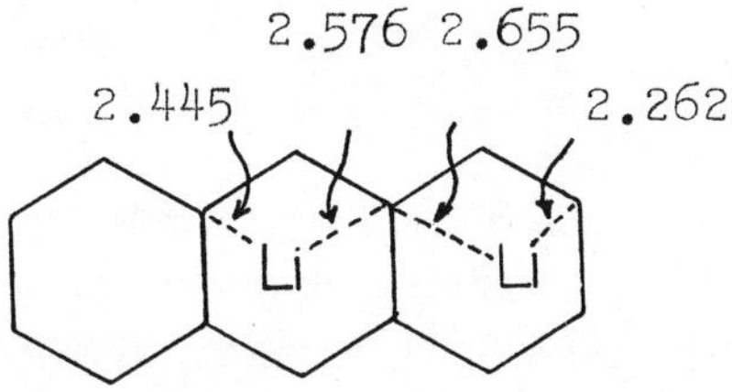

Journal of the American Chemical Society

Figure 38. Positions of the lithium atoms with respect to the anthracene moiety in [LiTMED]$_2$anthracene (30)

Figure 34 for naphthalene, the angle that the N—Li—N plane makes with a line passing through the midpoint of C_2—C_3 and C_7—C_8 should be the same in the two molecules. The actual values are 15° ± 3° for both the N—Li—N groups in anthracene, compared with 18° ± 3° in naphthalene.

The two highest HOMO's of the acenaphthalene dianion have the symmetry shown in Figure 39. There are three possible sites for the two lithium atoms:

1) over the center of the five-membered ring on either side of the acenaphthalene dianion,
2) positions similar to those in the naphthalene dianion, and
3) between the 5 and 6 carbon atoms.

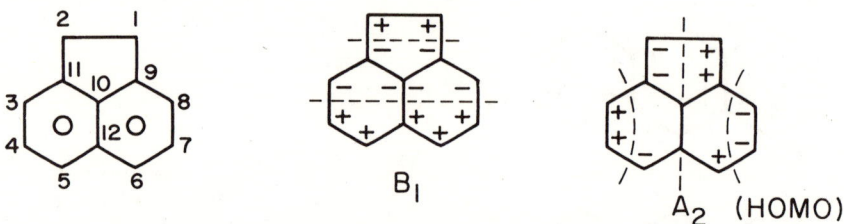

Figure 39. *Two HOMO's for the acenaphthalene dianion*

Hückel calculations place the largest negative charge on the 1,2 and 5,6 positions ($-0.28e^-$). The nearest nodal plane to the 1,2 position is at the center of the five-membered ring while the nearest nodal plane to the 5 and 6 carbon atoms places the lithium atom close to both those carbon atoms, as in the HOMO in the third position 3 listed. The first position is actually observed; it seems to be favored in that the second HOMO does not contain a node in position 3 but does for position 1. The N—Li—N plane should permit overlap between the 1-11 and 2-9 positions for the two N—Li—N fragments on either side of the five-membered plane. The angle of the N—Li—N plane with respect to the twofold axis of the molecule should therefore be between a maximum of 36° and a minimum of 10°. The observed value is 15°.

The ethylenic dianions of stilbene, tetraphenylethylene, and bifluorenylidene should have a HOMO that is antibonding with respect to the olefinic bond. In addition, the most basic carbon atoms are predicted to be the olefinic carbon atoms. The structure of [LiTMED] bifluorenylidene appears in Figure 40. The two lithium atoms lie on a twofold axis that bisects the C=C bond.

The molecular structure of the *trans*-stilbene dianion is similar to the N—Li—N fragments on the twofold axis on either side of the anion plane. Unfortunately, the stilbene dianion—but not the N—Li—N group

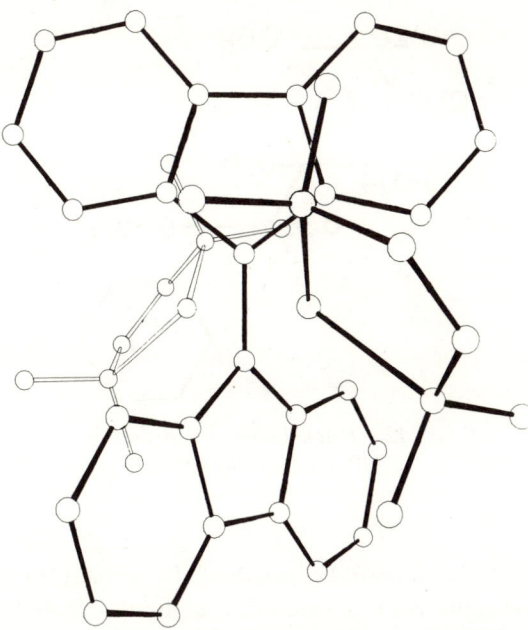

Figure 40. Molecular structure of [LiTMED]$_2$ bisfluorenylidene

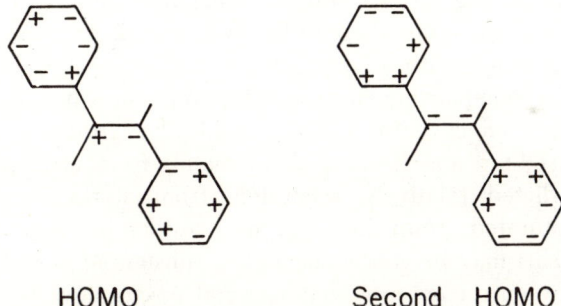

HOMO   Second HOMO

Figure 41. Symmetry of the two highest occupied molecular orbitals of trans-stilbene

—seems to be disordered in this structure, and we have not been able to interpret many of the structural details of the carbanion. The expected symmetries of the first and second HOMO's are shown in Figure 41; the Hückel charge distribution is given in Figure 42. In both the stilbene (135) and bifluorenylidene structures, the planar or nonplanar geometry solid-state configuration of the parent hydrocarbon is preserved, and in both structures the bisector of the N—Li—N bond is directed at the midpoint of the olefinic C—C bond.

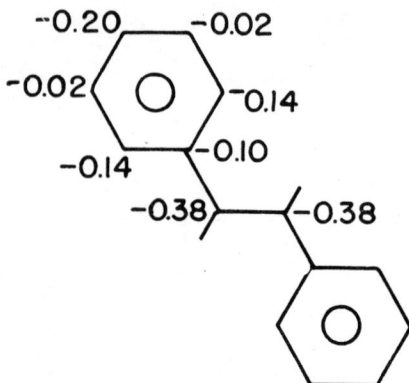

Figure 42. Hückel charge distribution for trans-*stilbene*

Summary

The complexes preferentially obtained by crystallizing N-chelated organolithium reagents from a solution containing excess bidentate tertiary amine base, TMED, are characterized by a 1:1 base-to-lithium ratio. A comparison of metal–base distances shows a dependency of the metal-N distance on the carbanion stability and perhaps to a lesser extent on the base strength of the tertiary amine. Chelation of a lithium atom by TMED results in the optimal use of the lone pair electrons on the nitrogen atoms. Metal atoms with a larger atomic radius are not able to use the nitrogen atom lone-pair electrons as effectively, and thus, the magnitude of the chelate effect with TMED should be greatly reduced. The distance of sodium and heavier alkali metals from unsaturated groups can be predicted relatively accurately from atomic radii; however, lithium–unsaturated group distances are considerably shorter than those predicted. This may be consistent with a substantial degree of covalent bonding between second-row elements and unsaturated organic species.

All the benzylic systems, RLiTMED, examined have an allylic geometry with respect to the metal atom. The orientation of the N—Li—N plane, the charge distribution inferred from INDO and NMR results, INDO overlap integrals and molecular orbital coefficients, and the allylic configuration itself, are compatible with a three-center bonding model that invokes the use of all lithium 2s and 2p orbitals. The allylic geometry observed in the solid state is also consistent with recent NMR studies of solution equilibria. The position of the metal atom should vary with the degree of solvation.

The position of the lithium atom in N-chelated organolithium compounds, $(TMEDLi)_2R$, also cannot be predicted from electrostatic considerations alone but, as in the benzylic derivatives, is consonant with some directed covalent bonding involving the HOMO of the carbanion. In the ethylenic derivatives studied the closest approach of the lithium atom to the organic group is along a perpendicular bisector of the olefinic bond, thus placing the lithium atom on a nodal surface of the HOMO of the carbanion.

To evaluate conclusively the hypotheses presented here, additional solution and solid-state studies are needed, particularly for trisolvated organolithium compounds, solvent-separated complexes, and unsaturated complexes of the heavier alkali metals. The results do, however, furnish the first glimpse of the stereochemistry of a chemically important class of compounds and should also provide working models for understanding the role of Group Ia metals in polymerization reactions and in their reactions with electrophilic reagents.

Acknowledgment

The author thanks all of his research associates who contributed so much to this work: W. Rhine, R. Zerger, M. Walczak, J. Brooks, J. Davis, D. Brauer, C. Johnson, and J. Toney.

Literature Cited

1. Addison, C. C., Davies, B. M., *J. Chem. Soc., A* (1969) 1822, 1827, 1831.
2. Schlosser, M., *J. Organometal. Chem.* (1967) **8**, 9.
3. Phillips Petroleum Co., British Patent **1,029,445** (1964).
4. Schlenk, W., Bergmann, E., *Ann.* (1928) **464**, 1.
5. Langer, Jr., A. W., *Trans. N. Y. Acad. Sci.* (1965) **27**, 741.
6. Eberhardt, G. G., Butte, W. A., *J. Org. Chem.* (1964) **29**, 2928.
7. Eberhardt, G. G., Davis, W. R., *J. Polym. Sci. Part A* (1965) **3**, 3753.
8. Eberhardt, G. G., *Organometal. Chem. Rev.* (1966) **1**, 491.
9. Rausch, M. D., Ciappenelli, D. J., *J. Organomet. Chem.* (1967) **10**, 127.
10. Slocum, D. W., Engelmann, T. R., Ernst, C., Jennings, C. A., Jones, W., Koonsuitsky, B., Lewis, J., Shenkin, P., *J. Chem. Ed.* (1969) **46**, 144.
11. Wittig, G., Pieper, G., Fuhrmann, G., *Ber.* (1940) **73**, 1193.
12. Jenny, E., Roberts, J. D., *Helv. Shem. Acta* (1955) **38**, 1748.
13. Gloss, G. L., *J. Amer. Chem. Soc.* (1962) **84**, 809.
14. Speziale, A. J., Bissing, D. E., *J. Amer. Chem. Soc.* (1963) **85**, 3878.
15. Jones, M. F., Trippett, S., *J. Chem. Soc. C* (1966) 1090.
16. Rhine, W., Walczak, M., Stucky, G. D., unpublished data.
17. Yasuda, H., Ph.D. Thesis, Osaka University, Japan (1971).
18. Szwarc, M., *Science* (1970) **170**, 23.
19. Langer, A. W., Jr., *Trans. N. Y. Acad. Sci.* (1965) **27**, 741.
20. Eberhardt, G. G., Butte, W. A., *J. Org. Chem.* (1964) **29**, 2928.
21. Eberhardt, G. G., *Organometal. Chem. Rev.* (1966) **1**, 491.
22. Rausch, M., Advan. Chem. Ser. (1973) **130**, 248.
23. Coates, G. E., Wade, K., "Organometallic Compounds," Methuen, London (1967).
24. Brooks, J. J., Ph.D. Thesis, University of Illinois (1971).
25. Rhine, W., Stucky, G. D., unpublished data.

26. Katz, T. J., Rosenberger, M., O'Hara, R. K., *J. Amer. Chem. Soc.* (1964) **86**, 249.
27. Katz, T. J., Heton, N., *J. Amer. Chem. Soc.* (1972) **94**, 3281.
28. Kershner, L. D., Gaidis, J. M., Freedman, H. H., *J. Amer. Chem. Soc.* (1972) **94**, 985.
29. Cruickshank, D. W. J., Sparks, R. A., *Proc. Roy. Soc., Ser. A* (1960) **258**, 270.
30. Brooks, J. J., Rhine, W., Stucky, G. D., *J. Amer. Chem. Soc.* (1972) **94**, 7346.
31. Pople, J. A., Beveridge, D. L., "Approximate Molecular Orbital Theory," McGraw-Hill, New York (1970).
32. Davis, J., Rhine, W., Brooks, J. J., Stucky, G. D., unpublished data.
33. Walsh, A. D., *J. Chem. Soc.* (1953) 2296, 2325.
34. Pearson, R. G., *Chem. Phys. Lett.* (1971) **10**, 31.
35. Fujimura, Y., Yamaguchi, H., Nakajima, T., *Bull. Chem. Soc. Jap.* (1972) **45**, 384.
36. Lehmkuhl, H., *Ann. Chem.* (1968) **719**, 20.
37. Brauer, D. J., Stucky, G. D., *J. Amer. Chem. Soc.* (1970) **92**, 3956.
38. Pedersen, L., Griffin, R. G., *Chem. Phys. Lett.* (1970) **5**, 373.
39. Canters, G. W., Corvaja, C., de Boer, E., *J. Chem. Phys.* (1971) **54**, 3026.
40. Hirota, N., *J. Amer. Chem. Soc.* (1968) **90**, 3603.
41. Carrington, A., Dravnieks, F., Symons, M. C. R., *J. Chem. Soc.* (1959) 947.
42. Rhine, W., Davis, J., Stucky, G. D., unpublished data.
43. Rhine, W., Stucky, G. D., unpublished data.
44. Herkstroeter, W. G., Hammond, G. S., *J. Amer. Chem. Soc.* (1966) **88**, 4769.
45. Saltiel, J., Hammond, G. S., *J. Amer. Chem. Soc.* (1963) **85**, 2515.
46. Hammond, G. S., et al., *J. Amer. Chem. Soc.* (1964) **86**, 3197.
47. Byline, A., *Chem. Phys. Lett.* (1968) **1**, 509.
48. Byline, A., Grabowski, Z. R., *Trans. Faraday Soc.* (1969) **65**, 458.
49. Lamola, A. A., *Tech. Org. Chem.* (1969) **14**, 17.
50. Wagner, P. J., *J. Amer. Chem. Soc.* (1967) **89**, 2820.
51. Garst, J. F., Pacifici, J. G., Zabslotny, E. R., *J. Amer. Chem. Soc.* (1966) **88**, 3872.
52. Eargle, D. H., *J. Amer. Chem. Soc.* (1971) **93**, 3859.
53. Garst, J. F., *J. Amer. Chem. Soc.* (1971) **93**, 6312.
54. Fenimore, C. P., *Acta Crystallogr.* (1948) **1**, 295.
55. Nyburg, S. C., *Acta Crystallogr.* (1954) **7**, 779.
56. Bailey, N. A., Hull, S. E., private communication.
57. Bailey, N. A., Hull, S. E., *Chem. Commun.* (1971) 960.
58. Patterman, S. P., Karle, I. L., Stucky, G. D., *J. Amer. Chem. Soc.* (1970) **92**, 1150.
59. Brooks, J. J., Stucky, G. D., *J. Amer. Chem. Soc.* (1972) **94**, 7333.
60. Lewis, G. N., Magel, T. M., Lipkin, D., *J. Amer. Chem. Soc.* (1942) **64**, 1774.
61. Hoffmann, R., Bissell, R., Farnum, D., *J. Phys. Chem.* (1969) **73**, 1789.
62. Morimoto, H., Lehmann, H., and Perutz, M. F., *Nature* (1971) **232**, 408.
63. Perutz, M. F., *Nature* (1970) **228**, 726.
64. Koh, B., *Biochim. Biophys. Acta* (1961) **48**, 527.
65. Vernon, L. P., Ke, B., Shaw, E. R., *Biochemistry* (1967) **6**, 2210.
66. Fajer, J., Felton, R. H., Dolphin, D., *Proc. Nat. Acad. Sci. U. S.* (1970) **67**, 813.
67. Dolphin, D., Forman, A., Borg, D. C., Fajer, J., Felton, R. H., *Proc. Nat. Acad. Sci. U. S.* (1971) **68**, 618.
68. Fajer, J., Borg, D. C., Forman, A., Dolphin, D., Felton, R. H., *J. Amer. Chem. Soc.* (1970) **92**, 3451.

69. Rundle, R. E., *J. Amer. Chem. Soc.* (1947) **69**, 1327.
70. Rundle, R. E., *J. Chem. Phys.* (1949) **17**, 671.
71. Drew, D. A., Haaland, A., *Chem. Commun.* (1972) 1300.
72. McPherson, A., Stucky, G. D., unpublished data.
73. Allegra, G., Perego, G., Immirzi, A., *Makromol. Chem.* (1963) **61**, 69.
74. Snow, A. I., Rundle, R. E., *Acta Crystallogr.* (1951) **4**, 348.
75. Rundle, R. E., Lewis, P. H., *J. Chem. Phys.* (1952) **20**, 132.
76. Drew, D. A., Haaland, A., *Chem. Commun.* (1971) 1551.
77. Morosin, B., Howatson, J., *J. Organometal. Chem.* (1971) **29**, 7.
78. Atwood, J. L., Stucky, G. D., *J. Amer. Chem. Soc.* (1969) **91**, 4426.
79. Zerger, R., Rhine, W., Stucky, G. D., unpublished data.
80. Zerger, R., Stucky, G. D., *Chem. Commun.* (1973) 44.
81. Kennard, O., Watson, D. G., "Molecular Structures and Dimensions," vols. 2 and 3, International Union of Crystallography, A. OOsthoek, Utrecht, Netherlands (1971).
82. Cotton, F. A., Wilkinson, G., "Advanced Inorganic Chemistry," Interscience, New York (1972).
83. Brooks, J. J., Rhine, W., Stucky, G. D., *J. Amer. Chem. Soc.* (1972) **94**, 7339.
84. Zerger, R., Stucky, G. D., unpublished data.
85. Dessy, R. E., *J. Amer. Chem. Soc.* (1966) **88**, 460.
86. Toney, J., Stucky, G. D., *J. Organometal. Chem.* (1970) **22**, 241.
87. Magnuson, V. R., Stucky, G. D., *Inorg. Chem.* (1969) **8**, 1427.
88. Haaland, A., *Chem. Commun.* (1971) 430.
89. Atwood, J., private communication.
90. Grant, D., Killean, R. C., Lawrence, J. L., *Acta Crystallogr. B* (1969) **25**, 377.
91. Davis, J., Stucky, G. D., unpublished data.
92. Dixon, J. A., *J. Amer. Chem. Soc.* (1965) **87**, 1379.
93. McKeever, L. D., private communication.
94. Yuki, H., *J. Organometal. Chem.* (1971) **32**, 1.
95. Crutzner, J. B., Lawlar, J. M., Jackman, L. M., private communication.
96. Sandel, V. R., private communication.
97. Canters, G. W., Corvaja, C., de Boer, E., *J. Chem. Phys.* (1971) **54**, 3026.
98. Takeshita, T., Hirota, N., *J. Amer. Chem. Soc.* (1971) **93**, 6421.
99. Pedersen, L., Griffin, R. G., *Chem. Phys. Lett.* (1970) **5**, 373.
100. Goldberg, I. B., Bolton, J. R., *J. Phys. Chem.* (1970) **74**, 1965.
101. McClelland, B. J., *Chem. Rev.* (1964) **64**, 301.
102. Hogan-Esch, T. E., Smid, J., *J. Amer. Chem. Soc.* (1969) **91**, 4580.
103. Johnson, C., Stucky, G. D., *J. Organometal. Chem.* (1972) **40**, C-11.
104. Almenningen, A., Bastiansen, O., Haaland, A., *J. Chem. Phys.* (1964) **40**, 3434.
105. Pauling, L., "The Nature of the Chemical Bond," Cornell, Ithaca (1960).
106. Matthews, J. D., Swallow, A. G., *Chem. Commun.* (1969) 882.
107. Wilford, J. B., Whitla, A., Powell, H. M., *J. Organomet. Chem.* (1967) **8**, 495.
108. Weiss, E., Fischer, E. O., *Z. Anorg. Allg. Chem.* (1956) **284**, 69.
109. Nesmeyanov, A. N., Gusev, A. I., Pasynskii, A. A., Anisimove, K. N., Kolobora, N. E., Struchkov, Y. T., *Chem. Commun.* (1968) 1365.
110. *Ibid.* (1969) 277.
111. Bush, M. A., Sim, G. A., Knox, G. R., Ahmad, M., Robertson, C. G., *Chem. Commun.* (1965) 74.
112. Churchill, M. R., Fennessey, J. P., *Inorg. Chem.* (1967) **6**, 1213.
113. Chaiwasie, S., Fenn, R. H., *Acta Crystallogr.* (1968) **B24**, 525.
114. Wilford, J. B., Powell, H. M., *J. Chem. Soc., A* (1969) 8.
115. Berndt, A. F., Marsh, R. E., *Acta Crystallogr.* (1962) **16**, 118.

116. Joshi, K. K., Mais, R. H. B., Nyman, F., Owston, P. G., Wood, A. M., *J. Chem. Soc., A* (1968) 318.
117. Dunitz, J. D., Orgel, L. E., Rich, A., *Acta Crystallogr.* (1956) **9**, 373.
118. Mills, O. S., Nice, J. P., *J. Organomet. Chem.* (1967) **9**, 339.
119. Hardgrove, G. L., Templeton, D. H., *Acta Crystallogr.* (1959) **12**, 28.
120. Dahl, L. F., Smith, D. L., *J. Amer. Chem. Soc.* (1961) **83**, 752.
121. Churchill, M. R., Mason, R., *Proc. Roy. Soc. Ser. A* (1966) **292**, 61.
122. Hedberg, L., Hedberg, K., *J. Chem. Phys.* (1970) **53**, 1228.
123. Minasyants, M. K., Struchkov, Y. T., *Zh. Strukt. Khim.* (1968) **9**, 481.
124. Delbaere, L. T. J., McBride, D. W., Ferguson, R. B., *Acta Crystallogr. Ser. B* (1970) **26**, 515.
125. Drew, D. A., Haaland, A., *Chem. Commun.* (1971) 1551.
126. McPherson, A., Stucky, G. D., unpublished data.
127. Corradini, P., Allegra, C., *J. Amer. Chem. Soc.* (1959) **81**, 2272.
128. Cotton, F. A., LaPrade, M. D., *J. Amer. Chem. Soc.* (1968) **90**, 5418.
129. Cox, R. H., *J. Phys. Chem.* (1969) **73**, 2649.
130. Hogen-Esch, T. E., Smid, J., *J. Amer. Chem. Soc.* (1966) **80**, 307.
131. Chan, L. L., Smid, J., *J. Amer. Chem. Soc.* (1967) **89**, 4549.
132. Ellingsen, T., Smid, J., *J. Phys. Chem.* (1969) **73**, 2719.
133. Schaeffer, T., Schneider, W. G., *Can. J. Chem.* (1963) **41**, 966.
134. Kunz, V., Nowachi, W., *Helv. Chim. Acta* (1967) **50**, 1052.
135. Robertson, J. M., Woodward, I., *Proc. Roy. Soc., Ser. A* (1937) **162**, 568.
136. Toney, J., Stucky, G. D., *Chem. Commun.* (1967) 1168.

RECEIVED April 2, 1973. Work supported by the National Science Foundation under Grants GH-33634 and GP-31016X.

4

Magnetic Resonance Studies of Polytertiary Amine Chelated Alkali Metal Compounds

M. T. MELCHIOR, L. P. KLEMANN, and A. W. LANGER, JR.

Esso Research and Engineering Co., Linden, N.J. 07036

> *Magnetic resonance experiments have provided considerable structural information on chelated alkali metal compounds. The combination of ^1H and ^7Li NMR experiments on chelated lithium halides (Chel · LiX) with ESR experiments on chelated sodium naphthalenide (Chel · $Na^+C_{10}H_8^-$) has shown that these systems are strongly-chelated, tight ion pairs which aggregate in aromatic hydrocarbon solvents. For a given chelating agent there is a strong correlation between the ^{23}Na hyperfine interaction in Chel · $Na^+C_{10}H_8^-$ and the ^7Li chemical shift in Chel · LiBr. Evidence of a stereospecific collision complex between Chel · LiBr and aromatic solvents is derived from ^1H NMR experiments.*

In the past 10 years nuclear magnetic resonance (NMR) and electron spin resonance (ESR) have proved to be valuable tools for studying polytertiaryamine chelated alkali metal compounds. Early work in these laboratories concerned ^1H NMR studies of diamine chelated lithium alkyl compounds, primarily butyllithium (LiBu), in paraffinic solvents. It was shown (1, 2) that chelation results in a marked upfield chemical shift of the α-CH_2 protons of LiBu. The magnitude of this shift correlates with the stability of the chelated LiBu (Chel · LiBu) and presumably reflects a polarization and subsequent loosening of the Li–C bond. There is also a smaller downfield chemical shift of the N–alkyl group protons of the chelating agent caused by a general transfer of electron density from chelating agent to lithium to alkyl group. This downfield shift of the chelating agent protons is a useful criterion for the presence of chelated lithium. For the prototype system TMED · LiBu (see Table I for chelating agents and abbreviations) in paraffinic solvent, this (normal) chelation shift is about 0.1 ppm downfield for N–CH_3 protons and essentially zero

Table I. Skeletal Structures of Polytertiaryamine Chelating Agents

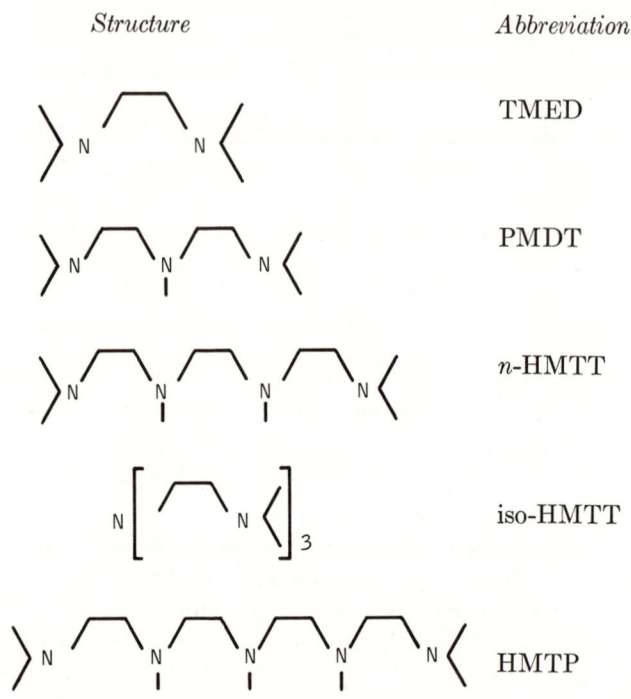

Structure	Abbreviation
	TMED
	PMDT
	n-HMTT
	iso-HMTT
	HMTP

for N–CH$_2$– protons. This is in distinct contrast to the behavior of TMED · LiBu in benzene where a large upfield shift of about 0.5 ppm is experienced by the N–CH$_2$– protons. (This statement takes into account a small correction caused by partial reaction of TMED · LiBu with solvent to form TMED · Liϕ.) This upfield methylene shift is evidence for a stereospecific solute–benzene collision complex (3) and provides an excellent probe of chelate structure. A similar upfield methylene shift has been noted for dimethoxyethane LiAlMe$_4$ in benzene (4).

The high reactivity of chelated lithium alkyl compounds severely limits structural study of pure compounds, particularly in aromatic solvents. Most of our more recent work on chelated lithium alkyl systems used ^{1}H and ^{7}Li NMR to observe various metalation reactions like the self-metalation or aging reaction of TMED · LiBu in heptane (1, 2). Much of our current insight into the structural features of chelated alkali metal systems comes from careful quantitative study of systems with relatively stable anions like resonance stabilized carbanions (5) and the systems described in this paper. We discuss magnetic resonance experiments on two systems: (a) chelated lithium halides Chel · LiX, examples of the recently discovered inorganic salt chelates (6), and (b) chelated sodium salts of naphthalene radical anion, Chel · $Na^+C_{10}H_8^-$. A large amount of data has been collected on these systems which provide stable models valuable for detailed structural study of chelated alkali metal compounds in general. Our discussion is concerned with the broad structural implications of the wide range of experimental results available.

Evidence is presented which shows that a chelated cation is a distinct, long-lived chemical species and that different chelated cations may coexist in solution as discrete observable species. Investigation of the anion–cation interaction shows that chelated salts in benzene exist as tight ion pairs down to the limit of spectrometer sensitivity. The effect of chelating agent on ion pair separation is considered. Finally we describe a series of experiments conducted in mixed solvents, the results of which reveal a stereospecific association of aromatic solvent molecules with a chelated lithium salt.

Experimental Procedures

The preparation of chelated lithium halides Chel · LiX has been described elsewhere (6). Solutions for ^{7}Li and ^{1}H NMR experiments were prepared from the solid complexes. Preparation of Chel · $Na^+C_{10}H_8^-$ was achieved by shaking a solution of the chelating agent and $C_{10}H_8$ with a strip of freshly cleaned sodium in a very simple cell joined at one end to an ESR sample tube. This technique is simple and allows the observation of successive increments of reduction. Its disadvantage is that it gives an unknown concentration of Chel · $Na^+C_{10}H_8^-$ in an excess of reactants. As noted below it succeeds only because of the unusually sluggish electron transfer in the hydrocarbon solvent. All samples were handled in a dry box.

ESR spectra were taken on a Bruker Scientific 418s X-band spectrometer equipped with Hall-probe stabilized magnetic field sweep. The 100 MHz proton NMR spectra were obtained on a standard Varian HA-100 spectrometer operating with internal lock using a solvent peak. All samples contained a trace of tetramethylsilane (TMS) as an internal reference. ^{7}Li NMR spectra were obtained at 23.3 MHz on a JEOL C60H NMR spectrometer equipped with a JNM-NS-100 Nuclear Single Sideband Unit. Samples for ^{7}Li NMR contained a capillary of 5.0M LiBr in methanol as external reference.

The deuterated chelating agent iso-HMTT-d_{18}, $N[CH_2CH_2N(CD_3)]_3$, was prepared from tris-(2-aminoethyl)amine, paraformaldehyde-d_2, formic acid-d_2 and D_2O *via* the Eschweiler-Clarke reaction (7).

Chelated Lithium Halides; ^{7}Li and ^{1}H NMR Results

The unusual properties of polytertiaryamine chelated lithium salts have been noted (6). The high solubility and conductivity of chelated lithium halides in benzene raise a number of important and interesting questions concerning the role of the aromatic solvent since these chelated salts are negligibly soluble in saturated hydrocarbons. Initial experiments focused on possible evidence for specific anion solvation by the aromatic solvent with little or no success. We have, however, obtained clear-cut evidence for a stereospecific association of solvent molecules localized at a chelated Li^+ cation. This interaction is manifested in the ^1H and ^7Li chemical shifts of the chelated salt Chel · LiX in benzene.

Table II shows the ^1H NMR data on $0.1M$ iso-HMTT and $0.1M$ iso-HMTT · LiBr in CH_2Cl_2 and in benzene. Of primary interest is the last entry in the table, Δ_{chel}, the chelation shift for iso-HMTT · LiBr relative to free iso-HMTT in each of the two solvents. The values of Δ_{chel} in CH_2Cl_2 are analogous to what is observed for lithium alkyl systems such as TMED · LiBu in paraffinic solvents. The small (~ 0.1 ppm) downfield shift of the $N-CH_3$ protons accompanied by essentially no shift of the $N-CH_2-$ protons is typical for lithium chelates in non-aromatic solvents and is what we have referred to as a normal chelation shift. Observation of this normal Δ_{chel} is evidence that iso-HMTT · LiBr does exist as the chelate in CH_2Cl_2.

Table II. Proton NMR Chemical Shifts for iso-HMTT, Free and Complexed with LiBr

	CH_2Cl_2		C_6H_6	
	CH_3	$CH_2{}^a$	CH_3	$CH_2{}^a$
iso-HMTT ($0.1M$)	−2.183	−2.442	−2.139	−2.541
iso-HMTT·LiBr ($0.1M$)	−2.319	−2.448	−2.240	−1.875
Δ_{chel}, (ppm)	−0.136	−0.006	−0.101	+0.666

^a Average shift for A_2B_2 pattern, all shifts in ppm downfield from TMS.

Table II shows that the chelation shift Δ_{chel} is quite different in benzene solution. Here again the behavior of iso-HMTT · LiBr is analogous to TMED · LiBu, with a large upfield shift of the $N-CH_2-$ protons. This large upfield methylene shift (large positive Δ_{chel}) in benzene is general for chelated lithium compounds and is seen as a manifestation of a stereospecific collision complex between benzene and the positive end of the molecular dipole moment of Chel · LiX. To study this unusual interaction we have carried out extensive experiments in mixed CH_2Cl_2–benzene solvents, some results of which are discussed below. In addition to studying the origin of this upfield methylene chelation shift in benzene, we have taken advantage of its existence in a number of ways described below.

The ^7Li NMR spectra of Chel · LiX have proved sensitive to various structural features. Figure 1 shows the ^7Li chemical shifts for PMDT ·

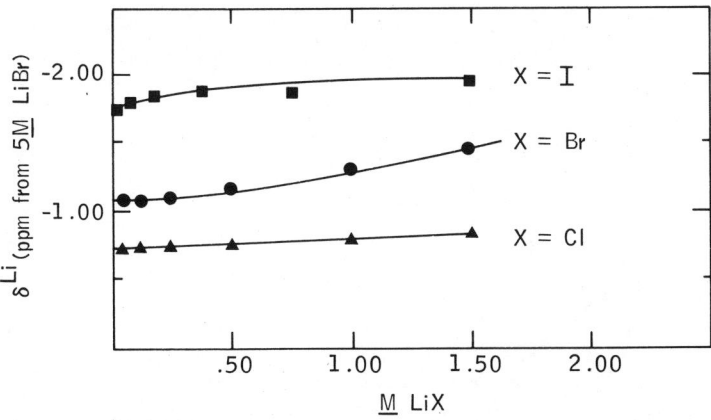

Figure 1. ^{7}Lithium NMR chemical shifts for PMDT · LiX, X = Cl, Br, I; room temperature in toluene

LiX (X = Cl, Br, or I) at room temperature in toluene. (All ^{7}Li chemical shifts are measured relative to an external standard of 5.0M LiBr in methanol.) Figure 1 shows that there is about a 1-ppm downfield shift in going from chloride to iodide in PMDT · LiX. Figure 2 shows the variation with concentration of ^{7}Li chemical shifts for Chel · LiBr in benzene for the five chelating agents studied. The limiting dilution chemical shifts for Chel · LiBr derived from Figure 2 are tabulated in Table III. In addition to these variations with anion and chelating agent, the ^{7}Li chemical shift is also influenced by stereospecific association of Chel · LiX with aromatic solvents to roughly the same degree as is the ^{1}H chemical shift of the chelating agent. The ^{7}Li chemical shift of 0.1M iso-HMTT · LiBr in benzene is about 0.6 ppm upfield from that of 0.1M iso-HMTT · LiBr in CH_2Cl_2. This may be compared with the ^{1}H NMR data in Table II. There is also concentration dependence of the ^{7}Li chemical shift although somewhat less pronounced than for ^{1}H. Clearly the ^{7}Li chemical shift of Chel · LiX is equally sensitive to the nature of the anion, chelating agent, and solvent, as well as concentration. This discourages attempts to intepret the magnitude of the observed ^{7}Li chemical shifts. In the following we will draw conclusions from the fact that ^{7}Li chemical shift differences exist in given situations without attempting to rationalize the magnitude of the shifts themselves.

Chelated Sodium Naphthalenide; ESR Results

Chelated sodium naphthalenide salts Chel · $Na^+C_{10}H_8^-$ have been prepared *via* sodium metal reduction of solutions of $C_{10}H_8$ and chelating agent in benzene. While $Na^+C_{10}H_8^-$ has been studied extensively in ether

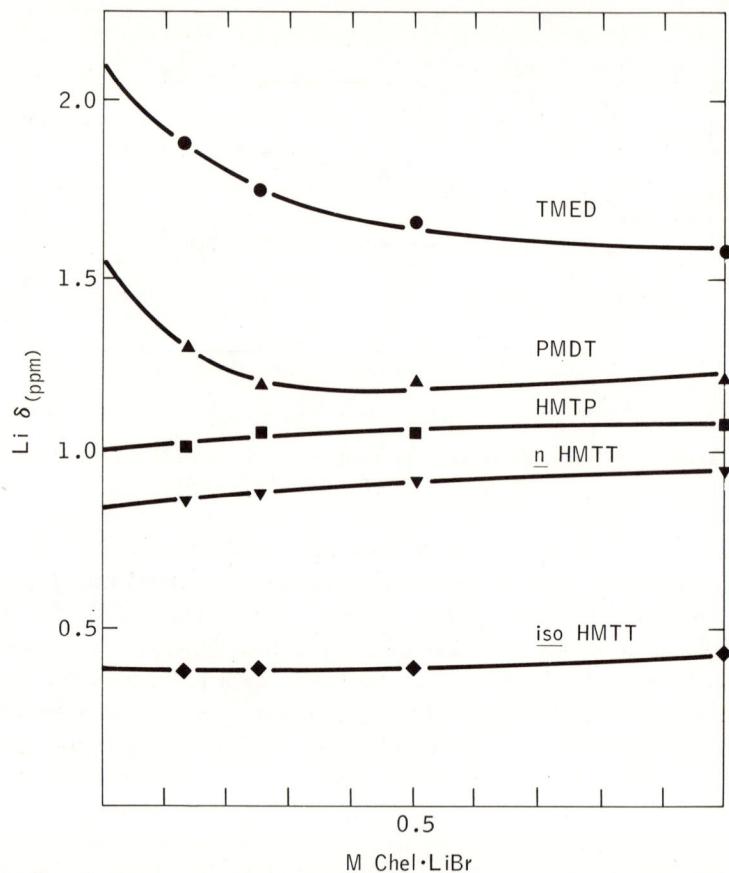

Figure 2. ⁷Lithium NMR chemical shifts for Chel · LiBr in benzene as a function of the concentration (moles/liter) of chelated salt. Chemical shifts are in ppm downfield from external 5M LiBr in methanol.

solvents (8, 9), this is the first study of such salts by ESR in hydrocarbon solvents. The observed spectra have a number of unusual features. A typical spectrum is shown in Figure 3, that of PMDT · Na⁺C$_{10}$H$_8$⁻ prepared by Na metal reduction of a benzene solution 0.002M in C$_{10}$H$_8$ and PMDT. Several features of this spectrum are worth noting:

(a) The spectrum in Figure 3 is characterized by narrow lines (< 0.1 G) despite the fact that the system contains an excess of the parent hydrocarbon C$_{10}$H$_8$. This indicates an unusually sluggish electron exchange between Chel · Na⁺C$_{10}$H$_8$⁻ and the parent C$_{10}$H$_8$ in the hydrocarbon solvent. (Preparation of Na⁺C$_{10}$H$_8$⁻ in an ether solvent such as THF using our technique produced only a single broad ESR line.) This observation is discussed below.

Table III. Observed ^{23}Na Hyperfine Interactions and Limiting Dilution ^{7}Li Chemical Shifts for Chelated Ion Pairs

Chelating Agent	Number of Nitrogens	a^{Na}, G^a	Li, ppm^b
TMED	2	1.80	2.08
PMDT	3	1.12	1.55
n-HMTT	4	0.89	1.00
HMTP	4^c	0.91	0.83
	5^c	0.20	
iso-HMTT	4	0.71	0.36

^{a} Observed ^{23}Na hf interaction for Chel • $N_a^+C_{10}H_8^-$ in benzene.
^{b} Observed limiting dilution ^{7}Li chemical shift for Chel • LiBr in benzene, in ppm downfield from $5M$ LiBr in methanol (see Figure 2).
^{c} See section on effect of chelating agent on ion pairing.

(b) The spectrum in Figure 3 is that of an ion pair evidenced by the superposition of a ^{23}Na hyperfine (hf) interaction on the 25-line pattern which would be observed for a free $C_{10}H_8^-$ anion. Figure 3 shows splitting of the central line and the outermost free ion line into quartets by the ^{23}Na hf coupling constant $a^{Na} = 1.12$ G.

(c) The ^{23}Na hf coupling constant in this and related systems is concentration independent and characteristic of the particular chelating agent. Observed vaues of a^{Na} are listed in Table III.

With two of the chelating agents, TMED and HMTP, time dependent ESR spectra were observed. The overall behavior of the TMED

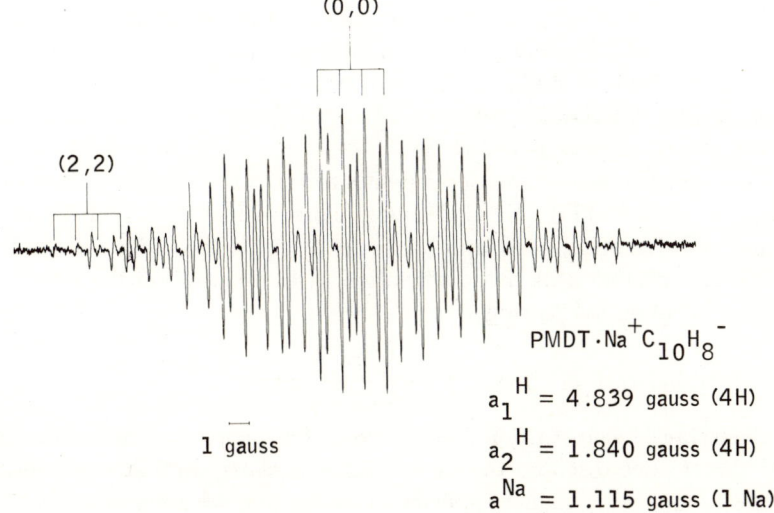

Figure 3. ESR spectrum of PMDT · Na$^+$C$_{10}$H$_8^-$ ion pairs in benzene. Spectrum shows ^{23}Na hf splitting of the 25-line pattern of the $C_{10}H_8^-$ anion. This splitting is shown for the central (0, 0) and the outermost (2, 2) line of the free anion.

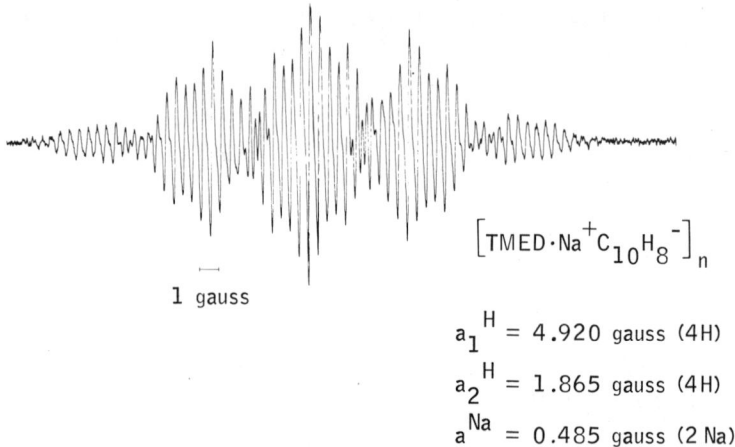

Figure 4. ESR spectrum of initial product of sodium reduction of $C_{10}H_8$ + TMED in benzene. Spectrum is that of an ion cluster and shows hf interaction with two equivalent ^{23}Na cations.

system is very complicated and is not discussed in detail. Two products are of interest—the initial product, and the product observed after long times in contact with sodium. The ESR spectrum of the initial product is shown in Figure 4. This spectrum shows hf interaction with two equivalent ^{23}Na nuclei, indicating initial formation of a triple ion or dimeric ion pair (TMED · $Na^+C_{10}H_8^-$)$_2$. The long contact time product is a typical tight ion-pair TMED · $Na^+C_{10}H_8^-$, data for which is included in Table III. These species are very similar to the two species observed when $Na^+C_{10}H_8^-$ is prepared in TMED as a solvent (10) even though the order in which the two species appear is inverted.

The second case of time dependent spectra is the observed decay of initially formed HMTP · $Na^+C_{10}H_8^-$ ion pairs with $a^{Na} = 0.91$ G, to a mixture of this species and another with $a^{Na} \leq 0.2$ G. We discuss a possible interpretation of this observation in the section on the effect of chelating agent on ion pairing.

General Structural Features of Chelated Salts

Chelating Agent–Cation Interaction. Our experiments have provided direct spectroscopic evidence that a chelated lithium or sodium cation is a distinct chemical species. Observation of a unique ^{23}Na hf interaction for each chelating agent in Chel · $Na^+C_{10}H_8^-$ by itself suggests that a chelated cation is a distinct entity. This point is made more convincing by the observation that ESR spectra of mixtures of PMDT · $Na^+C_{10}H_8^-$ and iso-HMTT · $Na^+C_{10}H_8^-$ are superpositions of the individual chelate spectra. Figure 5 shows the ^{23}Na hf splitting of the outermost proton hf component of $C_{10}H_8^-$ by two chemically distinct chelated cations, PMDT · Na$^+$ and iso-HMTT · Na$^+$, each with a ^{23}Na hf interaction identical to that observed in separate experiments. There is no evidence of exchange of Na$^+$ between chelating agents at room temperature.

The ESR spectrum in Figure 5 resulted from the reduction of a solution in which the ratio of PMDT to iso-HMTT was about 16:1. Nearly equal intensities of the ESR spectra of PMDT · Na$^+$C$_{10}$H$_8^-$ and iso-HMTT · Na$^+$C$_{10}$H$_8^-$ shows the strong preference of Na$^+$ cations for iso-HMTT over PMDT. Table IV shows the variation of the ESR intensity ratio [PMDT · Na$^+$C$_{10}$H$_8^-$]/[iso-HMTT · Na$^+$C$_{10}$H$_8^-$] with variations of the ratio of iso-HMTT to PMDT in the parent solution. Table IV also shows that the relative concentration of the two chelated ion-pairs are satisfactorily accounted for over a considerable concentration range by a simple equilibrium constant expression,

$$K_{eq} = \frac{[\text{PMDT} \cdot \text{Na}^+\text{C}_{10}\text{H}_8^-][\text{iso-HMTT}]}{[\text{iso-HMTT} \cdot \text{Na}^+\text{C}_{10}\text{H}_8^-][\text{PMDT}]} \qquad (1)$$

with $K_{eq} = 6.6 \pm 1.2$. This result corresponds to a free energy difference of about 1.6 kcal per mole in chelate strength. Clearly such measurements could form the basis for a quantitative scale of chelating ability.

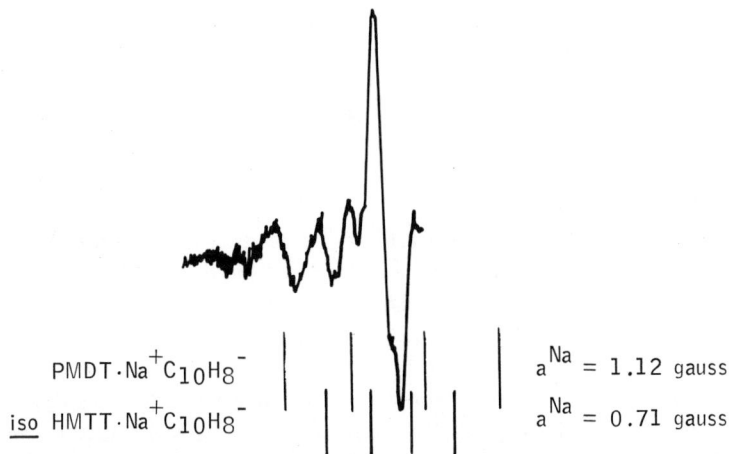

Figure 5. Low field portion of ESR spectrum of a mixture of iso-HMTT · Na$^+$C$_{10}$H$_8^-$ and PMDT · Na$^+$C$_{10}$H$_8^-$. The lines shown are the two quartets arising from ^{23}Na hf splitting of the end (2, 2) line of the $C_{10}H_8^-$ proton hf pattern for the two species. The strong line near the center of the pattern arises from ^{23}Na hf splitting of the next proton hf line (2, 1).

Table IV. Mixed iso-HMTT/PMDT Chelated $Na^+C_{10}H_8^-$; Relative Intensities and Ion-pair Equilibrium Constant

% iso-HMTT	ESR Intensity Ratio[a]	$K_{eq} \times 10^{2}$[a]
2	5.0	7.0
6	1.5	7.9
10	0.63	5.8
16	0.31	5.0
22	0.28	7.2

[a] Calculated from Equation 1. Concentrations of the free chelating agents calculated from starting concentrations assuming a total $Chel \cdot Na^+C_{10}H_8^-$ concentration of $10^{-3}M$ based on observed ESR intensities.

The ^{7}Li NMR spectra of Chel · LiBr in benzene and fluorobenzene provide equally strong, if not stronger, evidence of a discrete long-lived chelated cation. The data in Figure 2 and Table III show that the ^{7}Li NMR chemical shifts for Chel · LiBr at a given concentration in benzene are characteristic of the various chelating agents. Moreover Figure 6 shows that, at 4°C in fluorobenzene, separate ^{7}Li NMR peaks for PMDT · LiBr and iso-HMTT · LiBr can be resolved in a mixture of the two chelates. The ^{7}Li chemical shifts of the two chelates are the same as those observed in separate experiments. In this case, because of the relatively greater time scale of NMR, exchange of cations between chelate sites can be observed as a coalescence of the NMR peaks at the higher temperatures shown in Figure 6. The details of this exchange reaction are beyond the scope of the present discussion. We note only that a chelated Li$^+$ cation has a lifetime of about one second at $-10°C$ in this system.

Clearly this type of ^{7}Li NMR experiment has the potential for quantitative measurement of chelate strength. In this particular system, however, attempts to observe the equilibrium

$$\text{iso-HMTT} + \text{PMDT·LiBr} \rightleftarrows \text{iso-HMTT·LiBr} + \text{PMDT} \quad (2)$$

have failed because the equilibrium is displaced overwhelmingly to the right. From these experiments it has been estimated that the additional free energy of chelation for iso-HMTT relative to PMDT is at least 15 kcal/mole. Thus Li$^+$ cations show even more preference for iso-HMTT over PHDT than do Na$^+$ cations.

Further evidence of the strength of the chelating agent–cation interaction comes from the ^{1}H NMR spectra of chelated LiBr systems which have an excess of chelating agent. We have already noted the large upfield shift of the N–CH$_2$– protons of iso-HMTT · LiBr in benzene relative to free iso-HMTT (Table II). In several experiments distinct peaks for free and complexed iso-HMTT could be observed. In order to study this system without the interference of the N–CH$_3$ protons, selectively deuterated iso-HMTT-d_{18} was used ($> 99\%$ N–CD$_3$). Figure 7 shows the

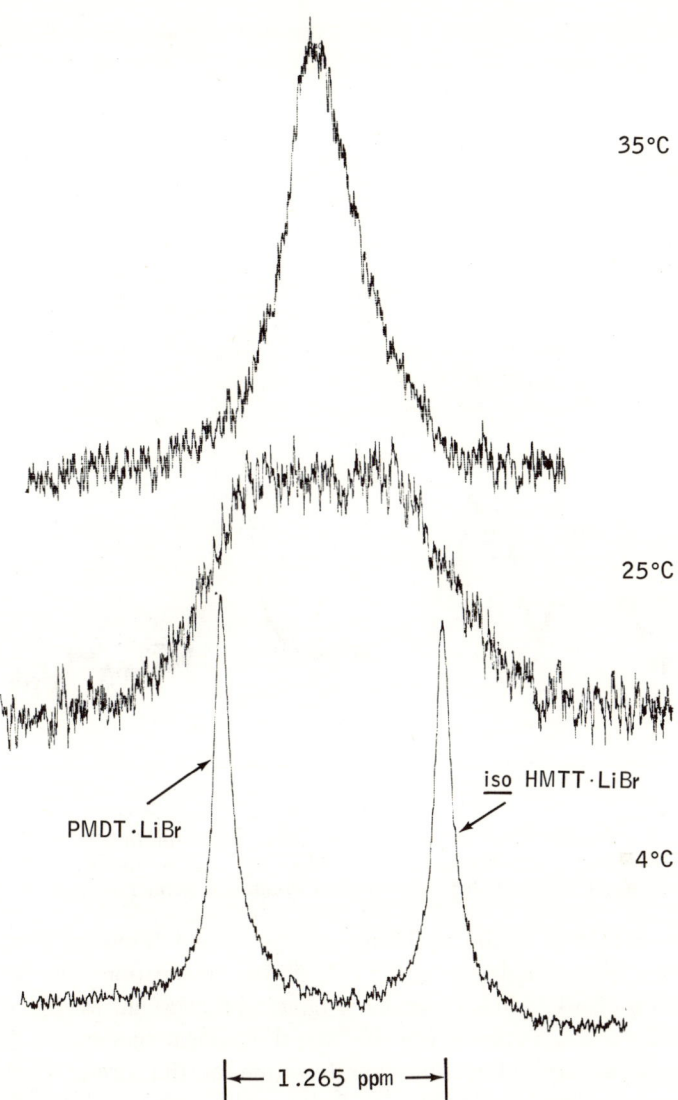

Figure 6. *^{7}Lithium NMR spectra of a mixture of 0.5M PMDT · LiBr and 0.5M iso-HMTT · LiBr in fluorobenzene at several temperatures*

^{1}H NMR spectrum of a solution 0.05M in iso-HMTT-d_{18} and 0.05M in iso-HMTT-d_{18} · LiBr in benzene at 5°C. Clearly resolved peaks from free and complexed iso-HMTT-d_{18} can be observed. These peaks coalesce at temperatures above 40°C as the exchange of LiBr between iso-HMTT molecules becomes sufficiently fast. The spectrum in Figure 7 is strong evidence for a well defined 1:1 chelate structure iso-HMTT · LiBr.

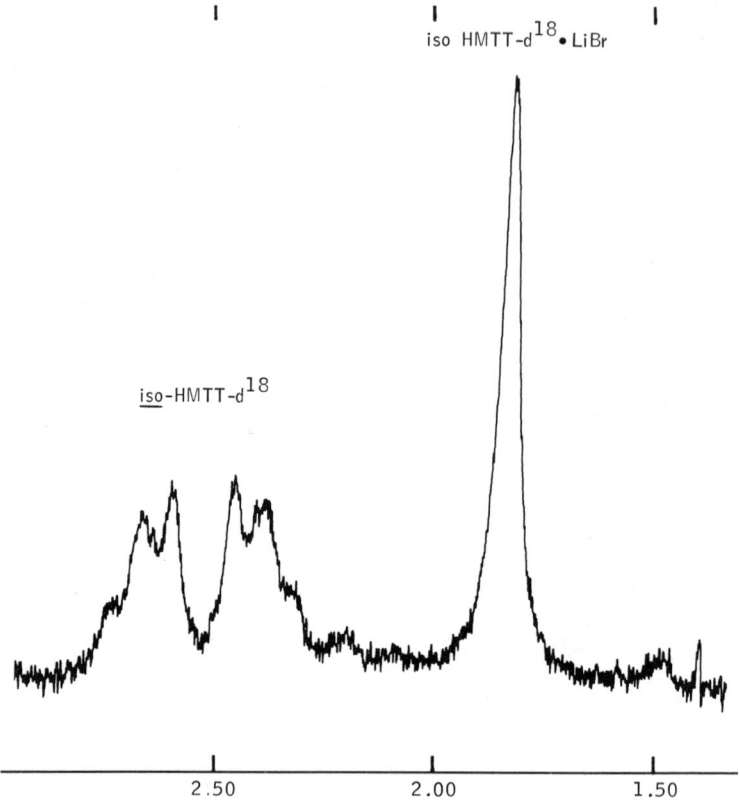

Figure 7. Proton NMR spectrum (100 MHz) of a solution of 0.05M in iso-HMTT-d_{18} and 0.05M in iso-HMTT-d_{18} · LiBr in benzene at 5°C. Chemical shifts are in ppm downfield from TMS.

We have looked at the chelating agent–cation interaction from the point of view of the chelating agent (^{1}H NMR), the cation (^{7}Li NMR), and the anion (ESR). In each case we have seen clear-cut evidence that a chelated Li^+ or Na^+ cation is a well defined chemical species.

We note a very interesting manifestation of the strong chelating agent–cation interaction, *i.e.* the ESR linewidths observed for Chel · $Na^+C_{10}H_8^-$ in the presence of excess $C_{10}H_8$. We consider two processes which can broaden the ESR lines, anion–neutral molecule electron transfer

$$C_{10}H_8 + Chel \cdot Na^+C_{10}H_8^- \underset{}{\overset{k_1}{\rightleftarrows}} Chel \cdot Na^+C_{10}H_8^- + C_{10}H_8, \quad (3)$$

and anion–anion spin–spin exchange

$$Chel \cdot Na^+C_{10}H_8^- + Chel \cdot Na^+C_{10}H_8^- \overset{k_2}{\rightleftarrows} Chel \cdot Na^+C_{10}H_8^- + Chel \cdot Na^+C_{10}H_8^- \quad (4)$$

In Equation 4 no exchange of ion partners need occur. We have not attempted to measure k_1 and k_2 in any of these systems but we can infer order of magnitude estimates. In the slow exchange limit (11) Equation 3 should give a line broadening independent of radical anion concentration

$$(W - W_0) \approx k_1 [C_{10}H_8], \tag{5}$$

where W and W_0 are linewidths (in Hz) in the presence and absence of exchange, respectively. In these systems we have never observed line broadening attributable to this process even with concentrations as high as $[C_{10}H_8] = 0.5M$. By assuming we could easily see a broadening of $W - W_0 \geq 0.05$ G (0.15 MHz) we obtained a conservative upper limit for the rate constant for electron transfer in Equation 3, $k_1 \leq 10^6 M^{-1}$ sec^{-1}. This upper limit is significantly below the range of values for electron transfer rate constants in ether solvents (12). At high concentrations ($\sim 10^{-2}M$) of Chel·Na$^+$C$_{10}$H$_8^-$ we have observed concentration dependent broadening which can be attributed to spin–spin exchange, Equation 4. In this case conservative estimates of the quantities involved lead to the conclusion $k_2 \geq 5 \times 10^8 M^{-1}$ sec^{-1}. This lower limit is comparable to spin–spin interaction rates ($\sim 10^9$) in ether solvents considered to be diffusion controlled (13). Presumably then the low value of k_1 in these systems is not caused by an unusually low frequency of anion–neutral molecule encounters. Concerted electron transfer as in Equation 3 requires a symmetric transition state in which the cation is shared equally by two $C_{10}H_8$ molecules (12, 14). With a tightly chelated cation steric considerations make such a transition state difficult to visualize. Alternative pathways involve either decomplexation

$$\text{Chel·Na}^+\text{C}_{10}\text{H}_8^- \rightleftarrows \text{Chel} + \text{Na}^+\text{C}_{10}\text{H}_8^-, \tag{6}$$

or ion-pair dissociation

$$\text{Chel·Na}^+\text{C}_{10}\text{H}_8^- \rightleftarrows \text{Chel·Na}^+ + \text{C}_{10}\text{H}_8^-, \tag{7}$$

either of which lead to high activation energies for the overall electron transfer process.

Effect of Chelating Agent on Ion Pairing. We have already noted that the observation of ^{23}Na hf interaction in the ESR spectra of Chel·Na$^+$C$_{10}$H$_8^-$ is direct evidence that these chelated salts exist as tight ion pairs in very dilute solution, at least on the ESR time scale. The available ^{7}Li NMR data on chelated lithium halides show that these systems are also tight ion pairs down to the limits of present spectrometer sensitivity. Figure 1 shows the ^{7}Li chemical shift of PMDT · LiX (X = Cl, Br, or I) as a function of concentration in benzene. The downfield chemical shift for the series Cl, Br, I perhaps is caused by increased covalency of the LiX bond but this is complicated by other factors. An adequate theory of the ^{7}Li NMR chemical shifts in chelated lithium salts will have to consider the combined effects of anion, chelating agent, solvent, and concentration or degree of aggregation. Our results indicate that these four factors are of comparable magnitude. The important feature of the data in Figure 1 for the present discussion is the absence of any trend toward a common free ion value for δ_{Li} at low concentrations. It seems clear from Figure 1 that there is no ion-pair dissociation at the concentrations used.

We now consider the role of the chelating agent in determining the ion-pair interaction in chelated salts. The ion-pair interaction is manifested in the ^{23}Na hf interactions observed for Chel · Na$^+$C$_{10}$H$_8^-$ given in Table III. It was noted above that two ion-pair species HMTP · Na$^+$C$_{10}$H$_8^-$ were observed. We postulate that the two species represent the linear pentadentate HMTP chelating Na$^+$ with either four or five nitrogens giving hf interactions of $a^{Na} = 0.91$ G and $a^{Na} \leq 0.2$ G, respectively. Providing we make this assumption concerning HMTP, the data in Table III show that the magnitude of a^{Na} decreases as the number of nitrogen bases increases. The order in which the magnitude of a^{Na} places the chelating agents parallels that observed for dc conductivity of Chel · Na$^+$C$_{10}$H$_8^-$ (15) and thermal stability of related lithium salts (6, 16). It is generally recognized that the ^{23}Na hf interaction in Na$^+$C$_{10}$H$_8^-$ is a very sensitive probe of the geometry of the ion pair (17). This hf interaction is a complicated function of the precise location of the Na$^+$ cation over the plane of the C$_{10}$H$_8^-$ anion (17). Nevertheless it is a reasonable first approximation that a decrease in a^{Na} reflects an increased ion pair separation. We conclude from this that as the chelating agent in Chel · Na$^+$C$_{10}$H$_8^-$ becomes more effective, the average ion pair separation increases across the series

TMED < PMDT < HMTP$_{IV}$ ≈ n-HMTT < iso-HMTT < HMTP$_V$,

in which HMTP appears twice as discussed above. The data in Table III show that the limiting dilution ^{7}Li chemical shift for Chel · LiBr is strongly correlated with a^{Na} for Chel · Na$^+$C$_{10}$H$_8^-$ for a given chelating agent. It is not obvious whether δ_{Li} reflects the trend in ion pair separation, the direct effect of the different chelating agents, or both. For the moment we are satisfied to note the correlation with other data.

We conclude that chelated salts are tight ion pairs in aromatic hydrocarbon solution with an anion–cation interaction which is a sensitive function of the chelating agent–cation interaction.

Stereospecific Interaction of Benzene with Lithium Chelates. As we have noted, the marked ^{1}H chemical shift of the chelating agent –CH$_2$–

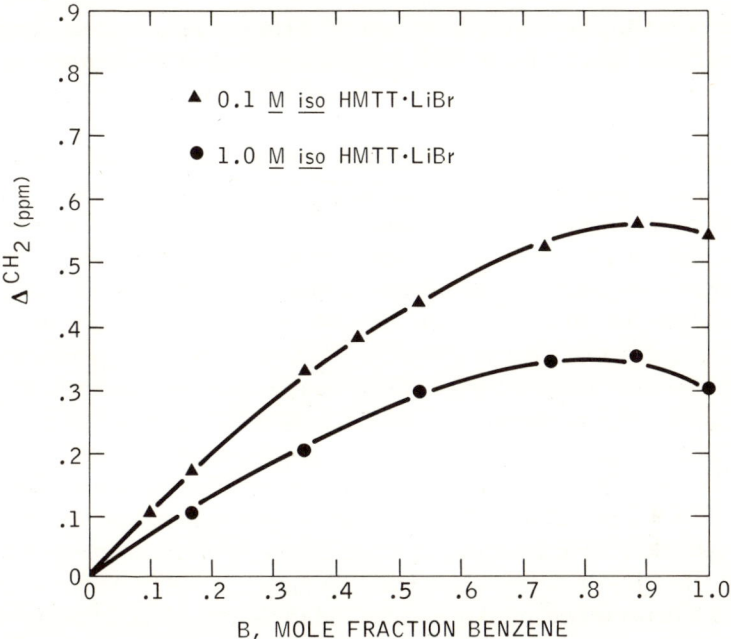

Figure 8. Proton chemical shifts of iso-HMTT · LiBr in mixed benzene–methylene chloride solution. The upfield methylene shift Δ^{CH_2} is defined as zero in pure methylene chloride (see Table II). Results are shown for two concentrations of chelated salt.

protons of Chel · LiX in aromatic solvents is evidence of a stereospecific collision complex between solvent and chelate. We have carried out extensive experiments to study this effect. The details of this work are beyond the scope of the present discussion but the basic conclusions will be described. Some of the results are shown in Figure 8 which is a plot of the average upfield chemical shift of the N–CH$_2$– protons of iso-HMTT · LiBr in benzene–methylene chloride mixtures. All chemical shifts are reported on a scale which places the chemical shift of iso-HMTT · LiBr in pure CH$_2$Cl$_2$ at zero. (The actual measurement of chemical shifts is always against a trace of tetramethylsilane as an internal reference.) In CH$_2$Cl$_2$ the ^{1}H shifts are concentration independent and show a normal chelation shift (*see* Table II) relative to free iso-HMTT, analogous to the behavior of lithium alkyls in non-aromatic hydrocarbon solvent. As the mole fraction of benzene in the mixed solvent is increased, Figure 8 shows that there is a large upfield shift of the N–CH$_2$ protons which depends on both the solvent composition and the total iso-HMTT · LiBr concentration. A series of dilutions at fixed solvent composition provided the data for the infinite dilution (c → o) chemical shift plot shown in Figure 9. The variation of the limiting dilution shift $\Delta_{c\to 0}$ (B) with B (the mole fraction of benzene in the solvent) is consistent with a simple equilibrium between monomeric iso-HMTT · LiBr and its complex or complexes with benzene, the observed shift $\Delta_{c\to 0}$ (B) being a weighted average over free and complexed species. The assumption that iso-HMTT · LiBr dissociates to monomeric species at low concentration is supported by our recent cryoscopic molecular weight measurements. These measurements confirm that dimers and higher aggregates predominate in the range 0.2–1.0M iso-HMTT · LiBr. From the shape of the limiting dilution shift plot in Figure 9, we have estimated that about 1/3 to 1/2 of the available complexation sites (assumed to be independent) are occupied in pure benzene solvent. On the other hand comparison of observed shifts with calculated ring current contributions (18) suggests an average of about one benzene per iso-HMTT · LiBr, indicating there may be several independent sites per iso-HMTT · LiBr. The calculation gives best agreement for a time averaged model which places the face of one benzene ring over one of the N–CH$_2$CH$_2$–N groups of iso-HMTT · LiBr.

The dependence of the chemical shifts on total iso-HMTT · LiBr concentration shown in Figure 8 is consistent with the formation of

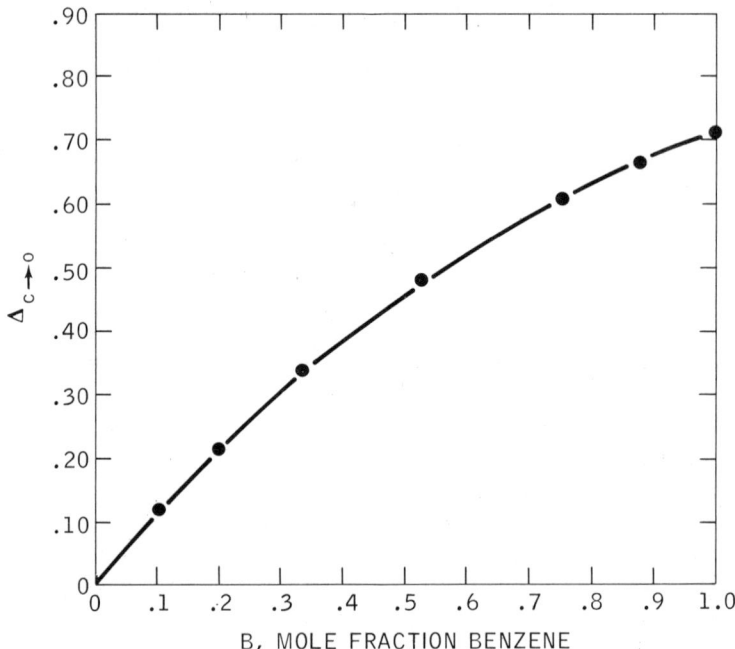

Figure 9. Proton NMR data for iso-HMTT · LiBr in mixed C_6H_6/CH_2Cl_2 solvent. The limiting dilution methylene shift $\Delta_{c\to 0}$ is plotted vs. the mole fraction benzene in the solvent.

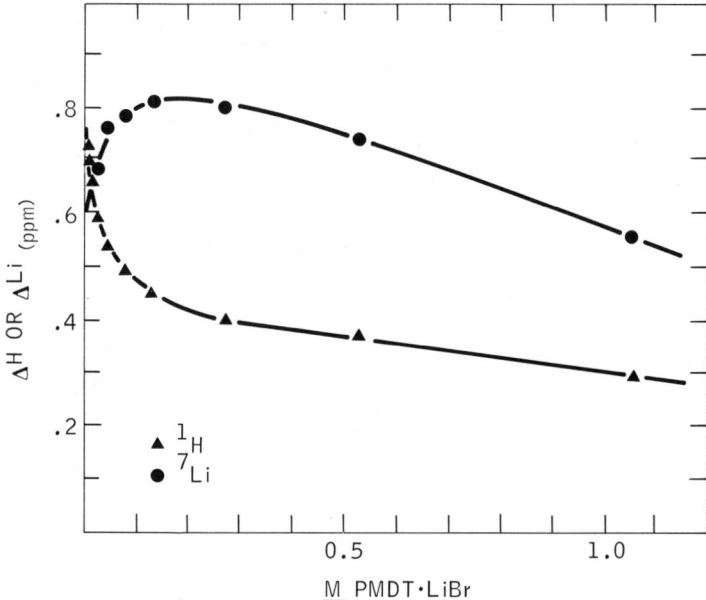

Figure 10. Methylene proton and ^{7}Li NMR chemical shifts of PMDT · LiBr in benzene as a function of PMDT · LiBr concentration. The chemical shifts are in ppm upfield from the corresponding values for PMDT · LiBr in CH_2Cl_2.

dimers and higher aggregates which do not form the stereospecific collision complex with benzene. The degree of aggregation is considerably reduced as the mole fraction of the more polar CH_2Cl_2 solvent is increased, leading to much less dependence of the chemical shift on total iso-HMTT · LiBr concentration.

The ^{7}Li chemical shifts for iso-HMTT · LiBr in this mixed solvent system are very strongly correlated with the N–CH$_2$ ^{1}H chemical shifts at higher concentrations of iso-HMTT · LiBr. At low concentrations, however, the upfield ^{7}Li chemical shift reaches a maximum and then decreases. This behavior is even more striking in the case of PMDT · LiBr in benzene shown in Figure 10. The divergent behavior of the ^{1}H and ^{7}Li chemical shifts at very low concentration of Chel · LiBr may be evidence that a second type of benzene solvation occurs in monomeric Chel · LiBr. We speculate that this may be evidence of benzene solvation in the vicinity of the anion.

Literature Cited

1. Langer, A. W., Amer. Chem. Soc., Div. Polym. Chem., Prepr., **7** (1), 132 (Phoenix, January, 1966).
2. Melchior, M. T., Langer, A. W., unpublished data.

3. Aroney, M. J., LeFevre, R. J. W., Vaughan, H. J., *J. Chem. Soc., A* (1970), 2224.
4. Ross, J. F., Oliver, J. P., *J. Organometal. Chem.* (1970) **22**, 503.
5. Forster, E. O., Langer, A. W., Amer. Chem. Soc., Div. Polym. Chem., Prepr. **13** (2), 656 (New York, August, 1972).
6. Langer, A. W., Whitney, T. A., U.S. Patent **3,734,963**.
7. Clarke, H. T., Gillespie, H. B., Weisshaus, S. Z., *J. Amer. Chem. Soc.* (1933) **55**, 4571.
8. Szwarc, M., *Accounts Chem. Res.* (1969) **2**, 87.
9. Hirota, N., *J. Amer. Chem. Soc.* (1968) **90**, 3603.
10. Reddoch, A. H., *J. Chem. Phys.* (1965) **43**, 3411.
11. Johnson, C. S., *Advan. Magn. Resonance* (1965) **1**, 33.
12. Hirota, N., Carraway, W., Schook, W., *J. Amer. Chem. Soc.* (1968) **90**, 3611.
13. Miller, T. A., Adams, R. N., *J. Amer. Chem. Soc.* (1966) **88**, 5713.
14. Sharp, J. H., Symons, M. C. R., "Ions and Ion Pairs in Organic Reactions," Vol. I, M. Szwarc, Ed., pp. 239–244, Interscience, New York, 1972.
15. Klemann, L. P., unpublished data.
16. Klemann, L. P., Whitney, T. A., Langer, A. W., Amer. Chem. Soc., Div. Polym. Prepr. **13** (2), 661 (New York, August, 1972).
17. Goldberg, I. B., Bolton, J. R., *J. Chem. Phys.* (1970) **74**, 1965.
18. Johnson, C. E., Bovey, F. A., *J. Chem. Phys.* (1958) **29**, 1012.

RECEIVED September 4, 1973.

5

Ac Conductivity of Some Organolithium Complexes in Aromatic Solvents

E. O. FORSTER and A. W. LANGER, JR.

Esso Research and Engineering Co., Linden, N.J. 07036

> *The nature of the electrical conductance of N-chelated aryl and aralkyl lithium compounds chelated with polyamine-type complexing agents has been studied in aromatic hydrocarbons ranging in concentration from 10^{-5} to 1 mole/liter between 10^2 and 10^7 Hz and between $-30°$ and $80°C$. The chelating agents included N,N,N',N'-tetramethylethylenediamine and N,N,N',N'',N''',N'''-hexamethyltriethylenetetraamine; the results obtained with these systems were compared with those obtained with tetra-n-amylammonium thiocyanate. Ion pairs representing dipoles contribute significantly to the conduction process. The drastic change in conductivity observed at concentrations greater than 10^{-2} mole/liter has been attributed to the formation of ion aggregates. The dielectric constant of one of these complexes has been determined from dilute solutions to be about 16. The behavior of these complexes is similar to that of $(n\text{-amyl})_4$-NCNS.*

The chemistry of organometallic compounds has received considerable attention recently. In particular, adducts of alkali metals with aliphatic and aromatic hydrocarbons have been studied in detail because of their general usefulness in organic synthesis. Until very recently the role played by the solvent and the metal on the formation and dissociation equilibria of these adducts was not clearly understood. Various studies (*1, 2*) indicated that the polarity of the C—Li bond could be increased by addition of electron-donor compounds as evidenced by an increase in electrical conductivity.

The significance of the increased ionic character of the carbon–metal bond in solvents with dielectric constants of less than four was not understood well. Kraus (*3*) and Fuoss (*4*) postulated earlier that dissolution of essentially ionic compounds such as quaternary ammonium salts in solvents of low dielectric constant should yield, in addition to some free ions, ion pairs and ionic aggregates. Comprehensive reviews of this subject are presented by Szwarc (5) and Blandamer and Fox (6). According to these authors three types of ion pairs can be encountered in solvents of low dielectric constant: contact pairs that are basically dipolar molecules, solvent shared ion pairs in which a cation is linked electrostatically through a solvent molecule to an anion, and solvent-separated ion pairs in which both ions are still linked electrostatically but separated by more than one solvent molecule (6). The relative contribution of these structures to physical properties of the solution can be deduced from conductivity or kinetic data. Of these two techniques conductivity studies have contributed considerably to the understanding of ion-pair equilibria (7). This does not mean that developments in other fields such as analysis of kinetic data, electron spin resonance (ESR) spectra, and ultrasonic relaxation data have had no impact on this field. These three fields have been very helpful in elucidating the structure of aqueous electrolytes where conductivity measurements are harder to perform.

Interest in N-chelated organolithium compounds stems from their remarkable reactivity which was first noted by Langer (8). From studies of various reactions and from NMR chemical shifts Langer concluded that the reactivity of these complexes in dilute solution is related to the increased ionic character of the Li—C bond (9). The question arose whether these complexes could be considered as some sort of stabilized ion pairs. Thus it seemed desirable to study the conductivities of these systems over a wide concentration range and to analyze the resulting data in the light of existing models and theories. This paper presents the results of a detailed study of the electrical conductance of aryl- and aralkyllithium compounds chelated with various polyamine-type complexing agents in aromatic hydrocarbons over a wide frequency and concentration range. The results are interpreted in the light of classical theories.

Experimental

The experimental details and the chemicals used have been described elsewhere (*10, 11, 12*). The instrumentation permitted measurements from 10 to 10^7 Hz from $-30°$ to $80°C$ at concentrations from 10^{-5} to 1.0 mole/liter. For comparison, studies were also carried out on tetra-*n*-amylammonium thiocyanate in aromatic solvents, a system that has been investigated in complete detail by Kenausis *et al.* (*13, 14*).

Results

The studies presented here involved several variables. First, the effect of chelating agent was studied as a function of both its structure and concentration with respect to the aralkyllithium compound. The role of solvent was then investigated for a given chelated system. With a knowledge of the influence of these variables on the electrical conductivity, one specific system was selected and its frequency and temperature dependence were studied as a function of concentration. These last results were then compared with those obtained with the quaternary ammonium salt system.

Effect of Chelating Agent. Before the role of the chelating agent can be properly determined, it is advisable to evaluate the electrical properties of the respective components alone, using the appropriate solvent (Table I).

Table I. Conductivities of Some Organolithium Compounds
$(25 \pm 0.1)\,°C$, 1 KHz

Compound	Solvent	Conc. mole/liter	Conductivity, $(ohm\ cm)^{-1}$
—	n-C_7H_{16}	—	$< 10^{-15}$
C_4H_9Li	n-C_7H_{16}	0.025	1.1×10^{-11}
TMED $(CH_3)_2N(CH_2)_2N(CH_3)_2$	—	—	1×10^{-11}
C_4H_9Li–TMED	n-C_7H_{16}	0.02	6×10^{-11}
C_6H_5Li–TMED	C_6H_6	0.05	3.8×10^{-11}
$(C_6H_5)_2CHLi$–TMED	C_6H_6	0.025	3.5×10^{-11}
$(C_6H_5)_2CHLi$–TMED	$C_6H_5CH_3$	0.025	3×10^{-11}
$(C_6H_5)_3CLi$–TMED	$C_6H_5CH_3$	0.025	4×10^{-11}

It was not possible to prepare butyllithium–TMED complexes in benzene (8) because the complex reacted with the solvent within one hour to produce phenyllithium complexes with TMED. Of the three aralkyl complexes tested the diphenylmethyllithium proved the most soluble. Therefore the diphenylmethyllithium complex was chosen to study further the effect of chelating agent type. The results are shown in Table II.

The HMTT complex at a 1:1 mole ratio produces nearly as conductive a solution as TMED at a 1:2 mole ratio. This suggests that four nitrogen atoms are probably required to produce the maximum coordination around the lithium atom to optimize the ionic character of the Li—C bond (9, 10). However, the efficiency of the coordination is probably higher with TMED because its two pairs of nitrogen atoms are not as sterically restricted in the alignment as are the four in HMTT. The TMED and HMTT complexes were selected for further studies.

Solvent Effects. The overall effect of solvent on the conductivity of $(C_6H_5)_2CHLi$–$(TMED)_2$ is shown in Figure 1. As the number of methyl groups around the benzene ring increases, conductivity decreases. The

Table II. Effect of Chelating Agents on Conductivity of $(C_6H_5)_2CHLi$–Chel in Toluene

(0.2M, 25°C, 1 KHz)

Chelating Agent	Ratio of RLi to Chelating Agent	Conductivity, $(ohm\ cm)^{-1}$
TMED[a]	1/1	1×10^{-5}
TMED	1/2	5×10^{-5}
PMDT[b]	1/1	2.6×10^{-8} [d]
HMTT[c]	1/1	2×10^{-5}

[a] N,N,N',N'-tetramethyl ethylene diamine
[b] N,N,N',N'',N''-pentamethyl diethylene triamine
[c] N,N,N',N'',N''',N'''-hexamethyl triethylene tetramine
[d] Complex does not go completely into solution.

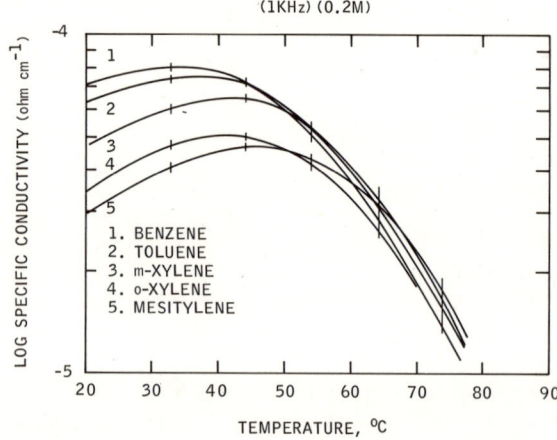

Figure 1. Effect of temperature on conductivity of $(C_6H_5)_2CHLi/2TMED$ in various solvents (1 KHz) (0.2M)

significance of the general shape of the temperature dependence curve of these conductivities is discussed elsewhere in this paper. Toluene was selected as solvent for subsequent studies because unlike benzene it has a considerably lower freezing point and its solution can be studied over a broader temperature range.

In those studies that use solvents other than the hydrocarbon corresponding to the carbanion, any metalation of the solvent would change the nature of the conductive species and complicate data interpretation. For example, the pK_a's of toluene (35) and diphenylmethane (ca. 33) are close enough so that toluene metalation by diphenylmethyllithium could be significant at very low concentrations of the lithium compound according to the equilibrium:

$(C_6H_5)_2CHLi \cdot Chel + C_6H_5CH_3 \rightleftarrows (C_6H_5)_2CH_2 + C_6H_5CH_2Li \cdot Chel$

No attempt was made to correct for this effect because early studies had indicated that similar systems were extremely slow in reaching equilibrium and in most cases the equilibrium contribution would not change the conclusions. These assumptions will be examined more critically in future work.

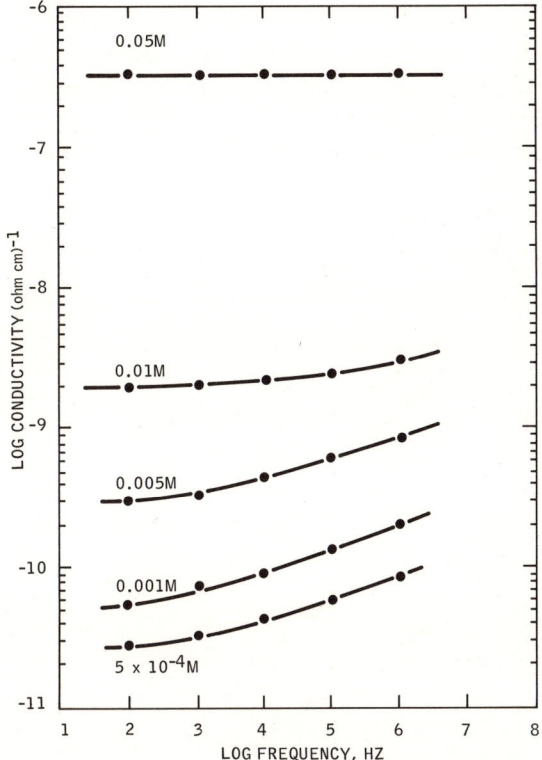

Figure 2. AC conductivity of $(C_6H_5)_2CHLi \cdot (TMED)_2$ in $C_6H_5CH_3$ at 25°C

Frequency and Temperature Effects. The frequency dependence of $(C_6H_5)_2CHLi-(TMED)_2$ as a function of concentration is shown in Figure 2. At the highest concentration studied (1M), the conductivity is essentially constant over six decades, changing from 3.9×10^{-4} (ohm cm)$^{-1}$ at 10 Hz to 4.6×10^{-4} (ohm cm)$^{-1}$ at 10^8 Hz. The temperature dependence of the conductivity of these solutions is alluded to in the preceding section, and it is shown in more detail in Figure 3 for both the TMED and HMTT complex. The unusual shape of the log conductivity vs. the reciprocal temperature curve disappears at concentrations greater than 0.2M and gives way to the familiar linear dependence. The HMTT complex appears to be more stable even at the lower concentrations. The activation energy for conductivity can be calculated via the Arrhenius equation to be about 1600 cal/mole for the HMTT complex while that for the TMED complex (at concentrations above 0.4M) is 1300 cal/mole. The peculiar shape of the conductivity vs. temperature curve shown for 0.2M solutions of the $(C_6H_5)_2CHLi-(TMED)_2$ complex suggests that the conductive species becomes unstable at higher temperatures which in turn implies that the forces holding them together are about as large as kT. The implications of these deductions are dealt with in a later section.

A convenient way to summarize the concentration dependence of this complex is shown in Figure 4, where the logarithm of the equivalent conductance of the complex as determined at 1 KHz is plotted as a function of the logarithm of concentration. For comparison a similar

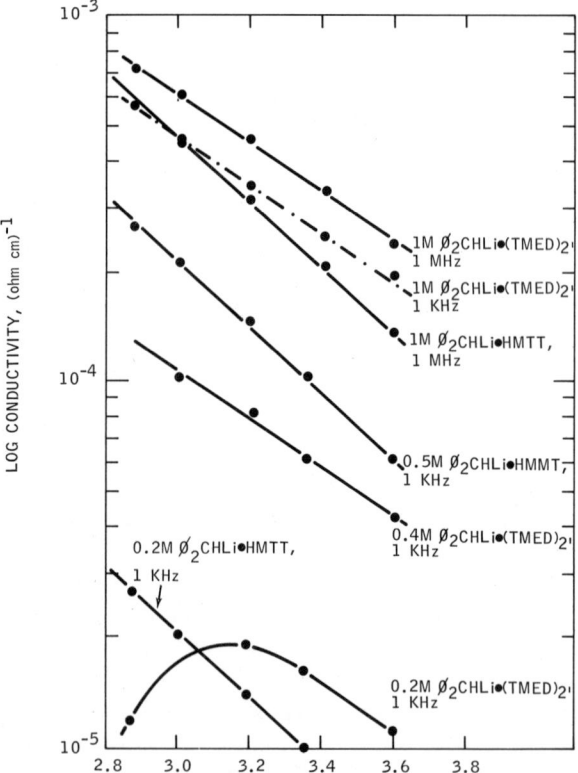

Figure 3. Temperature dependence of conductivity of $(C_6H_5)_2CHLi \cdot$ complexes in toluene

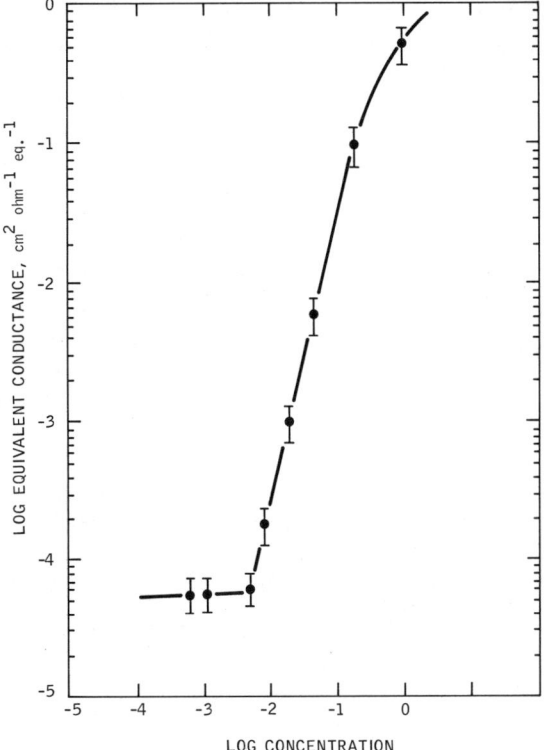

Figure 4. Equivalent conductance of $(C_6H_5)_2$-$CHLi(TMED)_2$ in toluene (25°C, 1 KHz)

plot is shown in Figure 5 for the well-studied quaternary ammonium salt, tetraamylammonium isocyanate in *p*-xylene.

Discussion

The data reported above indicate that the N-chelated aralkyl lithium complexes are quite conductive species, particularly in concentrated solutions. This is even more surprising since the aromatic solvent has a low dielectric constant and the absence of an inherent dipole in the solvent molecule seems to have little effect on the overall results. (The dielectric constant of TMED of 2.8 is certainly not going to contribute either.) Obviously, the solvent's dielectric constant is not the whole story.

From the results reported in the literature using low dielectric constant solvents such as dioxane (15), tetrahydrofuran and dimethoxyethane (16), or benzene (13, 14), the chemical makeup of the solvent, its molecular structure, or both, might well influence the final results. With the first three solvents the presence of oxygen atoms seems to be

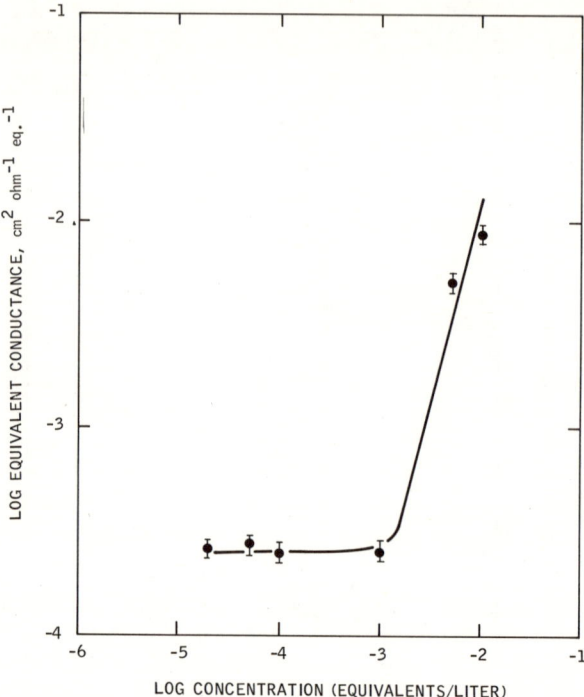

Figure 5. Equivalent conductance of $(n\text{-}amyl)_4N^+\text{-}CNS^-$ in p-xylene (25°C, 1 KHz)

important since they can act as electron donors or charge-transfer agents, thus stabilizing the charge separation within the complex. It is harder to comprehend the situation with aromatic solvents such as benzene or toluene. Apparently the molecular structure is important. These molecules have a high degree of symmetry, and small deformations lead to the formation of induced dipoles (*17*). Conversely, in the presence of a strongly polar solute molecule the nearest-neighbor solvent molecules are subject to induced polarization. Thus, molecules such as benzene and toluene, which are readily polarizable, will be very effective in "solvating" these solute diploles.

This process will increase the dielectric constant of the solution. Superimposed on this effect is the tendency of the solute to aggregate. This aspect has been recognized by many workers (*13, 14, 15, 17*). Dielectric measurements give information concerning the contribution of both processes, as shown in Table III, using the Onsager relationship (*12*).

The increase in the solute's apparent dielectric constant followed by a subsequent decrease at the highest concentration is similar to the observation reported by Kraus (*17*) on the abrupt decrease of the association

Table III. Dielectric Constant of $(C_6H_5)_2CHLi-(TMED)_2$ Solutions in Toluene

(10 KHz, 25°C)

Concentration, mole/liter	Dielectric constant	
	Solution	Solute
0	2.37	—
0.001	2.40	16
0.005	2.41	16
0.01	2.42	17
0.05	2.70	19
0.1	3.02	21
1.0	6.10	16.5

number of tetraamylammonium thiocyanate. This drop might be interpreted as suggesting that the ratio of solute to solvent of roughly 1 to 3 at the 1M level favors complete solvation of the solute, thus leaving few if any solute molecules associated. The question can then be raised as to the nature of the conductive species. The solution with an apparent dielectric constant of 6.1 is no longer unfavorable toward dissociation of solute molecules, and it is logical to visualize the existence of solvated ions in equilibrium with solvent-separated ion pairs. The picture is less definite on the other side of the concentration spectrum. It is not easy to gain a detailed understanding of the conductive species in very dilute solution. From the data in Figure 2 there appears to be some indication of a frequency-independent, ohmic-conduction region down to concentrations of about 0.005M, which would be attributable to ions. For lower concentrations the conductivity vs. frequency plot becomes quite nonlinear down to below 100 Hz, rendering the existence of a frequency-independent conduction mechanism questionable (18). In addition to these considerations it is appropriate to examine the frequency dependence of the dielectric constant in these very dilute solutions, as shown in Table IV.

Electrode polarization effects produced by the migration of ions to the electrode surface should be evident in the presence of ions (19). Such an ion layer can have considerable influence on the apparent capacitance, particularly for conductivity levels greater than 10^{-8} (ohm cm)$^{-1}$. At the low levels of conductivity prevailing in dilute solutions containing less than 10^{-3} mole/liter such polarization effects are not expected although the drop in dielectric constant for the 0.05M solution can be attributed to this effect. All this suggests that there are few if any ions present in these dilute systems. At the intermediate concentration levels, there might exist multiple ions while at the highest concentrations (above

Table IV. Frequency Dependence of the Dielectric Constant of $(C_6H_5)_2CHLi-(TMED)_2$ Solutions in Toluene (25°C)

Conc., mole/liter	Dielectric constant (Hz)					
	100	200	500	1000	5000	10000
0.05	18	10	4.6	4.1	3.1	2.7
0.001	2.42	2.45	2.40	2.40	2.40	2.40
0.0005	2.38	2.38	2.38	2.38	2.38	2.38

1 mole/liter) highly polarized, solvated ion pairs form the bulk of the solution (see Table III).

Reference has been made to results obtained with such ionic substances as quaternary ammonium salts (13, 14). The validity of these comparisons is supported by a comparison of Figure 4 with 5. Those two figures show plots of the equivalent conductance of $(C_6H_5)_2CHLi-(TMED)_2$ in toluene (Figure 4) and of $(n\text{-amyl})_4$ NCNS in p-xylene (Figure 5) as a function of concentration. The similarity of both curves is striking. The slight change in slope at high concentrations in Figure 4 is probably caused by viscosity effects (13). Both systems show a drastic change in slope near 10^{-2} mole/liter, yet one is an ionic substance in the solid state while the other is at best only partially ionic in character. It seems appropriate, therefore, to question the explanation offered by Kenausis et al. (13, 14) that only ions are responsible for the conduction in dilute solution. Indeed, from the frequency dependence of these dilute systems it seems reasonable to conclude that ion pairs representing dipoles might contribute to the overall conduction process by causing an increase in ac conductivity with frequency. On the other hand, the assignment of ion aggregates as the conductive species at concentrations above 10^{-2} mole/liter seems quite satisfactory in both cases. At very high concentrations (around 1 mole/liter and above) deaggregation apparently takes place, leading to an equilibrium between individual solvated ions and solvent-separated ion pairs.

Literature Cited

1. Shatonshtein, A. I., Petrov, E. S., *Usp. Khim.* (1967) **36**, 269.
2. Bagdasar'yarr, A. Kh. et al., *Dokl. Akad. Nauk. SSR* (1965) **162**, 1293.
3. Kraus, C. A., *J. Phys. Chem.* (1956) **60**, 129.
4. Fuoss, R. M., Aecascina, F., "Electrolytic Conductance," Chap. 16, Interscience, New York, 1959.
5. Szwarc, M., *Makromol. Chem.* (1965) **89**, 44.
6. Blandamer, M. J., Fox, M. F., *Chem. Rev.* (1970) **70**, 1.
7. Barthel, J., *Angew. Chem., Intern. Ed. Engl.* (1968) **7**, 260.
8. Langer, A. W., *Trans. N. Y. Acad. Sci.* (1965) **27**, 741.
9. Langer, A. W., *Amer. Chem. Soc., Div. Polym. Chem., Preprint*, **7** (1), 132 (Phoenix, Jan., 1966).

10. Forster, E. O., Langer, A. W. in "Phenomenes de Conduction dans les liquids isolants," p. 236, Colloques Intern. No. 179, Grenoble, Sept. 1968, CNRS Ed., Paris, 1970.
11. Forster, E. O., Langer, A. W., in "1969 Annual Report, Conference on Electrical Insulation and Dielectric Phenomena," *Nat. Acad. Sci. Publ.* **1764,** 87 (1970).
12. Forster, E. O., Langer, A. W., *Amer. Chem. Soc., Div. Polym. Chem., Preprint,* **13** (2), 656 (New York, August, 1972).
13. Kenausis, L. C., *et al., Proc. Natl. Acad. Sci.* (1962) **48,** 121.
14. Kenausis, L. C., *et al., Proc. Natl. Acad. Sci.* (1963) **49,** 141.
15. Kraus, C. A., *J. Phys. Chem.* (1956) **60,** 129.
16. Ellingsen, T., Smid, J., *J. Phys. Chem.* (1969) **73,** 2712.
17. Kraus, C. A., *J. Phys. Chem.* (1954) **58,** 673.
18. Forster, E. O., in "4th Intern. Symposium on Conduction and Breakdown Phenomena in Liquid Dielectrics," Dublin, 1972.
19. Hill, E. N., *et al.,* "Dielectric Properties and Molecular Behavior," p. 285, Van Nostrand-Reinhold, New York, 1969.

RECEIVED March 26, 1973.

6

Inorganic Complexes and Separation Processes

LAWRENCE P. KLEMANN, THOMAS A. WHITNEY, and ARTHUR W. LANGER, JR.

Corporate Research Laboratories, Esso Research and Engineering Co., Linden, N.J. 07036

> *Polyethylene polyamines are available as mixtures of isomeric and homologous compounds. By using inorganic compounds of Group IA and IIA metals, individual polyamines were selectively complexed from several multicomponent samples. Commercial polyamine mixtures containing primary and secondary nitrogens as well as their N-permethylated tertiary amine counterparts can be separated. The separation involves complex formation, isolation, and destabilization. Initial complexation and subsequent destabilization are sensitive to the lattice energy of the inorganic salt. Separation selectivities are generally high, and yields of purified polyamines, recovered by thermal dissociation of the complexes, are excellent. The tertiary amine chelates of sodium and lithium salts are unusual examples of hydrocarbon-soluble inorganic complexes. The NMR spectral and physical properties of these chelate compounds are examined.*

Initial work by Langer (1, 2, 3) on diamine chelates of organolithium reagents and the subsequent discovery of inorganic lithium salt chelates by Langer and Whitney (4) prompted a search in these laboratories for a convenient source of polydentate nitrogen bases. Polyethylene polyamines were preferred since these ligands give particularly stable five-membered chelate ring structures when complexed with metal ions.

However, the commercial oligomerization used to produce these polyamines gives multicomponent distillation fractions (5) rather than pure compounds. These fractions, recently studied by gas chromatography and mass spectroscopy, have proved to be mixtures of isomeric and homologous compounds (*6, 7*). The various distillation fractions are named for the major polyamine component contained in each—for example, ethylenediamine (EDA), diethylenetriamine (DETA), triethylenetetramine (TETA), and tetraethylenepentamine (TEPA).

A standard method used to separate single polyamine components from these fractions involves neutralization of the amines with a mineral acid and subsequent fractional crystallization of the salt adducts. Some purified polyethylene polyamines have been obtained from DETA (*8*), TETA (*9*), and TEPA (*10*), as well as from technical grade 1,2-diaminocyclohexane, DACH (*11*), using this procedure. Fractional crystallization of certain polyamine hydrates has also been reported (*12*). The difficulties associated with these methods—poor yields and low separation selectivities—appear in the literature, however (*13, 14*).

Another approach, involving polyamine complexes of transition metal salts, has been studied with specific triamine (*15*) and tetramine (*12*) mixtures. Here again, however, because of protic solvents, the results have been less than adequate to meet the difficult separation problem.

Our background with chelate complexes suggested the use of Group IA and IIA metal salts for selective polyamine complexation. The specificity of the interaction between alkali-metal and alkaline-earth salts and certain polyamines provides a sensitive technique for separating single polyamines from multicomponent samples. These separations, the factors that affect complex formation, and the unique properties of the polytertiary amine chelates of inorganic lithium compounds are discussed in this paper.

Polyamines

Group IA and IIA metal salts have been used to separate single polyamines from multicomponent samples. Three steps are generally involved in a separation: complex formation, complex isolation, and complex dissociation to recover the polyamine. It was not surprising to find that an interdependence of these steps exists so that factors favoring complex formation tend to make subsequent dissociation of the complex more difficult. A discussion of these factors appears later in this paper.

To avoid confusion in describing polyamine separations, Table I shows skeletal structural formulas of the isomeric and homologous polyethylene polyamines. Included with each formula is a derived abbreviation corresponding to the N-permethylated tertiary amine counterpart of the particular structure. The abbreviations are based on the homologous relationship of the N-permethylated structures to the first member of the series, N,N,N',N'-tetramethylethylenediamine, or TMED (Compound 1,

Table I. Skeletal Structures of n-Permethylated Tertiary Polyamines

Compound	Structure	Abbreviation
1	N∼N	TMED
2	N∼N∼N	PMDT
3	N∼N∼N∼N	n-HMTT
4	N(∼N)₃	iso-HMTT
5	N∼N◯N∼N	sym-c-TMTT
6	N◯N∼N∼N	unsym-c-TMTT
7	N∼N∼N∼N∼N	n-HMTP
8	N∼N∼N(∼N)₂	iso-HMTP
9	N∼N◯N∼N∼N	N,N'-c-PMPP
10	N◯N∼N∼N∼N	N-c-PMPP
11	N◯N∼N(∼N)₂	iso-c-PMPP
12	(cyclohexane with two N substituents, cis)	cis-TMCHD
13	(cyclohexane with two N substituents, trans)	trans-TMCHD

Table I). The next member is N,N,N',N'',N''-pentamethyldiethylenetriamine, or PMDT (Compound 2). Abbreviations for the higher homologs are derived accordingly, and isomeric differentiation is made where applicable.

The tetramethylcyclohexanediamine isomers, Compounds 12 and 13 in Table I, occurred admixed with at least three other components in the technical grade samples used. Except for the TMCHD isomers, some or all of the remaining compounds in Table I are found in differing amounts in the various N-permethylated polyethylene polyamine fractions (6, 7). Experimentally the N-permethylated amines were obtained via the Eschweiler-Clarke reaction (16).

Diamine Mixtures

The first practical separation was demonstrated on a sample of N-permethylated cyclohexanediamine (4). The sample, obtained by fractional distillation, contained 95.8% trans-TMCHD (Compound 13 in Table I) and at least four impurities. A heterogeneous mixture of this material with 0.9 equivalent of lithium bromide in heptane was agitated for several days to aid complex formation. The white, sparingly soluble complex that formed was isolated by filtration and was characterized as a 1:1 complex of the diamine and LiBr. The complex was dissociated in aqueous potassium hydroxide, and the polyamine was recovered by extraction into heptane. Gas chromatographic analysis of this extract showed the diamine to be 99.6% *trans*-TMCHD. One of the original impurities could no longer be detected while the remaining three totaled only 0.4%.

Data from a subsequent separation involving a grossly impure TMCHD mixture are summarized in Table II. As indicated there, the polyamine mixture used contained major amounts of *cis*- and *trans*-

Table II. Selective Complexation of *trans*-TMCHD

	Composition (wt %)	
Component[a]	Initial Mixture	LiBr Chelate
Unknown	1.7	—
cis-TMCHD	31.7	—
1,3-TMCHD	43[b]	—
trans-TMCHD	23[b]	100

[a] Listed in order of gas chromatographic elution.
[b] Complete peak resolution not possible.

TMCHD as well as an additional isomer identified as N,N,N',N'-tetramethyl-1,3-cyclohexanediamine (1,3-TMCHD). Contacting this mixture with about 10 mole % of lithium bromide gave a stoichiometric complex that contained 100% pure *trans*-TMCHD.

While the *trans*-TMCHD isomer complexes preferentially with LiBr in a competitive experiment such as the one just described, a 1:1 complex of LiBr with *cis*-TMCHD (Compound 12 in Table I) can be prepared by a stoichiometric reaction between the free ligand and LiBr. Comparison of the solid state dissociation temperature of *cis*-TMCHD · LiBr with that of *trans*-TMCHD · LiBr suggests that the latter is thermodynamically more stable. Whether the kinetics of complexation favor the formation of *trans*-TMCHD · LiBr from a mixture containing both *cis*- and *trans*-TMCHD and insufficient lithium bromide is not certain.

The structure of the cyclohexane ring imposes a particular geometry for the N—C—C—N moiety in *cis*- and *trans*-TMCHD. The two planes defined by the N—C—C—N unit intersect with a dihedral angle of about 60° in both *cis*- and *trans*-TMCHD. This picture is presented by Newman projection formulas in Figure 1. The stability differences for *cis*- and *trans*-TMCHD · LiBr are best rationalized by entropy considerations for the respective ligands. *cis*-TMCHD, having one —$N(CH_3)_2$ group axial and one equatorial, can invert to its superimposable mirror image. Such an inversion for *trans*-TMCHD, which has both —$N(CH_3)_2$ groups equatorial, is precluded. The subsequent smaller entropy loss for the trans isomer on complexation makes the *trans*-TMCHD chelate more stable than the *cis*-TMCHD chelate.

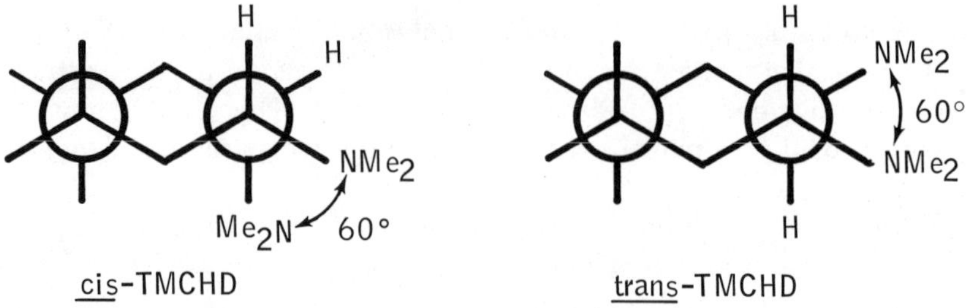

Figure 1. Newman projections with approximate N—C—C—N dihedral angles

A dihedral angle of 64.8° has been found for the chelating diamine unit in (ethylenediamine)$_2$ · LiBr (*17*). Similar puckering of the chelate ring has been found in TMED · LiC(C$_6$H$_5$)$_3$ (*18*). Since the N—C—C—N grouping has the potential for free rotation in ethylenediamine and TMED, the observed chelate geometries most probably approximate an energy minimum for the chelate ring containing a lithium ion. The close approximation of this N—C—C—N dihedral angle in the TMCHD isomers (*see* Figure 1) and related entropy considerations predict a greater thermal stability of TMCHD chelates relative to their TMED analogs. This has been confirmed experimentally with LiCl complexes of *trans*-TMCHD and TMED. The former is thermally stable at room temperature whereas TMED · LiCl can be prepared only at reduced temperatures.

The selective complexation of *cis*-TMCHD has been observed with MgCl$_2$ · 6H$_2$O. This salt was slurried with an 86% pure *trans*-TMCHD sample containing *cis*-TMCHD as the major contaminant. The insoluble *cis*-TMCHD · MgCl$_2$ complex was isolated by filtration, leaving a solution containing 96.2% pure *trans*-TMCHD. Attempts to explain this reversal of selectivity for magnesium chloride would require extensive speculation. Without more detailed information regarding the crystalline structures, coordinate geometries, and the role of the water of hydration in the respective complexes, any speculation would be meaningless.

Tetramine Mixtures

Our study of selective complexation was continued with samples of *N*-permethylated triethylenetetramines. These samples generally were found to contain four major components: the linear and branched acyclic isomers *n*-HMTT and *iso*-HMTT, respectively, and the cyclic piperazine derivatives, symmetrical and unsymmetrical TMTT. Unequivocal characterization of these four ligands was obtained by their unique chemical ionization mass spectra (*7*) and NMR spectra. The latter are summarized in Table III.

Table III. NMR Spectra of Tetramines[a]

Tetramine	Assignment	
	N—CH_2—	NCH_3
n-HMTT	7.53 m (12H)[b]	7.81 s (6H)
		7.86 s (12H)
iso-HMTT	7.48 m (12H)[c]	7.85 s (18H)
sym-c-TMTT	7.61 s (9H)	7.88 s (12H)
	7.55 s (7H)	
unsym-c-TMTT	7.58 m (16H)	7.82 s (3H)
		7.88 s (9H)

[a] Solutions about 10 wt % in benzene (internal TMS).
[b] Chemical shift (τ ppm); multiplicity (m = multiplet, s = singlet); relative intensity.
[c] A_2B_2 pattern.

Separations involving the N-permethylated tetramines were made using sodium iodide or lithium chloride as well as other alkali-metal salts (4). The results of several separations are given in Table IV. In that table, the compositions refer to the weight % of the tetramine components (Compounds 3–6) in each sample before the particular complexation. The salt used is included and so is the chelate isolated. The last column also gives the purities of the tetramines recovered from the chelates.

In the first experiment, sodium iodide gave a complex with the major polyamine component in the mixture, iso-HMTT. Purity of the iso-HMTT was increased to more than 99% by selective complexation. This separation was done by stirring a benzene solution of the crude polyamine

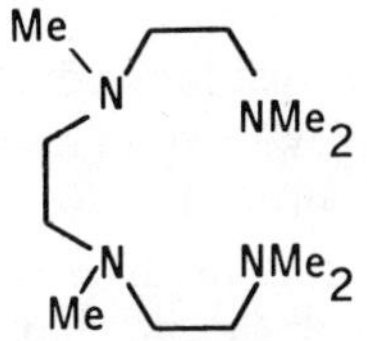

n-HMTT
(3)

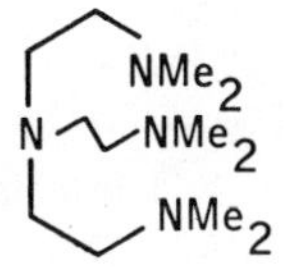

iso-HMTT
(4)

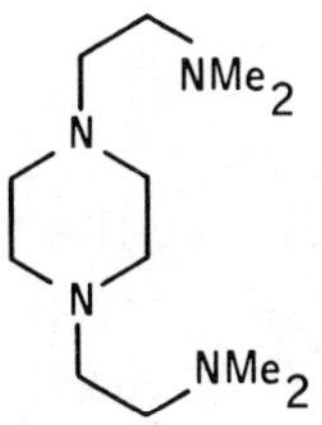

sym-c-TMTT
(5)

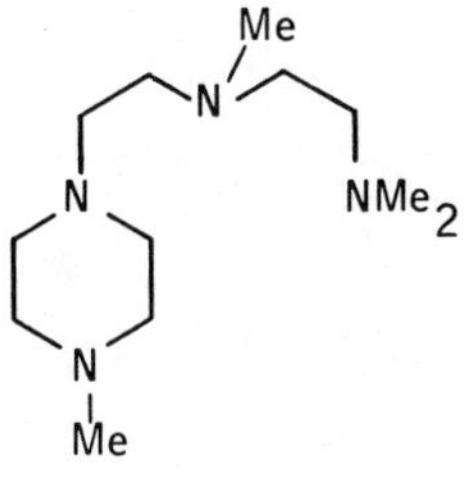

unsym-c-TMTT
(6)

sample with powdered sodium iodide and isolating the insoluble iso-HMTT · NaI by filtration.

An even more convincing demonstration of the complexation selectivity of sodium iodide for the branched tetramine, iso-HMTT, was obtained in the second experiment. The polyamine mixture in that instance contained 9.2% iso-HMTT along with 82% n-HMTT and 9.1% of the combined cyclic TMTT isomers. The sample also contained traces of TMED and PMDT (Compounds 1 and 2 in Table I); it represents a typical N-permethylated commercial TETA fraction. In this case, complexation with ca. 7.7 mole % sodium iodide suspended in benzene gave a polyamine chelate that still contained 97.3% iso-HMTT.

The reason for the high selectivity of the branched tetramine isomer in experiments with sodium iodide is not known for sure. However, complexes of both n- and iso-HMTT with sodium iodide were prepared, and their solubilities were measured in benzene and o-dichlorobenzene. In those experiments, iso-HMTT · NaI exhibited much lower solubility than did n-HMTT–NaI. These solubility differences may play an important role in the high degree of selectivity found between the acyclic tetramine isomers and sodium iodide.

In another experiment involving an initial tetramine mixture similar to that used in experiment 2 of Table IV, virtually all the iso-HMTT was removed by complexation with sodium iodide in pentane. Fractional vacuum distillation of the remaining tetramine mixture separated n-HMTT from the cyclic TMTT isomers. Gas-chromatographic analysis showed that the polyamine recovered this way was more than 99% n-HMTT.

The cyclic TMTT isomers obtained as the residue from a similar fractional distillation were separated by selective complexation with lithium chloride. The final entry in Table IV records the composition of the polyamine mixture that contained major amounts of the cyclic TMTT isomers. Contacting this mixture with excess LiCl (about 400 mole %)

Table IV. Separation of N-Permethylated Tetramines

Experiment	Composition (wt %)				Salt	Compound Isolated (Amine Purity, %)
	n-HMTT	iso-HMTT	sym-c-TMTT	unsym-c-TMTT		
1	19	81	—	—	NaI	iso-HMTT·NaI (>99)
2[a]	80.9	9.2	4.8	4.3	NaI	iso-HMTT·NaI (97.3)
3	2.2	—	54.2	43.6	LiCl	sym-c-TMTT·LiCl (97.2)

[a] Remaining 0.8% was a mixture of TMED and PMDT.

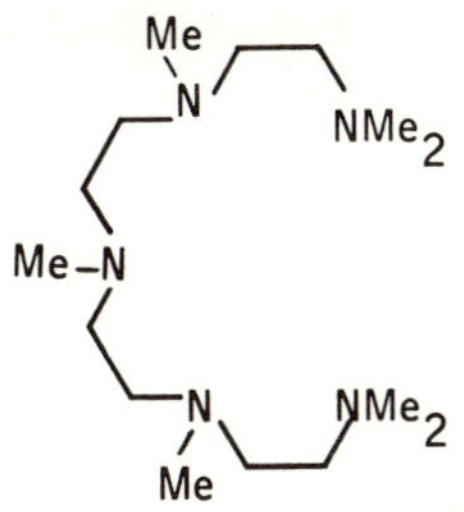

n-HMTP
(**7**)

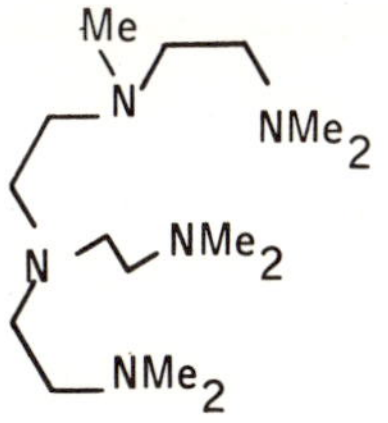

iso-HMTP
(**8**)

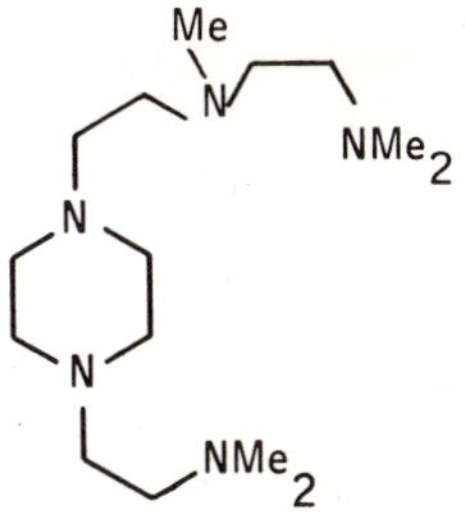

N,N'-c-PMPP
(**9**)

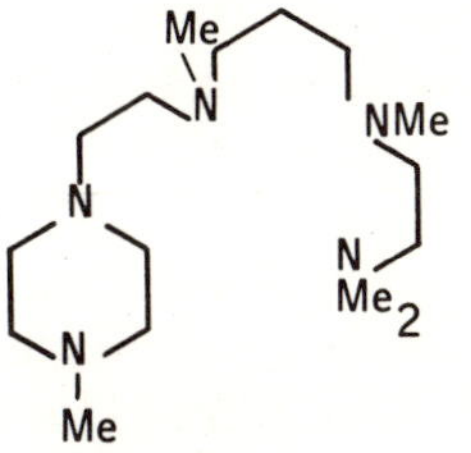

N-c-PMPP
(**10**)

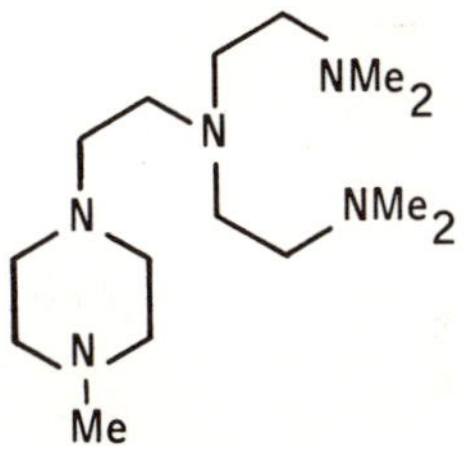

iso-c-PMPP
(**11**)

suspended in heptane produced an insoluble lithium chloride complex of the symmetrical-TMTT isomer. The *sym*-c-TMTT recovered from this complex was 97.5% pure. Subsequent analysis of that portion of the initial mixture that did not form a complex with LiCl showed its composition to be 4.1% n-HMTT and 95.9% *unsym*-c-TMTT. Therefore complexation of the cyclic TMTT isomers with excess LiCl demonstrates the high selectivity of the symmetrical isomer for adduct formation.

Pentamine Mixtures

Subsequent work revealed that chelate formation with Group IA and IIA metal salts could successfully separate N-permethylated TEPA samples. As mentioned, TEPA is the notation given to the commercial polyamine fraction containing mainly the isomeric tetraethylenepentamines (6). The N-permethylated pentamine components, characterized by NMR and chemical-ionization mass spectrometry (19), include the two acyclic isomers—Compound 7 and 8—and three derivatives of piperazine—Compounds 9 through 11. Of these, only three of the pentamines—n-HMTP (Compound 7), iso-HMTP (Compound 8), and N,N'-c-PMPP (Compound 10)—are present in amounts greater than 10% in a typical N-permethylated sample of TEPA. The amounts of these components found in one such sample are listed in Table V.

Table V. Sequential Separations from N-Permethylated TEPA

Component	Composition (wt %)		
	Initial Sample[a]	NaI Chelate	LiCl Chelate
n-HMTP	48.8	82.6	0.3
iso-HMTP	13.9	10.1	1.1
N,N'-c-PMPP	11.4	1.1	94.8
Others[b]	25.9	6.2	3.8

[a] Components listed are present in amounts greater than 10%.
[b] Eight other components detected by gas chromatography.

The data in Table V were obtained by sequential treatment of the initial sample with sodium iodide and lithium chloride. Complexation with sodium iodide was done in a heptane–benzene slurry. The sparingly soluble sodium iodide chelate was isolated by filtering the mixture. The remaining solution was concentrated, and the residue obtained was contacted with lithium chloride in pentane. After stirring this heterogeneous mixture, a solid lithium chloride chelate complex was isolated by filtration. Decomposition of the alkali-metal salt complexes followed by recovery and analysis of the polyamine components showed that the sodium iodide complex contained 82.6% n-HMTP while the LiCl complex contained 94.8% N,N'-c-PMPP. Table V shows that the initial polyamine sample contained 48.8 and 11.4% of these ligands, respectively.

If any analogy were to be drawn between experiments involving the isomeric tetradentate and pentadentate piperazine derivatives, it is that in both cases the more symmetrical ligands would complex with lithium chloride. Specifically, in the two studies, sym-c-TMTT (Compound 5) and N,N'-c-PMPP (Compound 9) ligated LiCl selectively.

An independent observation of considerable interest concerns the chemical-ionization mass spectral analysis of the isomeric TMTT and PMPP ligands. The spectra of pure TMTT and PMPP isomers were

Table VI. Relative Abundance of $[M + 1]^+$ in CH_4-CIMS Spectra of Cyclic Polyamine Bases[a]

Number of Nitrogens	Base	Relative abundance $[M + 1]^+$, %
4	sym-c-TMTT	34.4
4	unsym-c-TMTT	14.7
5	N,N'-c-PMPP	19.0
5	N'-s-PMPP	6.9
5	iso-c-PMPP	4.1

[a] $[M + 1]^+$ represents the amount of protonated base relative to total ion flux.

obtained by using methane as the reactant gas (7, 19). While positive-ion fragments accounted for an appreciable fraction of the total ion flux under these conditions, the protonated polyamine, $[M + 1]^+$, was found in each case. The relative abundance of this ion (Table VI) shows the preference of the protonated species from sym-c-TMTT and N,N'-c-PMPP in the respective isomeric series. The more symmetrical base seems to have the higher proton affinity under the analytical conditions used.

To study the complexation selectivity of sodium iodide for the linear pentamine isomer, n-HMTP (Compound 7), and not the branched isomer (as in the analogous tetramine separation), both n-HMTP · NaI and iso-HMTP · NaI were prepared and studied. No substantial difference in solubilities of the pentamine–sodium iodide complexes in benzene could be observed. However, the stability of n-HMTP · NaI in benzene seemed to be appreciably greater than that of iso-HMTP · NaI in the same solvent. The latter complex dissociated in benzene to give free iso-HMTP and insoluble sodium iodide at 40°C. The corresponding dissociation temperature for n-HMTP · NaI in benzene was 72°C. The complexation selectivity found by combining N-permethylated TEPA with sodium iodide may be rationalized if the above results indeed were to reflect a difference in the relative stabilities of the two complexes.

Proton NMR Spectra of Sodium Iodide Complexes

Since both n-HMTP · NaI and iso-HMTP · NaI are soluble in benzene, the ^1H NMR spectra of these compounds were recorded, together with the spectra of the corresponding uncomplexed ligands. These spectra, obtained in benzene using tetramethylsilane (TMS) as an internal standard, are compared in Figure 2. A clear differentiation between the N—CH_2 and N—CH_3 proton resonances is possible in the spectra of the uncomplexed ligands, n-HMTP and iso-HMTP. The higher field resonances in these spectra are assigned to the N—CH_3 groups while the resonances occurring at lower field are assigned to the N—CH_2 protons. Such differentiation becomes more difficult with the NaI complexes. Upon complex formation, an upfield shift of the N—CH_2 resonances and a smaller downfield shift of the N—CH_3 resonances from their respective spectral positions in the uncomplexed ligands occur.

With iso-HMTP · NaI, the bulk of the N—CH_2 signal is shifted upfield to the point where N-methylene and N-methyl resonances show considerable overlap. Peak assignment can tentatively be made with n-HMTP · NaI, however, and the magnitude of the chelation shift can be estmiated. In n-HMTP · NaI, the upfield shifts of the N—CH_2 protons average +0.13 ppm; the downfield shifts of the N—CH_3 resonances are

−0.06 and −0.03 ppm relative to their respective positions in the spectrum of free n-HMTP.

This shift phenomenon has been observed in benzene solution and has been discussed for TMED · LiC$_6$H$_5$ (20) and iso-HMTT · LiBr (21).

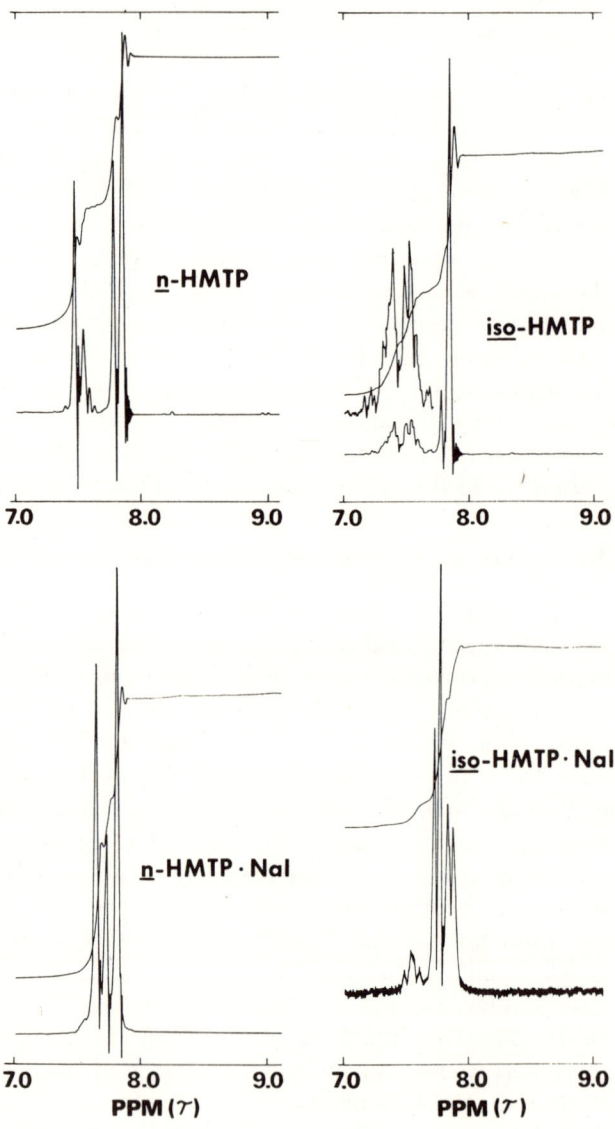

Figure 2. NMR spectra of pentamines and NaI chelates in benzene

An explanation offered for the upfield chelation shift envisions a loose association of benzene molecules about the positive end of the chelate complex dipole (21). A similar solvation model has been proposed for TMED·$ZnCl_2$ in mixtures of dioxane and benzene (22). According to the model, the N—CH_2 protons of the ligand would be near the benzene π-molecular orbitals, the associated diamagnetic anisotropy of which would induce an upfield shift in the N—CH_2 proton resonances.

Mixtures Containing Primary and Secondary Amines

Selective chelation is also effective in separating commercial polyamine fractions that have not been modified by prior N-permethylation. These polyamine mixtures therefore contain primary and secondary as well as tertiary amino groups. This can be seen, for example, in the structures of the major components found in triethylenetetramine (TETA) shown below with their common notations. Commercial TETA generally contains about 75% of the linear and about 8% of the branched tetramines, trien and tren, respectively. The remaining 17% or so is a mixture of dien and tet isomers. Data on separations involving such a mixture are given in Table VII. That table lists the ligand compositions of the NaBr and LiBr complexes isolated as functions of complexation time. These experiments were conducted at 0°C, and the complexes were isolated from the homogeneous solutions by precipitation with cold toluene followed immediately by suction filtration and drying.

Composition of the complexes varies with time. With NaBr, the complex isolated after 5 minutes was essentially pure tren·NaBr; after

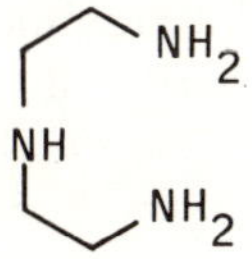

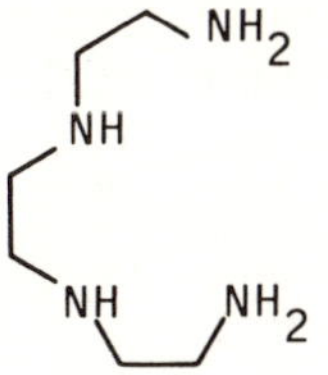

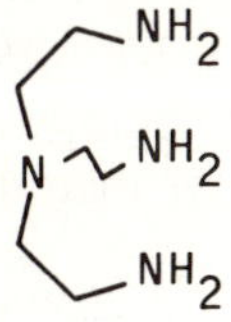

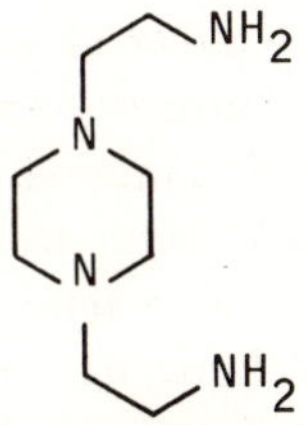

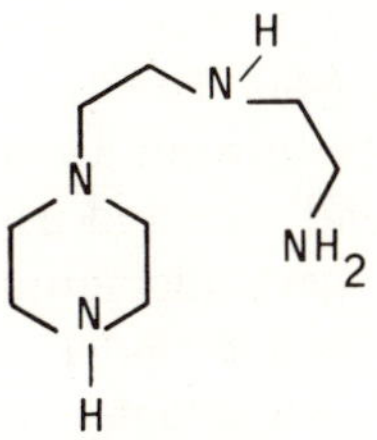

Table VII. Separation of Commercial TETA Mixture[a]

Time (min)	Ligand Composition (wt %)			
	NaBr complex		LiBr Complex	
	tren	trien	tren	trien
5	99.5	0.5	—	—
45	48.5	51.5	82.4	17.6
75	5.6	94.4	72.2	27.8

[a] Temperature 0°C.

75 minutes the complex isolated contained mainly the linear isomer trien. The results obtained with LiBr were similar directionally although the initial selectvity for tren · LiBr formation appeared to be lower, and a much longer time was necessary to form the trien complex. (After 12 days, the LiBr complex contained 76% trien.)

This was the first instance where the products of kinetic and thermodynamic complexation were observed. The mechanism of ligand exchange that leads to the more stable trien complex has not been studied, but no doubt it involves substantial ligand reorganization in the primary coordination sphere of the metal ion. Such reorganization should be inhibited by an increase in the charge/size ratio of the metal ion and is consistent with our observation of much slower ligand exchange with lithium bromide.

Complexation studies were continued at room temperature using tetramine mixtures in which the mole ratio of tren:trien was varied in (Table VIII). The salt used, the mole ratios of tren:trien:salt, and the diluent used (if any) are given in the table. In all cases selective complexation of the branched tetramine, tren, was observed. The final column in Table VIII gives the purity of the branched tetramine in the tren–salt complex that was isolated. Generally the purity of the tren was greater than 99%.

Salt Lattice Energy

Earlier work has shown that for lithium salts successful complexation was achieved only with salts having lattice energies of less than about 210 kcal/mole (4). To study further the effect of salt lattice energy on the salt's ability to form a complex, experiments were done with Group IA and IIA metal salts and tris-(2-aminoethyl)amine.

Salts of a particular group IA or IIA metal were stirred with tris-(2-aminoethyl)amine under a specific set of reaction conditions (25°C, 6 mmoles salt, 6.5 mmoles amine, 7 ml benzene, and 65 hours). The yield of the insoluble complex was determined after filtering and drying. Some salts formed complexes while others did not, depending on the salt's

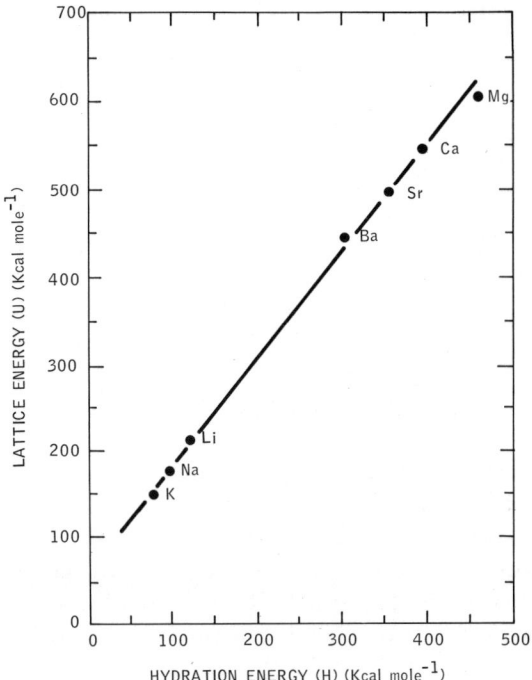

Figure 3. Plot of lattice energy vs. hydration energy for salts complexed by tris-(2-aminoethyl)-amine

lattice energy. The critical lattice energy for complexation, U_{crit}, was taken as the highest lattice energy represented by the salts of a given metal that successfully complexed with tris-(2-aminoethyl)amine. These critical lattice energies correlate linearly with the hydration energy of the respective alkali metal and alkaline earth cations (*see* Figure 3).

The above results indicate that complexation is favored for low-lattice-energy salts. Also, the lower the salt's lattice energy the more

Table VIII. Complexation of tren from Primary Tetramine Mixtures

Salt	Mole Ratio Reactants tren:trien:salt			Diluent	Amine Purity tren·salt Complex (%)
NaBr	1	4	1	—	99.5
NaBr	4	0.9	1	—	99.8
NaBr	1	1	0.9	EDA	99.2
LiBr	4	0.9	4	THF[a]	98.9

[a] Tetrahydrofuran.

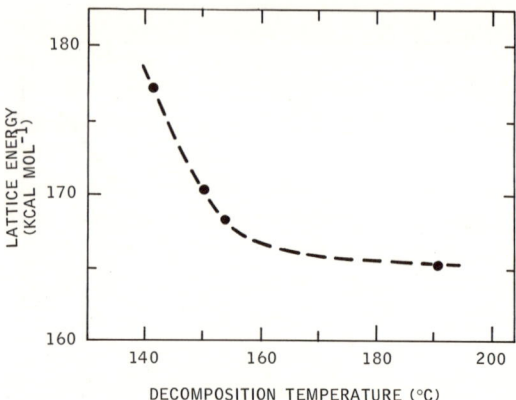

Figure 4. Effect of salt lattice energy on tren · salt decomposition temperature

stable the complex and the higher the temperature required to dissociate the complex thermally into the free polyamine and salt. Figure 4 presents this effect for sodium salt complexes of tren. A similar relationship exists for other polyamine complexes of Group IA and IIA metal salts.

Consequently, to aid thermal dissociation of the polyamine · salt complex, the salt should be chosen initially so that its lattice energy is near the critical lattice energy for the particular metal. Since the U_{crit} values determined with tris-(2-aminoethyl)amine (Figure 3) are not necessarily the same for another polyamine ligand, the choice of a metal salt for selective polyamine complexation requires some experimentation. The selectivities of various salts toward a multicomponent polyamine sample are best determined by several small-scale experiments using a variety of metal salts. Consideration must be given to the salt lattice energy which should be selected to optimize both the complexation and the dissociation. It is also possible, by varying the temperature used in the complexation, to enhance further the selectivity and flexibility of the selective chelation technique.

Complex Dissociation

Dissociation of the polyamine · salt adduct must be achieved to liberate the salt and recover the polyamine. With complexes of the N-permethylated polyethylene polyamines, it is convenient (on a small scale) to dissociate the complex by dissolving it in aqueous potassium hydroxide. The tertiary polyamine is generally extracted from this mixture without difficulty. Dissolution of the complexes containing primary polyamines likewise leads to dissociation. However, high solubility of the primary polyamines in water creates a recovery problem. With complexes containing primary amine functions, it is preferable, and generally easier in all cases, to recover the amines by thermal dissociation. Such a procedure obviates the use of polar, protic solvents for polyamine recovery and permits recovery of crystalline, anhydrous salts.

Thermal dissociation of tren · NaBr was studied, and the results are summarized in Table IX. In these experiments thermal dissociation was done by extracting the solid complex in a Soxhlet apparatus with a high-boiling hydrocarbon solvent. The solvent was Isopar G, an isoparaffinic distillation fraction boiling at 159° to 177°C. During the extraction the hot solvent bathes the complex and causes its dissociation. The insoluble salt is simultaneously separated from the polyamine which dissolves in the hot solvent. When the hydrocarbon-tren solution is cooled, a phase separation occurs since polyamines containing primary (and secondary) nitrogens have low solubilities in paraffins at ambient temperature (5).

Table IX shows that this treatment for four hours gave a quantitative recovery of NaBr and a 67.2% recovery of tren. The low yield of recovered tren results from the small but finite solubility of tren in Isopar G (about 0.9 gram/100 ml). This solvent, now saturated with tren, can be recycled to recover the polyamine from a fresh charge of tren · NaBr. The results of a dissociation experiment using recycled solvent appear in the second entry of Table IX. The data show that the recoveries of NaBr and tren by this procedure are both quantitative. The major advantages of this technique are high recovery yields and the avoidance of contacting the polyamines with water or other protic solvents that might adversely affect the purity of the recovered products.

The same type of dissociation technique may be used to recover N-permethylated polyamines from their metal–salt complexes. The solvent is the major factor to be considered in designing such a procedure. The solvent chosen should boil near or slightly above the thermal dissociation temperature of the particular polyamine · salt complex. With N-permethylated polyethylene polyamines, the final recovery of a polyamine from the extraction solvent is best done by distilling the hydrocarbon solution. Recovery by phase separation is not possible with the tertiary polyamines since these compounds generally are miscible with most hydrocarbon solvents.

Table IX. Thermal Dissociation of tren · NaBr and Recovery of Components[a]

Run	Extraction Time (hours)	% tren Recovered	% NaBr Recovered
1	4	67.2[b]	100
2	2[c]	100	100

[a] Hot solvent extraction of tren · NaBr by Isopar G (bp, 159°-177°C).
[b] Tren solubility in Isopar G equivalent to 0.9 gram/100 ml.
[c] Isopar G recovered from Run 1 was recycled.

Benzene-Soluble Inorganic Complexes

The N-permethylated polyethylene polyamine complexes of inorganic lithium salts show unusual properties. These compounds generally exhibit 1:1 stoichiometries of amine to lithium salt although $(TMED)_2$ $LiAlH_4$ (4) and TMED $(LiAlH_4)_2$ (23) have been reported. The solid lithium salt chelates show exceptional stability, as shown by the data in Table X for chelate complexes of lithium aluminum hydride. The thermal stabilities of the $LiAlH_4$ chelates equal or exceed the thermal stability of free $LiAlH_4$. With PMDT · $LiAlH_4$, the complete complex can actually be vacuum sublimed intact. PMDT · LiI can be handled similarly (4).

The high solubility of tertiary polyamine–alkali-metal salt chelates in benzene has been mentioned briefly. The data in Table XI illustrate the range of solubilities observed for LiBr and $LiNO_3$ chelates of bi-, tri-, tetra-, and pentadentate tertiary amine ligands. Benzene solutions containing several moles of chelate have been obtained.

Preliminary studies indicate substantial aggregation of the chelates at concentrations greater than about $0.1M$. Many of the inorganic lithium chelates exhibit high conductivities in benzene; they approach the electrical conductivity measured for lithium bromide in propylene carbonate. This observation has prompted application of chelate solutions as electrolyte systems (4).

In short the chemistry and properties of the inorganic lithium chelates should promote their uses in many applications. This is the case for the N-chelated organolithium reagents, as evidenced by the frequency of appearance of these systems in the literature. We hope that the polyamine-purification technique presented here will improve the availability of useful chelating ligands and stimulate additional research in this area.

Experimental

Elemental analyses were performed by the Analytical and Information Division of Esso Research and Engineering Co. NMR spectra were recorded on a Varian Associates A-60 spectrometer, and mass spectra were obtained on either a C.E.C. model 21-103 spectrometer or the Esso

Table X. Thermal Stabilities of $LiAlH_4$ Chelates

Chelate	Thermal Stability
TMED · $LiAlH_4$	Decomp. 125°C (1.3 torr)
$(TMED)_2$ · $LiAlH_4$	mp 118–120°C
PMDT · $LiAlH_4$	Sublimes 125°C (0.5 torr)[a]
n-HMTT · $LiAlH_4$	mp > 200°C

[a] Mp 150°-155°C.

Table XI. Solubilities of Chelated Lithium Salts in Benzene

Chelating Agent	Solubility (M at 25°C)	
	LiBr Chelate	$LiNO_3$ Chelate
TMED	1.6	0.3
PMDT	2.5	2.66
n-HMTT	0.3	>3[a]
iso-HMTT	>0.5	1.0
n-HMTP	>2	—

[a] Liquid chelate miscible with benzene.

chemical physics chemical ionization mass spectrometer (7, 24) operating with isobutane as the reactant gas. Compositions of amine mixtures were analyzed by gas-liquid chromatography. Analyses were made on a Varian Aerograph model 200 or a Perkin Elmer model 900 gas chromatograph. The columns used were 10' by 1/4" or 6' by 1/8" packed with 10-15% Carbowax 20M-KOH on 60/80 mesh Chromosorb W. Analyses were made between 100°–200°C, and peak areas were determined using an Infotronics model CRS 101 digital integrator.

The sources of amines included Geigy, Dow, Union Carbide, Jefferson Chemical, Ames Laboratories, and Matheson Coleman and Bell. Tertiary amines were generally obtained by N-permethylation of the corresponding technical-grade amine fraction using common techniques (16).

trans-**TMCHD.** A 3.4-gram sample of TMCHD (containing 95.8% of the trans isomer) was combined with 1.57 grams hydrated LiBr. The resulting thick slurry was diluted with 9 ml heptane, and the mixture agitated occasionally for 5 days. The solid chelate complex was isolated by filtration, washed with heptane, and decomposed in aqueous KOH. The recovered amine was 99.6% pure *trans*-TMCHD. Structural assignment of the cis and trans isomers was supported by comparison with authentic samples prepared by a diffcrent route (11). Elemental analysis of *trans*-TMCHD · LiBr gave 47.78% C, 9.25% H, 11.21% N, and 31.42% Br. (Theoretical: 46.71% C, 8.62% H, 10.89% N, and 31.08% Br.)

In another example, a sample of TMCHD (*ca.* 5 mmoles, containing 86% *trans*-TMCHD with the cis isomer as the major contaminant) was contacted for 2 days with 4.3 mmoles of $MgCl_2 \cdot 6H_2O$ in 2.3 ml of saturated hydrocarbon diluent. After removing solids by filtration, the amine content of the filtrate was 96.2% *trans*-TMCHD, showing that the cis isomer had been removed by complexation.

***n*-HMTT.** One mole of impure HMTT (composition: TMED, PMDT 0.8%; iso-HMTT, 9.2%; n-HMTT, 80.9%; c-TMTT isomers, 9.1%) in 1.5 liters n-pentane was stirred for 24 hours with 23 grams powdered NaI and then for an additional 24 hours with a second 23-gram charge of NaI. The solids were filtered off, and the solution contained less than 0.6% iso-HMTT. This solution was distilled (bp about 80°C/0.075 mm on a 15-plate Oldershaw column operating at a 20:1 reflux ratio) to give 58.6% n-HMTT of purity above 99%.

***iso*-HMTT.** A similar sample of HMTT (8.1 grams) was stirred with 0.405 gram NaI in 19 ml benzene. After 3 days, the solid chelate (0.701 gram) was isolated and decomposed in the normal way. The recovered polyamine was 97.27% iso-HMTT and 2.72% *n*-HMTT.

***sym*-TMTT and *unsym*-TMTT.** A 1.14 gram (about 5 mmole) sample of an amine mixture (composition: 2.2% *n*-HMTT, 54.2% *sym*-c-TMTT, and 43.6% *unsym*-c-TMTT) was combined with 0.85 gram (20 mmoles) LiCl and 8 ml heptane. After stirring for 2.5 days, the mixture was centrifuged. The supernatant liquid contained 4.1% *n*-HMTT and 95.9% *unsym*-c-TMTT. The solid residue was isolated by filtration and decomposed in aqueous KOH. The amine recovered was 2.5% *unsym*-c-TMTT and 97.5% *sym*-c-TMTT. Larger-scale runs gave similar results. The c-TMTT isomers were identified by their NMR spectra (benzene-TMS): *sym*-c-TMTT had one N—CH_3 singlet at τ 7.88; *unsym*-c-TMTT had two N—CH_3 singlets in a 1:3 ratio at τ 7.82 and 7.88, respectively.

***iso*-HMTT.** An HMTT sample containing 81% iso-HMTT and 19% *n*-HMTT (178 grams) was combined with 80.0 grams NaI and 2.5 liters benzene. After stirring for 28 hours, the solid chelate complex was isolated by filtration, washed with 200 ml of benzene, and dried. The first crop of chelate weighed 170 grams.

The filtrate was stirred for a week with an additional 25 grams NaI, and filtration gave 66 grams of a second crop of chelate complex. The total yield of chelate was 89% of theoretical. The amine in both crops of chelate was more than 99% iso-HMTT.

N-Permethylated TEPA. The components of an N-permethylated TEPA mixture containing five or more nitrogen atoms were separated by preparative gas-liquid chromatography using a Varian Aerograph Autoprep at 200°C with a 10' by 3/8" diameter, 15% 20M Carbowax KOH column. Molecular weights of these components were determined *via* the CIMS technique (7).

Sequential Separations. A 28.7-gram sample of crude N-permethylated TEPA was stirred with 9.75 grams NaI in 120 ml 5:1 heptane:benzene. After 41 hours the solid chelate complex A was isolated by filtration. The filtrate was concentrated, filtered, and 0.8 gram LiCl and 50 ml pentane added to the liquid residue. After stirring for 60 hours, filtration gave solid chelate complex B which was washed first with pentane, then with benzene, and dried.

The solid complexes were decomposed in aqueous KOH. The recovered amines were analyzed by gas-liquid chromatography and NMR (benzene-TMS). Complex A contained *n*-HMTP (82.6% pure); its NMR spectrum had two N—CH_3 singlets (2/3 ratio) at τ 7.79 and 7.85, respectively. Complex B contained N,N'-c-PMPP (94.8% pure); its NMR spectrum had three N—CH_3 singlets (1:2:2 ratio) at τ 7.80, 7.86, and 7.87, respectively. The original 11-component mixture contained 48.8% *n*-HMTP and 11.4% N,N'-c-PMPP.

tren · LiBr. A solution of LiBr (4.35 grams) in 20 ml tetrahydrofuran was mixed with 8.6 grams of an amine sample containing 4:1 tren:trien. The mixture was stirred overnight, and filtration gave 11.5 grams of a white, solid complex, melting with decomposition at 217°–220°C. This complex contained 98.9% tren and 1.1% trien. Elemental analysis of tren · LiBr gave 29.91% C, 7.92% H, and 22.60% N. (Theoretical: 30.92% C, 7.78% H, and 24.04% N.)

tren · NaBr. A solution of NaBr (0.55 gram) in 6 ml ethylenediamine was mixed with a 1.750-gram triethylenetetramine sample containing 1:1 tren:trien. The homogeneous solution was partially evaporated after 90 hours, and 15 ml benzene was added, producing a white precipitate. Filtration gave 1.01 grams of a white, solid complex, m.p. 139°–40°C. This complex contained 99.2% tren and 0.8% trien. Elemental analysis of tren · NaBr gave 28.75% C, 7.23% H, and 22.19% N. (Theoretical: 28.92% C, 7.28% H, and 22.49% N.)

Dissociation of tren · NaBr. Solid tren · NaBr (12.5 grams) was loaded in an Alundum extraction thimble which was then placed in a Soxhlet extractor. After extraction with 250 ml Isopar G (Enjay isoparaffinic solvent, bp 159°–177°C) for 2-8 hours, the solution was cooled to room temperature; a phase separation resulted. The lower liquid phase (containing pure tren and 1-3 wt % Isopar G) was removed. Solvent recycle gave quantitative recoveries of NaBr and polyamine.

tren and trien Method A, NaBr. A solution of NaBr (0.55 gram) in 2.2 ml ethylene diamine was mixed with 10.3 grams TETA at 0°C. After an appropriate time 20 ml cold toluene were added to precipitate the 1:1 polyamine–NaBr complexes. Ligand analyses of the latter were made after decomposition of the complexes in formic acid and formaldehyde (16). The complex contained 99.5% tren after about 5 minutes and 94.4% trien after 75 minutes.

tren and trien Method B, LiBr. A solution of LiBr (0.47 gram) in 10 ml tetrahydrofuran was mixed with 10.3 grams TETA. The polyamine–LiBr complexes were precipitated with 10 ml toluene and were analyzed as described above. The complex contained 80.2% tren after about 10 minutes and 72.2% tren after 75 minutes.

Acknowledgment

The experimental assistance of R. Santiago, E. Castillo, and W. Mykytka is gratefully acknowledged. The authors also wish to thank F. H. Field and T. M. Pugel for the CIMS data, and M. T. Melchior for helpful discussions of the NMR spectra.

Literature Cited

1. Langer, Jr., A. W., U.S. Patent **3,451,988** (1969).
2. Langer, Jr., A. W., *Trans. N.Y. Acad. Sci.* (1965) **27**, 741.
3. Langer, Jr., A. W., *Am. Chem. Soc., Div. Polymer Chem., Polymer Preprints* (1966) **7**(1), 132.
4. Langer, Jr., A. W., Whitney, T. A., U.S. Patent **3,734,963** (1973).
5. Yamashita, S., *Chem. Econ. and Eng. Rev.* (1971) **3**, 39.
6. Bergstedt, L., Widmark, G., *Acta Chem. Scand.* (1970) **24**, 2713.
7. Whitney, T. A., Klemann, L. P., Field, F. H., *Anal. Chem.* (1971) **43**, 1048.
8. Jonassen, H. B., LeBlanc, R. B., Meibohm, A. W., Rogan, R. M., *J. Amer. Chem. Soc.* (1950) **72**, 2430.
9. Jonassen, H. B., Strickland, G. T., *J. Amer. Chem. Soc.* (1958) **80**, 312.
10. Reilley, C. N., Vavoulis, A., *Anal. Chem.* (1959) **31**, 243.

11. Langer, Jr., A. W., Whitney, T. A., U.S. Ser. No. **240,870.**
12. Seitetsu Kagaku Kogyo Co., Jap. Patent Publ. No. **3364/71.**
13. Carr, J. D., Margerum, D. W., *J. Amer. Chem. Soc.* (1966) **88,** 1645.
14. Forsberg, J. H., Kubik, T. M., Moeller, T., Gucwa, K., *Inorg. Chem.* (1971) **10,** 2656.
15. Godfrey, N. B., U.S. Patent **3,038,904** (1962).
16. Clarke, H. T., Gillespie, H. B., Weisshaus, S. Z., *J. Amer. Chem. Soc.* (1933) **55,** 4571.
17. Durant, F., Piret, P., Van Meerssche, M., *Acta Crystallogr.* (1967) **23,** 780.
18. Brooks, J. J., Stucky, G. D., *J. Amer. Chem. Soc.* (1972) **94,** 7333.
19. Whitney, T. A., Klemann, L. P., Field, F. H., unpublished data.
20. Langer, Jr., A. W., Naegele, W., Melchior, M. T., unpublished data.
21. Melchior, M. T., Klemann, L. P., Whitney, T. A., Langer, Jr., A. W., *Am. Chem. Soc., Div. Polymer Chem., Polymer Preprints* (1972) **13** (2), 649.
22. Aroney, M. J., LeFèvre, R. J. W., Vaughan, H. J., *J. Chem. Soc., A* (1970) 2224.
23. Dilts, J. A., Ashby, E. C., *Inorg. Chem.* (1970) **9,** 855.
24. Munson, M. S. B., Field, F. H., *J. Amer. Chem. Soc.* (1966) **88,** 2621.

RECEIVED May 15, 1973.

7

Polymerizations Using N-Chelated Alkali Metal Catalysts

C. W. KAMIENSKI

Lithium Corp. of America, Bessemer City, N. C. 28016

>   *A discussion of the uses of alkali metal derivatives chelated with tertiary diamines as polymerization iniatators is presented, based mainly on the literature. Alkyllithium reagents are the main catalyst types covered. Chain elongation of alkyllithiums can be carried out under atmospheric or only slightly elevated pressure (80-90 psig) and mild temperatures to give polyethylenes having average molecular weights of about 300 to 3000. Products with relatively narrow molecular-weight distributions are obtained. Polymerizations are of the living type, and chain lengths are roughly controlled by polymer solubility. Butadiene is polymerized with* n-butyllithium–TMEDA *in hydrocarbon solution at rates about 10 times higher than with* n-butyllithium *alone. The resulting polybutadienes possess a high vinyl or 1,2-microstructure (80% or higher). Both block and random copolymers can be prepared from ethylene and either butadiene, styrene, or acrylonitrile.*

Work in the early 1960's by A. W. Langer and G. G. Eberhardt showed that monomeric one-to-one complexes of alkyllithiums, such as n-butyllithium, with certain diamines, such as N,N,N',N'-tetramethylethylenediamine and sparteine, drastically increase the polymerization rates of various monomers such as ethylene and 1,3-butadiene. A chelated structure has been proposed to account for this rate increase in which a monomeric or unassociated, more highly charge-separated complex functions as the initiating agent. These complexes are correctly considered to be initiating agents when they are used to produce "living" polymers, but they are catalysts when extensive chain transfer occurs as in telomerizations. The terms "catalyst" and "cocatalyst" have been in general use for this system and are used for consistency, recognizing the exceptions for living systems.

Types of Polymerizations

Ethylene. Langer (*1, 2, 3, 4*) was the first investigator to synthesize a plastics-range polyethylene by an ethylene growth reaction using an N-chelated alkyllithium compound. He found that catalyst aging led to metalation of the complexed chelating diamine by the alkyllithium compound and also to increased polymer molecular weights. Thus, n-butyllithium–TMEDA on aging for 40 days in heptane at room temperature polymerized ethylene to a highly crystalline polyethylene of molecular weight 142,000 containing built-in diamine cocatalyst end groups. In the absence of this aging phenomenon much lower molecular weights were obtained. The molecular weights in this low range could be controlled by varying the temperature because of the occurrence of chain-transfer reactions back to cocatalyst during polymerization. Lower-molecular-weight ethylene polymers were obtained at elevated temperatures (Table I).

Table I. Ethylene Polymerization—Effect of Polymerization Temperature (*1*)[a]

Temp (°C)	*Molecular weight*
85	11,700
70	23,000
50	32,500
30	42,500

[a] 2 Mmoles BuLi · TMEDA, 150 ml heptane, 1000 psig, 4 hrs.

Eberhardt (*5*) was able to obtain very low molecular-weight polyethylenes (1300-1400) by polymerizing ethylene at 110°C under pressure in n-octane using n-BuLi–TMEDA (or sparteine) complexes. At this high temperature the catalyst is rather unstable, presumably leading eventually to inactive lithium dialkylamide. Eberhardt postulated chain transfer to ethylene during polymerization to account for the formation of more than one mole of polymer per mole of catalyst and for the presence of terminal unsaturation in the polymer. This mechanism, as well as chain transfer to chelating diamine, would be expected to be operative at intermediate temperatures since there is no reason to suppose that Langer's aging effect does not occur during polymerization as well. A third type of chain-transfer reaction was proposed by Langer (*4*); in this reaction a cyclopentane end group forms in the polymer accompanied by chain transfer to ethylene, forming a molecule of ethyllithium.

Table II. Production of Straight-Chain Polyethylene Waxes (6)

N Type	Initial BuLi Conc. (M)	N/BuLi	C_2H_4 Pressure (psig)	Temp. (°C)	$\overline{M}n$	Moles of Polymer per Atom of Li
Spart.	0.03	2	400	100	1390	—
Spart.	0.03	2	800	100	1360	—
TED	0.06	12	850	105	1700	5.7
Spart.	0.03	4	800	115	1900	7.3
TMEDA	0.04	4	600	120	2000	2.8

The high polymerization temperatures (100°–130°C) used by Eberhardt (6) produced useful polyethylene waxes in the 1000-3000 molecular-weight range, albeit in low yield (Table II). These waxes have a low melt viscosity and are promoted as coating compositions for paper products such as drinking cups and as components of floor polishes.

Yields of wax obtained are generally from six to seven moles per atom of lithium. These waxy products possess terminal unsaturation as expected from a chain transfer to ethylene. Thus, although ethylene pressure affected the polymerization rate, it equally affected the chain-transfer rate and the molecular weight of the resulting polymers thus remained essentially unchanged with a change in ethylene pressure. The reaction temperature seems to affect polymerization more than chain transfer since a slight but significant increase in molecular weight occurs with increasing temperature. Waxes were not produced below 100°C. Apparently insolubility of the polymer–lithium species in the saturated hydrocarbon medium and the resulting product heterogeneity led to polymerization but not chain-transfer with a resulting much higher molecular weight and much poorer catalyst utilization.

Insolubility of the polymer–lithium species in saturated hydrocarbons at low reaction temperatures was demonstrated by Smith (7) and Screttas (8). These workers found that long-chain alkyllithiums of narrow composition that did not suffer extensive chain-transfer reactions were produced. They thus could be converted to useful derivatives such as fatty alcohols and acids or could be used as initiators for polymerization of other monomers. Precipitation of the long-chain alkyllithium with no further occurrence of polymerization after reaching a certain chain length was also regarded as the cause for the relatively narrow molecular-weight distributions obtained. Low ethylene pressures were used to avoid chain growth from any metalated TMEDA which is slowly formed during the polymerization. Using an n-butyllithium · TMEDA complex Smith obtained polyethylenes having melting points of 114° to about 125°C, with each product melting sharply (within a degree or so). Conditions used

Table III. Chain Elongation of

Run No.	Init. BuLi Conc. (M)	BuLi/Cocat.	C_2H_4 pressure (psig)
1	0.8	10[c]	Atm
2	0.2	5[c]	Atm
3	0.5	5[c]	Atm
4	0.2	5[d]	200
5	0.2	2.5[e]	600
6	0.4	0.25[f]	580

[a] Based on weight of carbonated product obtained ($\overline{X}n$ = moles of ethylene reacted/moles of butyllithium).
[b] Based on determined equivalent weight of lithium salt of carbonated product.
[c] N,N,N',N'-Tetramethylethylenediamine.

were generally: ethylene pressure, 30–150 psig; reaction temperature, room temperature to 60°C; reaction time, 1–24 hours. The molecular weights of the products obtained were about 2000. The lower temperatures and pressures gave products with molecular weights of less than 2000; the higher temperatures and pressures gave products having molecular weights higher than 2000. Thus, anywhere from 30 to 130 units of ethylene could be incorporated into a new alkyllithium compound by varying the reaction conditions.

Screttas obtained products having somewhat lower molecular weights using no elevated ethylene pressures but essentially depending on absorption of the gas by the chelated alkyllithium in solution. TMEDA was the only Lewis base that functioned efficiently at atmospheric pressure (Table III). Pressure was required when tetrahydrofuran, 2-dimethylaminoethoxyethane, or 2-dimethylaminoethyltetrahydrofuran were used.

Table IV. Preparation of

Experiment Number	Catalyst (eq.)	Cocatalyst (ml)	Feed Rate (l/min)	Reaction Temp.
7	DiLi-1 (.033)[a]	TMEDA (5)	1.0	40
8	DiLi-1 (.033)	Am. Et. (5)[b]	1.7	110
9	DiLi-1 (.034)	TMEDA (5)	7.5	110
10	DiLi-1 (.034)	TMEDA (5)	7.5	110
11	DiLi-1 (.033)	TMEDA (5)	2.5	104
12	DiLi-1 (.033)	TMEDA (10)	2.5	110
13	DiLi-1 (.033)	TMEDA (2.5)	2.5	110
14	DiLi-1 (.033)	TMEDA (5)	10	110
15	DiLi-1 (.033)	TMEDA (5)	1.7	60
16	DiLi-1 (.034)	Am. Et. (5)	1.7	110

[a] DiLi-1 = Dilithioisoprene produced, as shown in Example 1 of U.S. Patent 3,388,178.

n-Butyllithium with Ethylene

T, °C	Rx Time (hrs)	$\overline{X}_n{}^a$	$\overline{X}_n{}^b$
40	24	8.9	6.9
65	8	13.6	16.1
60	8	11.8	13.5
75	0.75	15.4	18.8
120	2	53	—
70	16	28	—

[d] 2-Dimethylaminoethoxyethane.
[e] 2-Dimethylaminoethyltetrahydrofuran.
[f] Tetrahydrofuran.

In addition, considerably longer chains were obtained with the latter cocatalysts than with TMEDA. Narrow molecular-weight distributions were found in these products—for example, the polyethylene from Run 5 had an $\overline{M}_w/\overline{M}_n$ of 1.23 while that from Run 6 had a value of 1.40. The densities of these products were 0.969 and 0.965, respectively. Their melting points were fairly sharp as indicated by differential thermal analysis. The first four runs showed good utilization of carbon–lithium initiator with little if any chain transfer taking place during the reaction. Discrepancies in the figures of the last two columns may be the result of incomplete drying, incomplete carbonation, or loss of active carbon–lithium bonds *via* cleavage of the cocatalyst during reaction.

Monoolefins higher than ethylene do not undergo a polymerization reaction of this type. Instead such compounds prefer to function as chain-transfer agents as do alkylaromatic compounds (5).

Liquid Polybutadienes (9)

Yield, grams	Viscosity, Poise (T,°C)	Mol Wt	Yield lb/eq. Catalyst
800	163 (60)	2,200	52
720	22 (22)	782	47
857	22.6 (25)	1,070	57
795	1,066 (22)	1,600	53
786	82 (23)	1,200	51
797	5.3 (23)	685	52
700	143 (24)	1,300	46
633	455 (23)	1,450	41
990	202 (58)	2,400	65
584	22 (22)	850	38

[b] Am. Et. = dimethylamino-2-ethoxyethane.

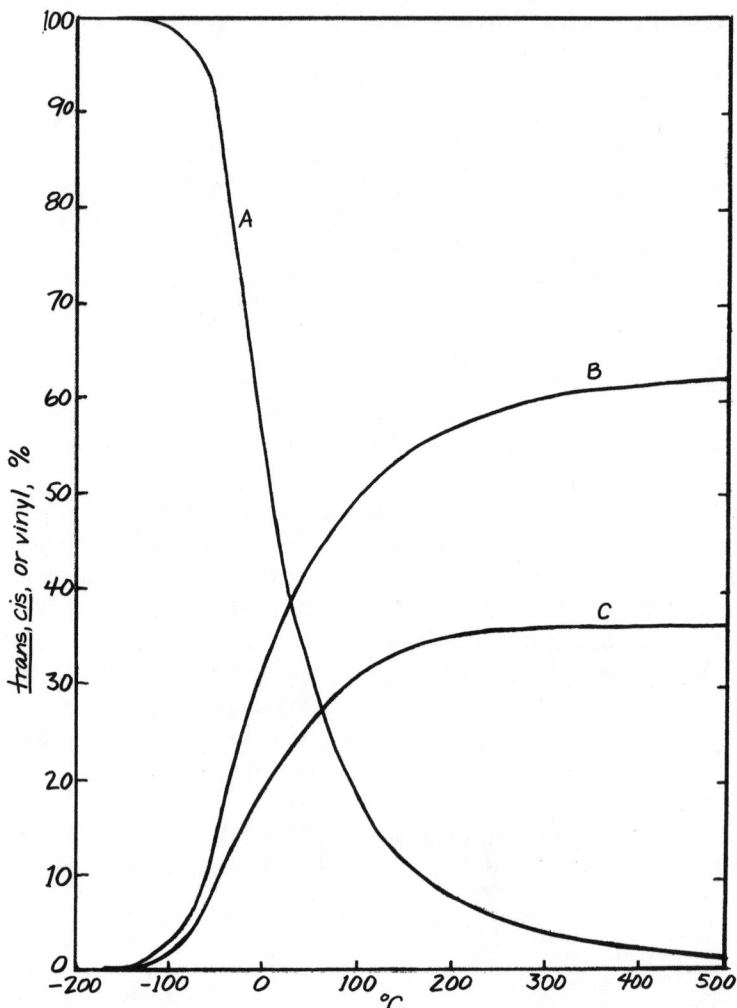

Figure 1. Temperature vs. % trans, cis, or vinyl for polybutadiene made in presence of 1.0 phm of THF (10)
A. vinyl; B. trans; C. cis

Conjugated Dienes and Other Monomers. Alkyllithiums such as n-butyllithium—and even the growing polyethylene carbon–lithium bond complexed with chelating diamines such as TMEDA—are effective initiators for the polymerization of conjugated dienes such as 1,3-butadiene and isoprene. A polybutadiene of high 1,2-content can be produced from butadiene in hydrocarbon solvents using these N-chelated organolithium catalysts.

The much greater stability of the growing polydienyllithium chain end compared with the growing polyethylenyllithium chain end has been demonstrated in a recent patent to McElroy and Merkley (9). Table IV shows that the yields of liquid polybutadienes produced per equivalent of catalyst at 100°C are about the same as those produced at 40° or 60°C although the rates of chain transfer to toluene solvent are significantly affected.

The important influence of temperature on the microstructure of organolithium-initiated polymerizations of butadiene, both with and with-

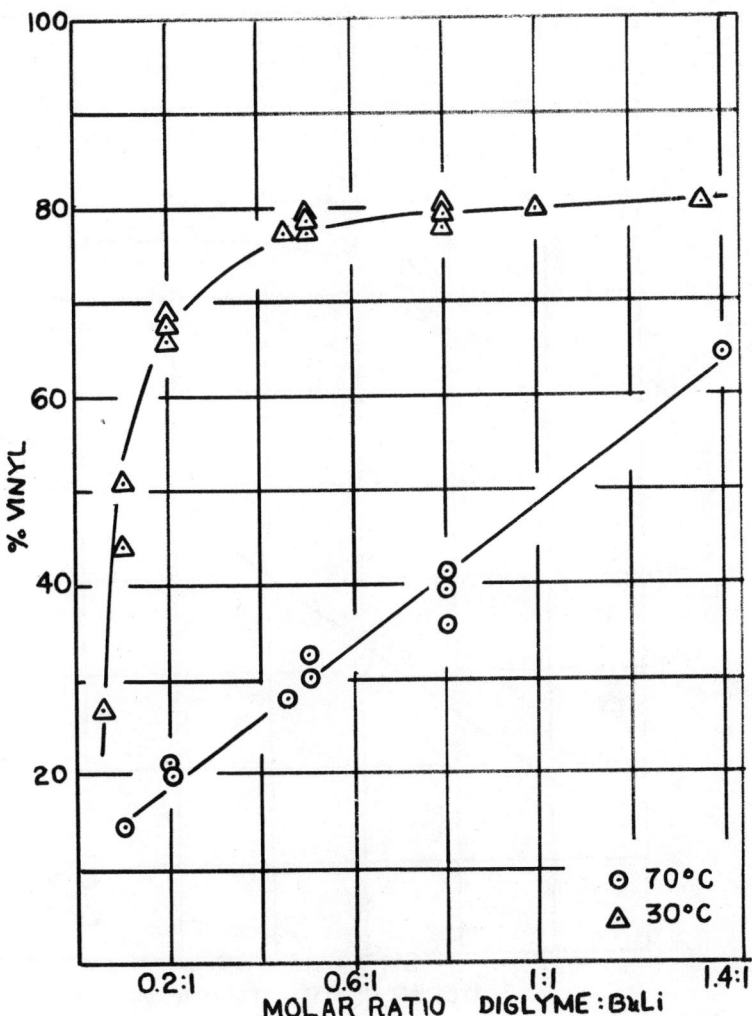

Figure 2. Effect of the molar ratio of diglyme to n-butyllithium on vinyl microstructure of polybutadiene

out chelated diamines such as TMEDA, has received little attention until recently. Uraneck (*10*) studied the effect of temperature on the microstructure of polybutadienes produced in the absence of such polar modifiers. He discovered that there is little effect as the temperature is varied between 0° and 100°C. However, he found a considerable temperature effect in the presence of polar modifiers such as THF (Figure 1), the 1,2 or vinyl content decreasing with an increase in temperatures. More recently, Antkowiak and co-workers (*11*) more completely studied the

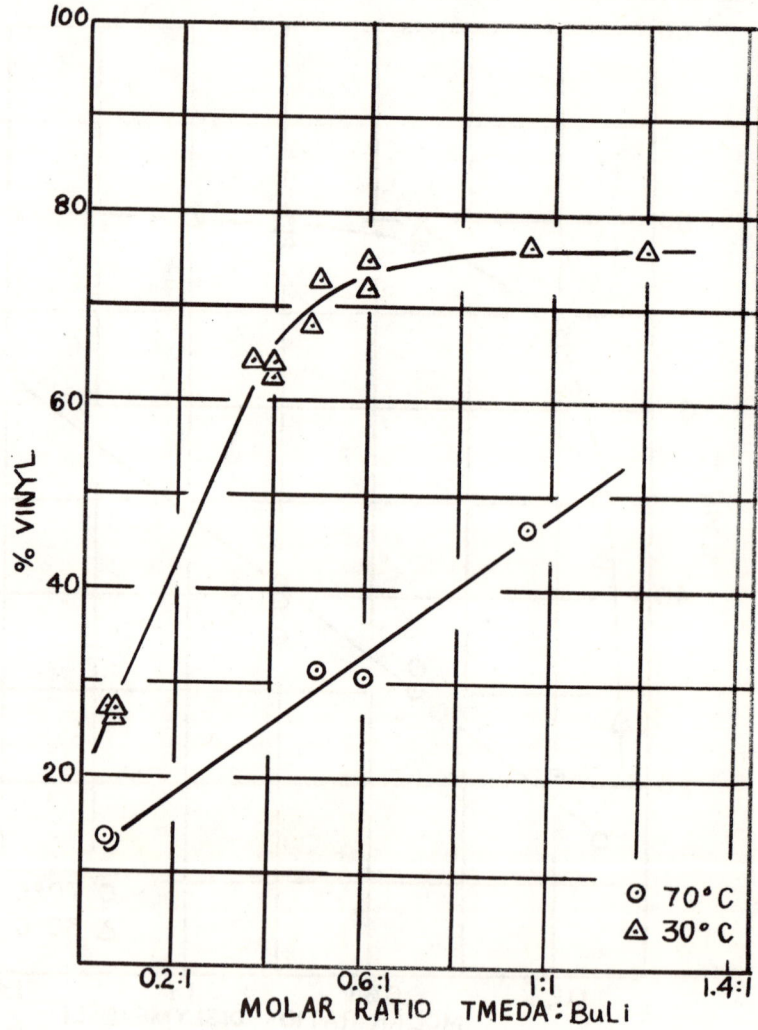

Figure 3. Effect of the molar ratio of N,N,N',N'-tetramethylethylenediamine to n-butyllithium on vinyl microstructure of polybutadiene

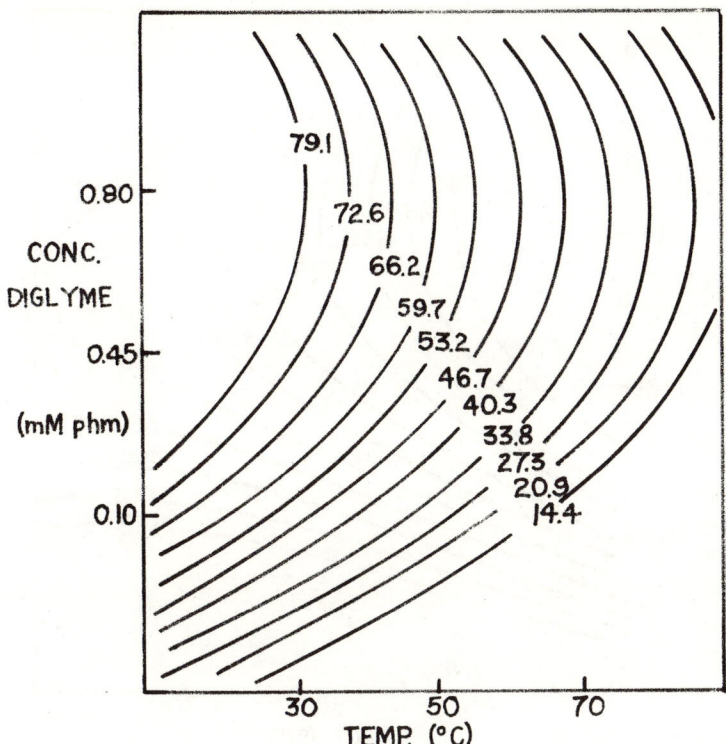

Figure 4. Effect of diglyme concentration and reaction temperature on vinyl microstructure of polybutadiene

effect of temperature on polar-modified alkyllithium polymerizations and copolymerizations of butadiene. These workers also showed that bi- and polydentate polar modifiers such as TMEDA and diglyme have a much greater influence on microstructure than do monodentate Lewis bases such as THF, diethyl ether, or triethylamine. At the same time temperature exerts a great effect on microstructure in the presence of these more powerful modifiers, especially at modifier:n-butyllithium ratios of about 0.2:1.0 (Figures 2 and 3). Those figures show that the effects of diglyme and TMEDA are almost identical and indicate that their mode of action is the same. Antkowiak and his group constructed contour plots using these two modifiers to allow calculation of the conditions for any desired vinyl content (*see* Figures 4 and 5) in a polybutadiene of about 200M. At higher temperatures there is a smaller change in the per cent of vinyl with increases in modifier concentration (modifier loses effectiveness at higher temperatures).

Langer (4) showed that a sharp decrease in both polymerization rate and vinyl content of the polybutadiene is obtained as the number

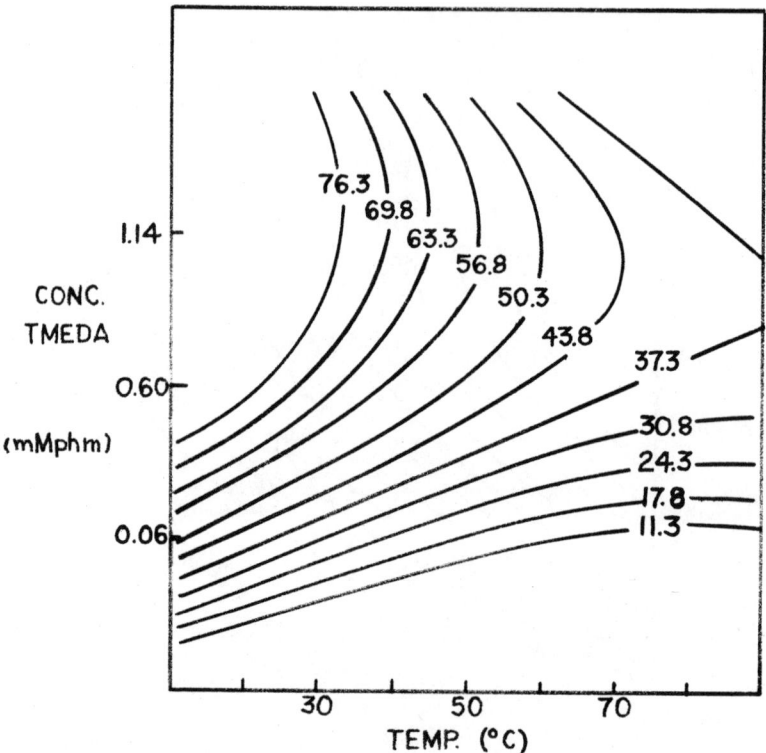

Figure 5. Effect of N,N,N',N'-tetramethylethylenediamine concentration and reaction temperature on vinyl microstructure of polybutadiene

of carbon atoms separating the nitrogen atoms in the chelating diamine is varied away from two or three (Table V). He also showed that increasing substituent size on the nitrogen atoms of the diamine decreases the polymerization rates but does not affect microstructure significantly at low temperatures. In addition, increases in the number of nitrogens in the chelating amine from two to three to four cause no change in vinyl microstructure (although a decrease in polymerization rate occurs presumably because of a greater stability of the chelate structure and greater steric hindrance to monomer approach).

Hay and co-workers (12) studied the kinetics of the polymerization of butadiene at low temperatures by the n-butyllithium–TMEDA initiator system (catalyst:cocatalyst ratios less than 0.5) and concluded that the polymerization was of the "living" type, giving polymers with predictable molecular weights and narrow molecular-weight distribution. n-Butyllithium · TMEDA ratios greater than 0.5 showed much lower initiator efficiencies. The authors postulated high polymerization rates resulting from strong solvation of lithium ions by TMEDA.

Table V. Polymerization of Butadiene (2)[a]

No. of C Atoms Separating N's in Diamine	Polymer Produced grams/grams BuLi in 2 hr at 25°C	% 1,2 or Vinyl Microstructure
1	104	25
2	>750	80
3	544	80
4	88	25
5	106	25

[a] Conditions: 0.002M BuLi-tetramethylated diamine (1:1), 2M C_7H_6 in n-C_5, 25°C.

As pointed out by Langer (and verified by Hay, Antkowiak, and Uraneck), both chelate type and temperature effects indicate that specific lithium solvation rather than a general solvent effect are operable here. Also, this high degree of lithium solvation results in propagation from a truly anionic reacting site located mainly at the secondary carbon of the terminal, active, delocalized allyl carbanion in the growing polymer.

Table VI. Effect of Diglyme on Vinyl Content and Initial Styrene at Different Temperatures (11)[a]

Diglyme/ BuLi	Initial Styrene, wt %			% 1,2(Bd = 100)		
	32°C	High Temp.	(°C)	32°C	High Temp.	(°C)
0.1:1	13.8	11.2	(63)	32-36	18-20	(63)
0.5:1	25.0	15.7	(66)	72-75	40-44	(66)
1:1	25.7	21.2	(66)	76-77	54-56	(66)
2:1	26.6	26.6	(66)	77-79	66-69	(60)
3:1	27.3	27.3	(66)	77-80	70-71	(57)

[a] Conditions: monomer, 65/35 butadiene/styrene, 15% solids; catalyst, 0.6mm phm BuLi.

Langer (13) has also disclosed the use of alkyllithium and dialkylmagnesium tertiary diamine complexes as catalysts for copolymerization of ethylene and other monomers such as butadiene, styrene, and acrylonitrile to form block polymers. Examples are given in which polybutadienyllithium initiates a polyethylene block, as well as *vice-versa*. Random copolymers of these two were also prepared, and other investigators have used not only tertiary diamines but hexamethylphosphoramide (14) and tetramethylurea (15) as nitrogenous base cocatalysts in such polymerizations. Antkowiak and co-workers (11) showed the similarity of action of diglyme and TMEDA in copolymerizations of styrene and

Table VII. Effect of TMEDA on Vinyl Content and Initial Styrene at Different Temperatures (11)[a]

TMEDA/BuLi	Initial Styrene, wt %		% 1,2 (Bd = 100)	
	32°C	66°C	32°C	66°C
0.1:1	12.1	9.0	25-27	14-16
0.4:1	21.3	16.1	52-55	26-28
1:1	28.8	26.0	64-66	50-51
2:1	29.2	29.2	67-72	58-61

[a] Conditions: monomer, 65/35 butadiene/styrene; 15% solids; catalyst, 0.6 mm phm BuLi.

butadiene (Tables VI and VII). They showed that as the ratio of polar modifier to BuLi was increased from 0.1 to 3, the amount of styrene initially incorporated also increased to almost the limiting concentration of styrene. Again, as with butadiene, increasing temperatures decreased the vinyl content as well as the initial styrene uptake. With no polar modifier present styrene was not incorporated into the polymer until most of the butadiene had polymerized (Figure 6). The effect of polar solvents on the structure of dienes and the composition of copolymers

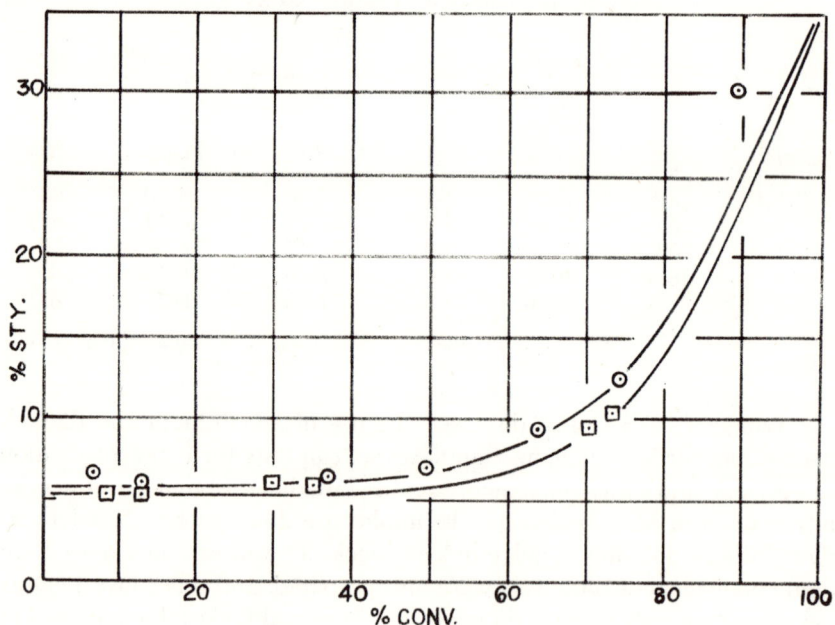

Figure 6. Effect of degree of conversion on the incorporation of styrene into a butadiene–styrene copolymer with no polar modifier present (□ = 32°C, ○ = 66°C)

has been reviewed for other catalyst systems by Morton (*16*) and by Hsieh and Glaze (*17*).

Like ethylene, butadiene (and styrene) can be polymerized with a metalated chelating diamine initiator resulting from aging of an *n*-butyllithium–TMEDA complex (*3, 18*). The resulting high-vinyl polybutadiene possesses nitrogen-containing end groups corresponding to TMEDA in the initial catalyst. In the absence of excess tertiary diamine, the monolithiated TMEDA is stable in aliphatic hydrocarbon solutions and does not significantly promote chain-transfer reactions during the polymerization of butadiene or styrene. Random copolymers with styrene or block copolymers with acrylonitrile can also be produced with the lithiated amine catalyst. Anderson and co-workers (*19*) found that an aminoether, *N*-methylmorpholine, can be similarly metalated with alkyllithium to give copolymers with alternating units of ethylene and α-methylstyrene. Triethylenediamine can also be used, and nitrogen appears in the polymers.

Other lithium compounds that also function as polymerization catalysts when combined with chelating diamines are lithium dialkylphosphides (*20*), used to polymerize 1,3-butadiene, and lithium chloride (*21*), used to polymerize *p*-vinylbenzamide.

Although alkyllithiums are used mainly in these *N*-chelated complexes, other alkali metal alkyls may also be used. For example, organosodium reagents have been solubilized in hydrocarbon solvents by chelating tertiary diamines and used as polymerization catalysts by workers at Borg-Warner Corp. (*22*).

Literature Cited

1. Langer, A. W., Jr., *Trans. N.Y. Acad. Sci.* (1965) **27**, 741.
2. Langer, A. W., Jr., *Am. Chem. Soc. Div. Polymer Chem. Preprints* (1967) 132.
3. Langer, A. W., Jr., U.S. Patent **3,451,988** (1969).
4. Langer, A. W., Jr., First Akron Summit Polymer Conference, Preprint (1970).
5. Eberhardt, G. G., Davis, W. R., *J. Polym. Sci., Part A* (1965) **3**, 3753.
6. Eberhardt, G. G., U.S. Patent **3,567,703** (1971).
7. Smith, W. N., Jr., U.S. Patent **3,579,492** (1971).
8. Screttas, C. G., unpublished data.
9. McElroy, B. J., Merkley, J. H., U.S. Patent **3,678,121** (1972).
10. Uraneck, C. A., *J. Polym. Sci., Part A-1* (1971) **9**, 2273.
11. Antkowiak, T. A., et al., *J. Polym. Sci., Part A-1* (1972) **10**, 1319.
12. Hay, J. N., et al., *Faraday Trans.* (1972) **1**, 1.
13. Langer, A. W., Jr., U.S. Patent **3,450,795** (1969).
14. Van de Castle, J. F., U.S. Patent **3,207,742** (1965).
15. Wofford, C. F., U.S. Patent **3,418,394** (1968).
16. Morton, M., Advan. Chem. Ser. (1966) **52**, 4.

17. Hsieh, H., Glaze, W., *Rubber Chem. Tech.* (1970) **43,** 22.
18. Langer, A. W., Jr., U.S. Patent **3,536,679** (1970).
19. Anderson, W. S., Levin, S. H., U.S. Patent **3,290,277** (1966).
20. Shell International Research, Brit. Patent **1,218,914** (1971).
21. Asahara, T., *J. Polym. Sci., Part A* (1969) **7,** 679.
22. Borg-Warner Corp., Brit. Patent **1,175,322** (1969).

RECEIVED February 12, 1973.

8

Metalation and Grafting by Anionic Techniques

ADEL F. HALASA

The Firestone Tire & Rubber Co., Akron Ohio 44317

Elastomers made by anionic initiators were metalated by organolithium reagents activated with chelating diamines. Polybutadiene and polyisoprene were metalated with n-BuLi · TMEDA. The lithiated polymers were used as sites for grafting of various monomers that can be polymerized anionically. Grafting and catalyst efficiencies were determined. Useful block copolymers using this system of grafting were made. The number of active sites were determined from the molecular weight of the block styrene bound to the rubber. The grafting efficiencies were determined from the amount of unbound polystyrene formed during polymerization.

This work deals mainly with anionic graft copolymers, their mode of preparation, and their characterization. The procedure used is one in which anions are generated on the backbone of a preformed polymer, and are used as sites for grafting of various monomers that can be polymerized anionically. The reagent used for generating sites on the polymer backbone is n-BuLi–N,N,N',N'-tetramethylethylenediamine (TMEDA). Catalyst efficiency determined by the site generation, as well as the efficiency of each site to initiate polymerization of grafts, is reported.

Discussion and Results

There are several types of reactions by which graft copolymers can be produced: (1) free-radical, (2) cationic, (3) condensation, and (4) anionic. Free-radical grafting is an old art. It suffers from the fact that control of the grafting position is difficult. Transfer of the radical to monomer gives large amounts of homopolymer. With unsaturated polymers, gelation and cross-linking caused by coupling reactions or propagation are often encountered. Cationic methods are also difficult to control and often lead to cross-linking. Only in those cases in which cation formation is controlled can even limited success be achieved. Since the reagents used to form the sites for grafting are usually the same as those used to cross link or cyclize polymers, this method of grafting is usually undesirable.

However the more recent work of Kennedy (1) seems to have circumvented these difficulties. He was able to produce well-characterized graft copolymers. However, his approach was limited to elastomers with a low per cent of unsaturation along the polymer backbone; otherwise, gelation ensues.

Condensation polymerization can be used to produce graft copolymers. However, like the other techniques, it suffers from self-condensation of the backbone, ring formation, and cyclization, which ultimately leads to gelation.

Before discussing the methods and the procedures of obtaining anionic graft copolymers, the new terminology used in this technique needs to be defined. First, grafting efficiency is a measure of the amount of grafted monomer compared with the total amount of monomer polyerized. For example, take styrene as the monomer to be grafted on a metalated polybutadiene. In this case, grafting efficiency is the amount of grafted styrene divided by the total amount of polymerized styrene found in the system:

$$\% \text{ Grafting efficiency} = \frac{\text{styrene grafted}}{\text{styrene grafted} + \text{homopolystyrene}} \times 100$$

Incomplete metalation would leave unreacted butyllithium available to initiate styrene and form homopolystyrene. This would result in poor grafting efficiency. Similarly, a chain-transfer process to monomer would also give poor grafting efficiency.

The second aspect to consider is catalyst efficiency of the metalating reagent. This is a measure of how many of the anions added to the system actually initiate chains. It is the expected molecular weight ($\overline{M}_n$) of the graft (determined from the moles of monomer and moles of the metalating agent added) divided by the experimentally determined number average molecular weight ($\overline{M}_n$). The found ($\overline{M}_n$) can be determined by isolating the styrene block after oxidative degradation of the polybutadiene backbone and determining its molecular weight.

$$\% \text{ Catalyst efficiency} = \frac{\overline{M}n \text{ calculated}}{\overline{M}n \text{ found}} \times 100$$

Any process that destroys the anionic sites or otherwise prevents initiation or propagation results in grafts of higher than calculated molecular weights. Impurities having active hydrogen are the usual cause of reduced catalyst efficiency.

The overall effectiveness of the method, therefore, is defined as the product of grafting efficiency and catalyst efficiency (overall effectiveness of the grafting process = catalyst efficiency × grafting efficiency). Catalyst efficiency and grafting efficiency play important roles in determining the overall effectiveness of metalation reagents in anionic graft copolymers. The structure of the copolymer formed depends on the number of sites that initiate polymerization. This is important in obtaining the desired physical properties.

Organometallic compounds and alkoxides add to activated double bonds and to functional groups such as ketones, esters, nitriles, and aldehydes. Many workers have taken advantage of this tendency and have attempted to prepare graft polymers by this method (2). A detailed description of these techniques and others like it is given in a recent review by Heller (3).

This report is limited to the most recent work on chelating diamines with organolithium compounds and their application to metalation and grafting.

Anion Generation on the Polymer Backbone

The approach to synthesis of anionic graft copolymers described here is to create anions on the polymer backbone and use these anions as sites for grafting onto the backbone. The advantages of this method are that it can give a controlled number of grafted side chains; it minimizes homopolymer formation; it provides narrow molecular-weight distribution of the graft; and it permits preparation of different types of graft copolymers. Disadvantage of the method is that it is applicable only to hydrocarbon polymers containing active hydrogens such as allylic or benzylic hydrogen, or exchangeable functional groups such as halides, or both.

The discovery of the powerful metalating agent, n-BuLi–N,N,N',N'-tetramethylethylenediamine, opened a new chapter in anionic grafting. This complex has been reported to metalate toluene and benzene within a few minutes to give quantitative yields of benzyllithium and phenyllithium, respectively (4). It also has been reported to polylithiate aromatic compounds (24, 25).

Several workers have used this complex to metalate hydrocarbon polymers. Plate and co-workers (5), for example, metalated polystyrene with n-BuLi · TMEDA and monitored butane evolution by gas chromatography. They reported 40% catalyst efficiency. They did not report the grafting efficiency or the overall effectiveness of this metalating reagent.

Chalk, Hay, and Hoogenboom (6, 7) using the same complex, reported lithiating poly(2,6-dimethyl-1,4-phenylene) ether and poly(2,6-diphenyl-1,4-phenylene) ether. The lithiation was done both at room temperature over a long time and at reflux for a shorter time. They reported catalyst efficiency of 17% as determined by the lithium content in the polymer. They attributed the low level of lithiation to the attack on THF by the metalating complex.

Table I. Change in Intrinsic Viscosity and M_n with Increasing Metalation Levels

Polymer	n-BuLi, m moles 100 grams Polymer	T, °C	Time, hr	$[\eta]dl/$gram[a]	$\overline{Mn}$ Values[b] Before Met	After Met
Polybutadiene	5.0	50	4		83,000	80,000
	10.0	50	4		90,000	60,000
	20.0	50	4		72,000	38,000
	30.0	50	4		87,000	30,000
				2.40[c]		
	4.0	70	2	2.0		
	8.0	70	2	1.60		
	16.0	70	2	1.20		
Polyisoprene				2.0[c]		
	8	80	4.8	1.5		
	16	80	4.8	1.05		
	24	80	4.8	0.77		

[a] At 25°C in toluene.
[b] GPC values.
[c] Initial $[\eta]$.

The same workers (8) also found, from the addition of vinyl monomers to the lithiated poly(2,6-dimethyl)- and poly(2,6-diphenyl-1,4-phenylene) ethers, that the grafting efficiency was very low. However, when the styrene was added over four hours to the lithiated polymer, the grafting efficiency was very high. This was determined by a quenching reaction with chlorotrimethylsilane, after which the $SiMe_3$ group was found on the end of the polystyrene graft. However, when the styrene was added rapidly to the lithiated polymer and the reaction quenched with chlorotrimethylsilane, the silyl group was found on the polyether aromatic group. This suggests that the initiation rate of styrene by the lithiated polyether is very slow compared with the propagation. Thus, rapid addition of styrene resulted in relatively few lithium atoms having the chance to start grafts. On slow addition, however, most anionic sites participate in initiating grafts.

This method of determining grafting efficiency proved successful since SiMe₃ groups on polystyrene appear at 10.1 to 10.3 τ by NMR while the SiMe₃ groups on the polyphenylene aromatic ethers appear at 10.4 τ.

Chalk and his associates reported 60 to 90% catalyst and grafting efficiencies. Using the above reagents, they were able to prepare graft polymers of styrene and methyl acrylate to poly(2,6-dimethyl)- and poly(2,6-diphenyl-1,4-phenylene) ether. In our study, n-BuLi · TMEDA was used as a lithiating agent for metalating polybutadiene, polyisoprene, and copolymers of o- and p-chlorostyrene with 1,3-butadiene elastomers. While this work was in progress, Minoura was doing similar metalation work (9, 10). The quantity of metalation reagent used, however, was quite different. He used extremely large amounts of metalating agents and probably had low efficiency.

We have attempted to determine grafting efficiency and catalyst efficiency. This was done by determining the per cent homopolymer, size of the grafted chain, and the number of the grafted chains. Polybutadiene and polyisoprene were metalated with the n-BuLi · TMEDA at several metalation levels (12); similar work has been reported on this subject elsewhere (12–16). After metalation, the polymer was hydrolyzed and compared with the original polymer. The $\overline{M}_n$ of the resulting polybutadiene was lowered drastically (Table I). The mechanism of this molecular-weight modification reaction is not clear at this time, but we feel that it involves scission at the vinyl or isopropenyl sites in the polybutadiene or polyisoprene, respectively. Grafting efficiency was determined by injecting freshly distilled styrene into the premetalated rubber, allowing the styrene to react for five hours, and then determining the amount of homopolystyrene either by acetone extraction or by gel permeation chromatography (GPC). Lithium atoms attached to the chain act as sites for initiating the polymerization of styrene grafts. Typical results for grafting efficiency are shown in Table II.

Table II. Efficiency of Grafting

Polymer	n-BuLi, m moles/ 100 grams Polymer	TMEDA, m moles/ 100 grams Polymer	Metalation Time, hrs	% Styrene Added	% Styrene as Grafting Efficiency
Polybutadiene	6	7.2	16	22.8	65.4
	10	12.0	16	28.6	66.7
	20	25	23	29.8	95.0
Polyisoprene	2.5	3.3	4	19.7	69.5
	5.3	6.7	4	16.3	75.0
	24.0	30.0	20	29.5	96.8

Some 5 to 25% homopolystyrene is generally observed even after a long metalation time. Several explanations for this observation can be proposed. There may be chain transfer because of transmetalation of unreacted styrene by the metalated polybutadiene, incomplete metalation of the polybutadiene because of an equilibrium between metalated polybutadiene and metalated TMEDA, or the presence of low-molecular-weight impurities capable of initiating polymerization. Since TMEDA is itself metalated by n-BuLi under the reaction conditions used, it is conceivable that metalated TMEDA and metalated polybutadiene are in equilibrium and are both initiating polymerization. If that is the case then the homopolystyrene should contain nitrogen.

The rate of the metalation reaction was followed by the disappearance of n-BuLi. All the n-BuLi was consumed in two hours. This was determined by quenching the metalation reaction with chlorotrimethylsilane and following the disappearance of trimethylbutylsilane by gas chromatography. The analysis for trimethylbutylsilane and the determination of homopolystyrene by acetone extraction are good methods for determining whether the polymer was completely metalated. The silylation reaction followed by gas-chromatographic analysis indicates whether TMEDA metalation occurred.

The metalated polymer can be used to initiate formation of graft polymers with very high grafting efficiency. The number of grafting sites is controlled by the amount of metalating agent used while the length of the grafted chain is controlled by the ratio of monomer to active sites. Catalyst efficiency is a measure of the number of active sites that initiate polymerization of the added monomer.

The length of the grafted chain is determined by fragmentation of the polybutadiene portion of the grafted copolymer with OsO_4/tert-butylperoxide oxidation and examination of the polystyrene residue by GPC. The results are given in Table III.

The results in that table are consistent with the picture of chain metalation since the data demonstrate the formation of multiple polystyrene blocks. The grafted polystyrene recovered has a rather broad molecular-weight distribution skewed toward the low-molecular-weight range. The high value of catalyst efficiency as determined in polymers 1 and 3 could result from either limitations in the analysis of molecular-weight distribution or from some form of chain-transfer mechanism.

Overall effectiveness of the metalating reagent in this metalation reaction is very high. This suggests that n-BuLi · TMEDA is a very effective metalating agent. It suggests too that the anionic sites introduced are also efficient in grafting-added monomer because the initiation and the propagation rates of these sites are about the same. The data in

Table III. Polybutadiene-Grafted Styrene Structures

	Polymer 1	Polymer 2	Polymer 3
n-BuLi (catalyst), mmoles per 100 grams polymer	1.20	0.7	0.7
Metalation: n-BuLi	6.0	6.0	6.0
TMEDA mmoles/100 grams polymer	12.0	6.0	6.0
Molecular weight graft copolymer, osmotic pressure	103,000	94,000	103,000
% Homopolystyrene	0.0	0.0	0.0
Graft styrene, %	30.9	24.8	32.2
Catalyst efficiency, %	100	100	100
Molecular weight Styrene calculated for 1 block	31,800	23,300	33,500
Molecular weight calculated for total Li	4,420	3,480	5,000
Molecular weight found Styrene by GPC	3,453	3,647	4,070
Overall effectiveness, %	128	95	122

Table III indicate that steric hindrance or penultimate effects are at a minimum. (While this volume was in preparation a presentation was made on the same subject by Cha (*17*). Although his methods of analyses were different from ours, his results and conclusions are the same.)

A high-impact polystyrene that has much better optical clarity than that obtained by usual blending or grafting techniques can be prepared by our technique. Polymers containing 90–95% styrene grafted to polybutadiene rubber by use of 12 mmole RLi–TMEDA/100 gram polymer showed quite good optical clarity.

The raw graft copolymers of butadiene and styrene, as well as isoprene and styrene, are tough and elastomeric. This is attributed to their having the structural elements characteristic of SBS block copolymers.

Polybutadiene having a comb-type structure was prepared by adding additional butadiene to metalated polybutadiene. The grafted portion of the polymer, however, has a predominantly vinyl structure because of the presence of TMEDA.

Another method of generating anions on the backbone chain is to have replaceable functional groups that exchange with organolithium compounds at moderate temperatures without modifying or cross-linking the resulting elastomer. Metal–halogen exchange is well known in simple organic compounds (*18, 19*). The reaction of organolithium compounds with halogenated polyethylene has been disclosed (*20*). However, the products were not well characterized.

We have found that copolymers of *o*- or *p*-chlorostyrene with butadiene can undergo metal-halogen exchange with n-BuLi in the presence of TMEDA at moderate temperatures. This way the amount of halogen and subsequent metalation on the polymer are controlled. The halogen attached to the aromatic ring and its reactions are not complicated to any great extent by side reactions. Such reactions are common with polymers in which the halogen is attached to aliphatic groups. The interchange is effected with a complex of an alkyl derivative of the alkali metal and the aliphatic chelating diamine.

The copolymers of o- and p-chlorostyrene with butadiene can be prepared by anionic initiators (21). Since the reactivity ratios of 1,3-butadiene or o- and p-chlorostyrene are close to unity, the resulting copolymers have a constant composition.

Table IV. Metal-Halogen Interchange

	Polymer 1	Polymer 2	Polymer 3	Polymer 4
Polymerization:				
n-BuLi initiator; mmoles phm	2.1	2.1	2.1	2.1
Exchange system:				
n-BuLi; mmoles phm	1.6	1.6	3.2	4.8
TMEDA: mmoles phm	—	1.6	3.2	4.8
time, hrs	12	12	12	12
temperature, °C	25	25	25	25
Total n-BuLi, mmoles phm	3.7	3.7	5.3	6.9
Theoretical % styrene in copolymer	(24)	(32)	(39)	(44)
Homopolystyrene in grafted polymer:				
% by acetone extraction	92.0	None	None	None
% grafting efficiency	12	100	100	100
calculated Mn	6,486	8,648	7,358	6,376
found Mn		8,000	9,500	10,000
catalyst efficiency		108	77	64

A copolymer of 1,3-butadiene and o-chlorostyrene was made with an anionic initiator at 50°C in a hydrocarbon solvent (21). The copolymer contained 3 to 5% o-chlorostyrene. This copolymer was subjected to a metal-halogen exchange reaction. The results of the exchange reactions are shown in Table IV. No homopolystyrene was found by acetone extraction of the graft polymer, suggesting that catalyst efficiency (exchange efficiency) is very high.

In addition to being a synthetic route to unusual graft copolymers, the metalation technique offers a way to add functional groups to the chain by reactions characteristic of organolithium compounds. Hydroxyl or carboxyl groups, for instance, can be added by treating the metalated polyisoprene or polybutadiene (22) solution with ethylene oxide or CO_2, respectively. The lithium alkoxide and carboxylic salt obtained (23) in those reactions are highly associated and form a swollen gel almost instantly after exposure of the metalated polymer to ethylene oxide or CO_2. The rubbery gel forms a surface skin that makes it very difficult to mix the reactants well enough to get complete reaction. The reaction is best carried out in thin films or sprays rather than by addition of reagent to the solutions. Treatment of this salt with excess methanol, however, returns the product to a fluid state where work-up can be accomplished.

Acknowledgment

The author acknowledges the assistance of the Research Analytical and Polymer Structure Divisions of Firestone Central Research Laboratories for polymer analyses and thanks The Firestone Tire & Rubber Co. for permission to publish this work.

Literature Cited

1. Kennedy, J. P., *Macromolecular Preprint 1, Int. Congr. Pure Appl. Chem.*, Boston, Mass. (1971).
2. Webb, F. J., U.S. Patent **3,627,837** (1972).
3. Heller, J., *Polym. Eng. Sci.* (1971) **11**, 6.
4. Rausch, M. D., Ciappenelli, D. J., *J. Organometal. Chem.* (1967) **10**, 127.
5. Plate, N. A., Jampolskaya, M. A., Davydova, S. L., Kargin, V. A., *J. Polym. Sci. C* (1969) 547.
6. Hay, A. S., Chalk, A. J., French Patent **1,586,729** (1967); U.S. Patent **3,402,144** (1968).
7. Chalk, A. J., Hay, A. S., *J. Polym. Sci., A-1* (1969) **7**, 691.
8. Chalk, A. J., Hoogeboom, T. J., *J. Polym. Sci., A-1* (1971) **9**, 3679.
9. Minoura, Y., Shina, K., Harada, H., *J. Polym. Sci., Part A-1* (1968) **6**, 559.
10. Minoura, Y., Harada, H., *J. Polym. Sci., Part A-1* (1969) **7**, 3.
11. Tate, D. P., Halasa, A. F., Webb, F. J., Koch, R. W., Oberster, A. E., *J. Polym. Sci., Part A-1* (1971) **9**, 139.
12. Dunlop Co. Ltd., French Patent **1,571,456** (1967).
13. Dunlop Co. Ltd., French Patent **1,566,853** (1967).
14. Borg-Warner Corp., Brit. Patent **1,172,477** (1967).
15. Sun Oil Co., Brit. Patent **1,121,195** (1968).
16. Sun Oil Co., French Patent **1,583,793** (1968).
17. Cha, C. Y., "Abstracts of Papers," 161st National Meeting, ACS, March, 1971, POLY 46.
18. Leavitt, F. C., U.S. Patent **3,234,193** (1966).
19. Husemann, E., German Patent **1,226,304** (1966).
20. Plate, N. A., Daydova, S. L., Jampolskaya, M. A., Mukhitdinova, M. A., Kargin, V. A., *Vysokomol. Soedin.* (1966) **8**, 1562.
21. Halasa, A. F., Adams, H. E., Hunter, C. J., *J. Polym. Sci., Part A-1* (1971) **9**, 677.
22. Nylor, F. E., U.S. Patent **3,382,225** (1968).
23. Koch, R. W., U.S. Patent **3,598,793** (1971).
24. Halasa, A. F., Tate, D. P., *J. Organometal. Chem.* (1970) **24**, 769.
25. Halasa, A. F., *J. Organometal. Chem.* (1971) **31**, 369.

RECEIVED February 12, 1973.

9

Telomerization Reactions Involving Amine-Chelated Lithium Catalysts

W. A. BUTTE and G. G. EBERHARDT

Sun Research and Development Co., Marcus Hook, Pa. 19061

Complexes of organolithium compounds with certain diamines are remarkably active in the metalation of unsaturated hydrocarbons and addition to ethylene. These complexes form a new class of initiators for the telomerization of ethylene with aromatic hydrocarbons and olefins. The distribution of the products is in accordance with a mechanism involving competitive transmetalation and addition to ethylene. By proper selection of the telogen and regulation of the ethylene pressure, it is possible to influence the nature of the telomeric products and produce phenylalkanes, polyalkylbenzenes, and long-chain olefins with a variable average degree of telomerization. The assistance of the amine in accelerating the reaction is attributed to the formation of a coordination complex with lithium which facilitates ionization of the carbon–lithium bond.

In 1955 Pines and Schaap (1) discovered that toluene was alkylated by ethylene in the presence of sodium or potassium metal or, more specifically, their organometallic derivatives. This reaction requires a high temperature (about 200°C) and considerable olefin pressure; the organometallic catalyst is essentially insoluble in the reaction medium. The catalyst cycle—for example, in the side-chain ethylation of toluene—involves a benzyl carbanion which adds to ethylene to form a primary alkyl carbanion. The latter immediately abstracts a proton from the excess toluene reactant to form n-propylbenzene and to reform the energetically-favored benzylic anion in a catalytic cycle.

On further conversion of the n-propylbenzene, additional benzylic hydrogens are ethylated. The high rate of transmetalation involving the primary aliphatic organosodium or potassium intermediates and the alkylaromatic hydrocarbon prevents formation of telomeric products with these catalysts even at higher ethylene pressures.

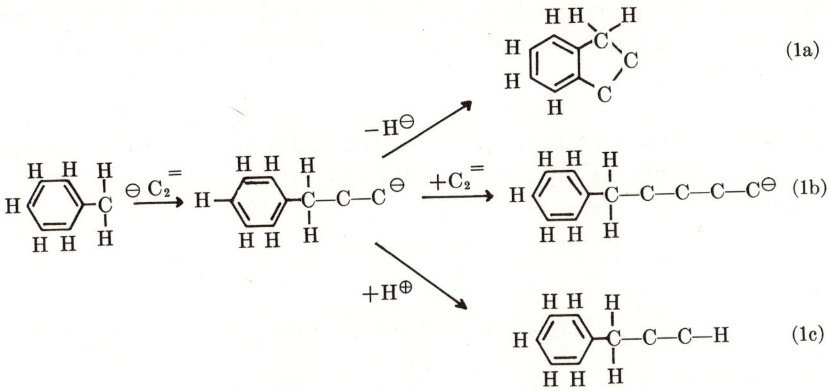

The carbanionic ethylene adduct can also undergo a cyclization reaction to indanes with elimination of a hydride ion (Reaction 1a). This reaction is more pronounced with organopotassium or some complex organopotassium catalysts (2). High-molecular-weight growth products are not obtained from either sodium- or potassium-derived catalysts.

An initiator system has been found independently in the laboratories of Esso and Sun Oil (3, 4). This system promotes transmetalation and chain propagation reactions at comparable rates so that a telomerization reaction of ethylene with aromatic hydrocarbons is realized under relatively mild operating conditions.

Nature of the Initiator

The initiator consists of an organolithium–amine complex. The organolithium component could be a commercially available material such as *n*-butyllithium in hexane solution. The amine component should be free of reactive hydrogen, including aromatic, allylic, and benzylic protons, and it is therefore limited to tertiary aliphatic amines. Maximum catalyst activity is obtained with chelating-type diamines like tetramethylethylenediamine (TMEDA) and bridgehead-type amines such as triethylenediamine (TEDA). Sparteine, a diamine containing tertiary bridgehead nitrogen atoms, forms an exceptionally stable and highly reactive catalyst.

The initiator may be preformed or generated *in situ* simply by combining the organolithium compound with the amine. Despite its high reactivity, the resulting complex can be easily handled as a hydrocarbon solution. Impurities such as water, air, and carbon dioxide must be rigidly excluded because of their rapid reaction with organolithium compounds.

Hydrogen acts as a poison because lithium hydride resulting from hydrogenolysis of the lithium–carbon bond is inactive as a catalyst component.

All of the evidence (3, 4) strongly suggests that the function of the amine is that of associating with the lithium in the reaction mixture. As a result the carbon–lithium bond is modified and a more reactive incipient carbanion results. A dynamic equilibrium exists between the lithium complex and the dissociated species. The instability constant, K_i, is a measure of the position of the equilibrium.

$$\text{RLi} \cdot n \text{ Amine} \overset{K_i}{\rightleftarrows} \text{RLi} + n \text{ Amine} \qquad (2)$$

With TEDA, K_i is relatively large. Therefore the degree of association and thus the reaction rate is materially improved by introducing an excess of the amine. With TMEDA, K_i is small so there is little advantage to using greater than equimolar amounts of amine.

Transmetalation

In contrast with the low reactivity observed with n-butyllithium solutions in hydrocarbons (5) or ether (6), the n-butyllithium–amine adducts rapidly metalate unsaturated hydrocarbons including even simple olefins. In this respect they surpass even organosodium compounds and will probably prove to be of considerable synthetic value. Addition of a hexane solution containing equimolar amounts of n-butyllithium and an amine to excess toluene at ordinary temperatures results in rapid formation of a benzyllithium–amine complex that in some cases separates as a yellow crystalline solid.

$$C_4H_9Li \cdot \text{Amine} + C_6H_5CH_3 \rightarrow C_6H_5CH_2Li \cdot \text{Amine} + C_4H_{10} \qquad (3)$$

The effecitveness of various amines as promoters for transmetalations can be readily discerned by comparing the amounts of benzyllithium formed under identical conditions (Table I). The marked influence of TEDA on the reactivity of n-butyllithium can be ascribed to the excellent donor characteristics of the bridgehead nitrogen atom (7) that disrupts the n-butyllithium aggregates (8). Furthermore the coordination of lithium by the amine polarizes the carbon–lithium bond thereby easing lithium–hydrogen interchange.

Table I also shows that TMEDA has a much greater effect on the reactivity of butyllithium than does either triethylamine or N,N'-dimethylpiperazine. To explain this difference it is suggested that the reactive species contain two coordinated amine groups. The unusual reactivity with TMEDA is then related to the favorable entropy change usually

Table I. Extent of Metalation of Toluene[a]

Amine	Amine/BuLi, molar ratio	Extent of metalation, %[b]
Et_3N	3	2
$MeN(CH_2CH_2)_2NMe$	1	5
$Me_2NCH_2NMe_2$	1	6
$Me_2NCH_2CH_2NMe_2$	1	77
$Me_2NCH_2CH_2CH(Me)NMe_2$	1	60
Sparteine	1	90
$N(CH_2CH_2)_3N$	1	43
$N(CH_2CH_2)_3N$	2	66

[a] One hour at 60°C, 1.0M butyllithium.
[b] Percent of theory based on butyllithium.

associated with chelate formation. Support for this interpretation comes from the predictable effect of ring size upon the relative reactivity of the adducts and from the high reactivity noted for the rigid bidentate complex formed with sparteine.

Telomerization of Ethylene with Aromatic Hydrocarbons

Ethylenation of n-butyllithium, phenyllithium, and benzylic lithium compounds does not occur at low temperature and ordinary pressure (9). Under more rigorous conditions, telomerization of ethylene in aromatic hydrocarbons proceeds vigorously in the presence of an organolithium compound and an amine. Although n-butyllithium is introduced initially, rapid transmetalation occurs to the more acidic aromatic hydrocarbon (telogen) which subsequently adds to ethylene (taxogen) and initiates the carbanionic polymerization of ethylene. This polymerization proceeds to modest molecular weight, but it is terminated by transmetalation back to the aromatic hydrocarbon which initiates another chain to complete the catalytic cycle.

$$n\text{-BuLi} + ArH \longrightarrow n\text{-BuH} + ArLi \tag{4}$$

$$ArLi + C_2H_4 \xrightarrow{k_1} ArCH_2CH_2Li \tag{5}$$

$$Ar(CH_2CH_2)Li + (n-1) C_2H_4 \xrightarrow{k_2} Ar(CH_2CH_2)_nLi \tag{6}$$

$$Ar(CH_2CH_2)_nLi + ArH \xrightarrow{k_3} Ar(CH_2CH_2)_nH + ArLi \tag{7}$$

Since the propagation reaction (Reaction 6) and the transfer reaction (Reaction 7) are competitive, the resulting product is a mixture of molecular weights governed by a simple statistical distribution shown in

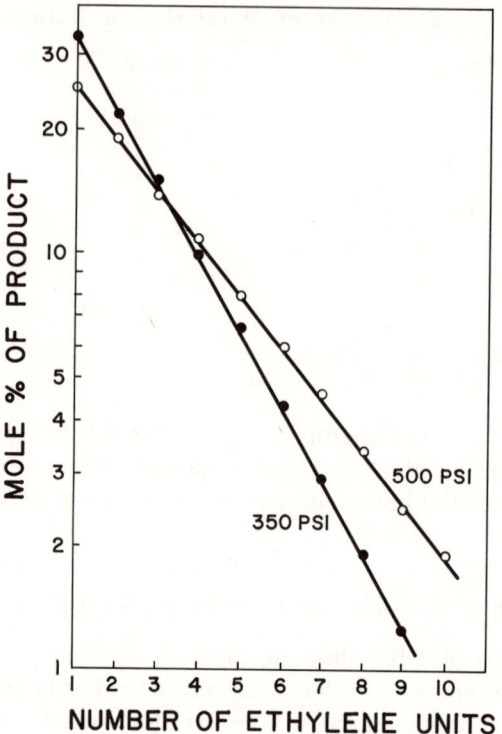

Figure 1. Distribution of telomeric products obtained from ethylene and benzene at 110°C

Figure 1. The average degree of polymerization of the product, $\bar{n}$, is a function of the relative rates of ethylene consumption and transfer. By invoking the steady-state assumption, it can be shown that the average degree of polymerization is governed by the competitive rates of propagation and transfer.

$$\bar{n} = \frac{k_2(C_2H_4)}{k_3(ArH)} + 1 = \frac{1}{\beta} + 1 \tag{8}$$

For convenience, the ratio of transfer and propagation rates is expressed as β since this ratio is related to the mole fraction, X_n, of product with degree of polymerization, n (Equation 9). This relationship may be modified to allow calculating β from the amount of product at two successive values of n, as shown in Equation 10.

$$X_n = \frac{\beta}{(1 + \beta)^n} \tag{9}$$

$$\beta = \frac{X_n}{X_{n+1}} - 1 \tag{10}$$

Equation 6 implies that the rate of ethylene consumption during telomerization is a function of its concentration but is essentially independent of the telogen concentration present in large excess. Furthermore, according to Equation 8, the β value may be varied by changing the pressure or the nature of the telogen. This is indeed the case, as seen from the average molecular weight of telomers obtained from reactions carried out at various ethylene pressures and with a number of aromatic telogens. Table II shows that the average degree of polymerization rises as the pressure is increased. Consequently the chain length—that is, the average molecular weight of the product—can be regulated by the proper

Table II. Influence of Pressure on Molecular Weight[a]

Telogen	Pressure, psig	$\bar{n}$	Average molecular weight
$C_6H_5CH_3$	100	1.4	131
	300	2.0	148
	500	2.7	168
	800	4.0	204
C_6H_6	100	1.6	123
	300	2.6	151
	500	3.9	187
	800	5.5	232

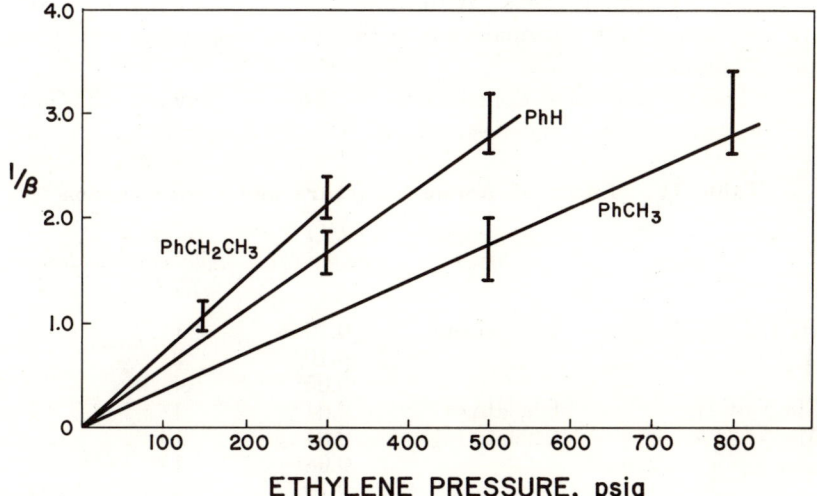

Figure 2. Influence of pressure on product distribution with three aromatic telogens at 100°C

Table III. Effect of Telogen on Telomer Structure[a]

Telogen	Telomer	n̄	Average molecular weight
Toluene	$C_6H_5CH_2(CH_2CH_2)_nH$	2.7	168
Xylene	$CH_3C_6H_4CH_2(CH_2CH_2)_nH$	3.0	190
Ethylbenzene	$C_6H_5CH(CH_3)(CH_2CH_2)_nH$	4.4	230
Benzene	$C_6H_5(CH_2CH_2)_nH$	3.9	187
Isobutylene	$CH_2C(CH_3)CH_2(CH_2CH_2)_nH$	8.1	380

[a] Conditions: 100°C, 0.05M n-butyllithium, TMEDA.

selection of operating conditions. The relationship of β to ethylene pressure for three telogens is shown in Figure 2.

The influence of the telogen upon molecular weight of the telomer is shown in Table III. Since, under equivalent conditions, the molecular weight depends only on the transmetalation rate, the hydrocarbons can be ranked in order of their decreasing kinetic acidity: toluene > xylene > benzene > ethylbenzene. The high reactivity of benzene is surprising but consistent with the facile metalation of benzene with the butyllithium amine complex noted elsewhere.

Reaction Rate

The rate of telomerization of ethylene in toluene is, as expected, directly proportional to the RLi–TMEDA concentration at 0.04M to 0.10M (Table IV). Thus solubility of the catalyst is not a limiting factor at these concentration levels. With other amines, the influence of structure and concentration is analogous to that discussed in connection with transmetalation.

A nearly first-order dependence is observed between the initial reaction rate and the ethylene pressure. A small deviation occurs which

Table IV. Effect of Amine Structure and Concentration

Amine	Structural type[a]	RLi molarity	Amine mole/mole RLi	Rate[b]
$N(CH_2CH_2)_3N$	Bridgehead	0.10	1	12
		0.10	2	19
		0.05	4	32
$(Me_2N)_2CH_2$	Chelate (4)	0.04	1	12
$(Me_2NCH_2)_2$	Chelate (5)	0.04	1	126
		0.08	1	128
		0.10	2	130
$(Me_2NCH_2)_2CH_2$	Chelate (6)	0.04	1	25

[a] Number in parentheses indicates ring size of chelate.
[b] Moles C_2=/mole RLi/hr over first hour at 105°C and 500 psig.

is explicable in terms of nonideal behavior of ethylene. The observed dependence indicates that the initial addition of ethylene to the carbanion derived from the telogen is the rate-controlling step.

The telomerization rates vary significantly with the nature of the telogen. The order of increasing rate—toluene < benzene < ethylbenzene—is consistent with the postulate that the primary addition of the telogen-carbanion to ethylene is rate determining. As the relative acidities of the telogens decrease, the relative reactivities of the derived carbanions increase since the same factors that promote metalation stabilize the resulting carbanion.

There is a very marked increase in the catalytic rates, following the order triethylamine < trimethylamine < triethylenediamine < TMEDA (3). The monodentate amine ligands are preferentially used in excess to contribute to the complex stability by mass law action. The bidentate ligands are effective in a molar equivalent. Successive replacement of the methyl groups of the TMEDA by ethyl groups leads to a decline in the rates, apparently because of steric hindrance. The coordination of lithium with an amine probably increases polarization of the carbon–lithium bond. This results in a decrease of the activation energy in an ionic addition reaction with ethylene, and at the same time it increases the base strength of the carbanion moiety so that it readily undergoes a protophilic transmetalation, a reaction that an alkyllithium would not undergo alone.

This striking effect of a strong donor base on the catalytic activity of an organolithium compound contrasts with the reverse effect of donor bases on the growth reaction of ethylene with trialkylaluminum compounds. The catalytic activity of the latter is connected with the electron deficient nature of the uncoordinated, monomeric, trialkylaluminum species (10). These facts point to a difference in the mechanism of ethylene addition between the amine-coordinated organolithium catalyst and the trialkylaluminum compounds.

Product Structure

The primary products of the telomerization reaction are *n*-alkyl aromatics. More than one product may be formed when two or more different types of hydrogen are available on the aromatic telogen for transmetalation. Thus the primary products from benzene in which all hydrogens are equivalent are exclusively the even-numbered homologous 1-phenyl-*n*-alkanes. With toluene, 90 mole % of the primary product stems from transfer to the benzylic carbon atom while the remainder results from nuclear attack. Thus 90% of the primary product consists of homologous odd-numbered 1-phenyl-*n*-alkanes and 10% of homologous even-numbered, methylphenyl-*n*-alkanes. Ethylbenzene is attacked at the benzylic and nuclear positions to give two series of products—60% 2-phenyl-n-alkanes and 40% ethylphenyl-n-alkanes. The amount of product stemming from nuclear metalation is consistent with results on carbonation of the metalated alkylbenzenes (11).

As the reaction products accumulate they compete with the telogen during the transmetalation step and thus give rise to other sets of homologous products. This effect may be minimized by maintaining a large excess of telogen throughout the course of the reaction. With benzene as telogen, secondary attack (of $C_6H_5CH_2CH_2^-$ on $C_6H_5CH_2CH_3$) is more likely than with toluene or ethylbenzene as telogens since the benzylic hydrogens of the primary product (1-phenylalkanes) compete more effectively with the solely aromatic hydrogens of the telogen during the transmetalation. At 10 and 50% benzene conversion, about 5 to 10% and 25% of the product, respectively, stems from these secondary reactions. Toluene and ethylbenzene at equivalent extents of conversion give rise to about half the level of secondary reaction products obtained from benzene.

Since the primary reaction products always have more possible metalation sites than does the starting telogen, the secondary products are necessarily more complex. Whereas benzene produces only 1-phenylalkanes, the ethylbenzene derived from it may undergo secondary attack to form homologous ethylphenyl-n-alkanes and 2-phenyl-n-alkanes. Similarly, the primary product, n-butylbenzene, may lead to homologous n-butylphenyl-n-alkanes and 4-phenyl-n-alkanes. The secondary reaction products from toluene and ethylbenzene are correspondingly more complex. The product patterns are shown in Figure 3.

Aromatic Telomer Waxes

With aromatic telogen in short supply, the chain transfer reaction is minimized, and predominantly waxlike products are formed even at moderate ethylene pressures (12). The molecular weight distribution of these waxes and their pertinent physical properties vary considerably, depending upon the extent of the reaction. With benzene as telogen, at 700–800 psi of ethylene, slightly more than 50% of the product consists of a high-melting, low-viscosity telomer. The lower-melting material can be recycled to give an additional 20–25% of high-melting product.

Table V shows the melting points, viscosities, and average molecular weights of similar fractions not altered significantly by the character of the telogen. This behavior is in line with the proposal that these products are low- to moderate-molecular-weight polymethylenes varying only in the nature of the end group. On the other hand the molecular-weight distribution markedly depends upon the relative acidity of the

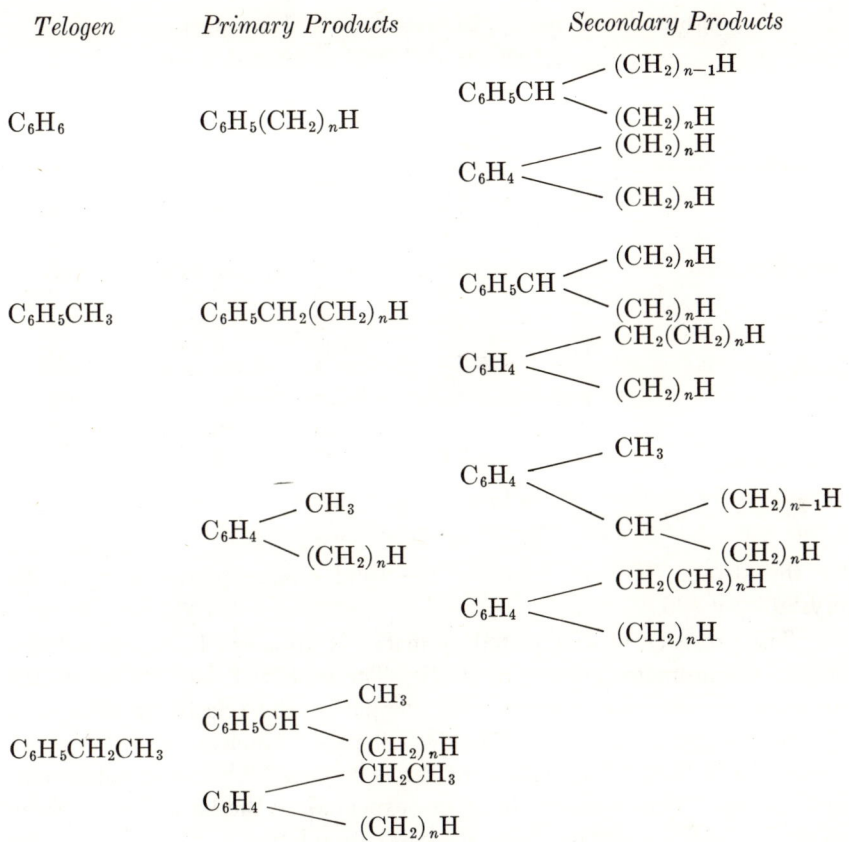

Figure 3. Structure of telomerization products

telogen. While toluene forms only a 26% yield of product within the two high-molecular-weight fractions, ethylbenzene and benzene produce 41 and 79% yields, respectively. These results agree with previous findings that the molecular weight is a function of the capacity of the telogen to effect chain transfer.

Complete characterization of the telomeric products is very hard to achieve because of the similarity among the several structural forms and because of the very broad molecular-weight distribution of homologous products. Nevertheless enough analytical data have been accumulated to permit a reasonably accurate appraisal. The crystalline birefringence, high melting point and XRD pattern indicate that the product contains an essentially unbranched carbon chain like that of polymethylene. This conclusion was confirmed by IR analysis in the 1400 and 2900 cm^{-1} regions indicating that the saturated portions were normal paraffins and

Table V. Product Distribution and Properties of Aromatic Telomer Wax

Telogen	Fraction	Wt %	Tm, °C	[n], decil/gram[a]	M, grams/mole
Toluene	A	57	[b]	0.013	340
	B	17	35	0.025	434
	C	13	83	0.043	590
	D	13	107	0.126	1050
Ethylbenzene	A	37	[b]	0.019	347
	B	22	35	0.029	459
	C	19	83	0.068	760
	D	22	109	0.087	870
Benzene	A	35	[b]	0.014	349
	B	11	35	—	435
	C	18	85	0.047	620
	D	36	106	0.079	820

[a] 1% solution in tetralin at 135°C.
[b] Oil at room temperature.

by the strong 730 cm^{-1} band of the solid characteristic of n-paraffin crystallinity (13).

The presence of dialkylated aromatics is apparent from an examination of the aromatic protons by NMR. The extent of dialkylation on the aromatic nucleus was estimated by comparing the relative intensities of the proton signals above and below 7.0 ppm. Nuclear dialkylation increases from 18 to 100% within the series benzene, toluene, ethylbenzene, and cumene. This is exactly the order expected on the basis that substitution of a benzylic carbon atom decreases its relative acidity (14) so that substitution upon the nucleus becomes more prominent.

These products are unusual in that they exhibit melting points higher than that of petroleum wax and melt viscosities lower than that of conventional polyethylene wax. This combination of desirable properties is difficult to attain since it is characteristic of a rather narrow molecular-weight region. The unique combination of high melting point and low viscosity provides numerous opportunities for product development and improvement. Materials of this kind are useful as modifiers for polyethylene and wax. An aromatic end group in the telomer waxes suggests the possibility of introducing functional moieties to make specialty products.

Poly-n-Alkyl Aromatics

The reaction of ethylene with a polymethylbenzene telogen typifies the formation of a product containing several alkyl chains (12). The multifunctional nature of a telogen, such as mesitylene, eases the introduction of more than one alkyl chain by way of the secondary reactions already outlined. Since the length of the alkyl chain is proportional to the ethylene-telogen ratio, this parameter is regulated by selecting the proper ethylene pressure. The number of alkyl chains can also be regulated since this parameter is related to the extent of conversion.

As is characteristic of a telomerization reaction, the product contains a broad molecular-weight range that can be separated into component fractions by distillation. An unusually large amount of intermediate-molecular-weight material results in this case because of the consecutive introduction of several alkyl groups. Under these circumstances the products pyramid. For example, the trialkylbenzene substituted with five-, seven-, and nine-carbon chains may result from any of three intermediates containing one-, five-, and nine-carbon chains; or one-, seven-, and nine-carbon chains.

Table VI. Major Components of a Light Distillate Telomer

E.T. min[a]	% of total fraction	Parent mass	$1,3,5\text{-}R_1R_2R_3C_6H_3$[b]		
			R_1	R_2	R_3
3.3	10	148	Me	Me	n-Pr
6.5	10	176	Me	n-Pr	n-Pr
7.5	20	176	Me	Me	n-Am
10.0	15	204	Me	n-Pr	n-Am
10.7	15	204	Me	Me	n-Hep
12.9	8	232	—	—	—
13.5	6	232	—	—	—

[a] Elution time: SE54, 6-foot, 7.5°/min from 90°C, 60 ml He/min.
[b] Assigned from MS, NMR, and IR of individual component.

The mass spectrum of the product obtained from mesitylene indicates that only homologous alkylbenzenes are formed. The presence of homologous 1,3,5-tri-n-alkylbenzes of odd carbon chain number has been confirmed by NMR and IR spectra of low-boiling components isolated with preparative gas chromatography (Table VI). Because of the large number of isomers possible, the isolation of single compounds from the higher molecular weight fractions is a formidable task which was not attempted.

Telomerization of Ethylene with Olefins

That even vinylic and allylic positions can be metalated by lithium from an amine complex is evident from the nature of the ethylene telomerization products formed in the presence of higher olefins (15). In the presence of excess butene, telomers are formed that vary in structure depending on the nature of the butene isomer (Table VII). Telomeric

Table VII. Telomerization of Ethylene with Olefinic Telogens[a]

Telogen	Product Structure	Molecular weight
None (ethylene)	Unbranched polyethylene wax	1300
1-Butene	3-Methyl-1-olefins and normal olefins (33:67)	360
2-Butene	3-Methyl-1-olefins and normal 2-olefins (60:40)	280 280
Isobutylene	2-Methyl-1-olefins	280

[a] Conditions: 110°C, 400 psig, excess telogen

olefins obtained from isobutylene are homologs of 2-methyl-1-pentene. The main products from butene-1 and -2 are homologs of hexene-2 and 3-methylpentene-1, respectively.

Addition of ethylene to an allylic anion is not involved since 1- and 2-butene would give rise to a common intermediate and identical products would result. Preferential metalation of the terminal vinylic position of 1-butene also seems unlikely since products derived from vinylic lithium intermediates are not obtained with isobutylene. To explain the product structure, ethylene insertion into the allylic carbon–lithium bond may possibly involve both a four-centered and six-centered transition state (Reactions 11a and 11b) with the latter predominating in butene-2. The exclusive formation of 2-methylpentene-1 and its homologs from the telomerization with isobutylene is easily understood since this product would result from both reaction pathways.

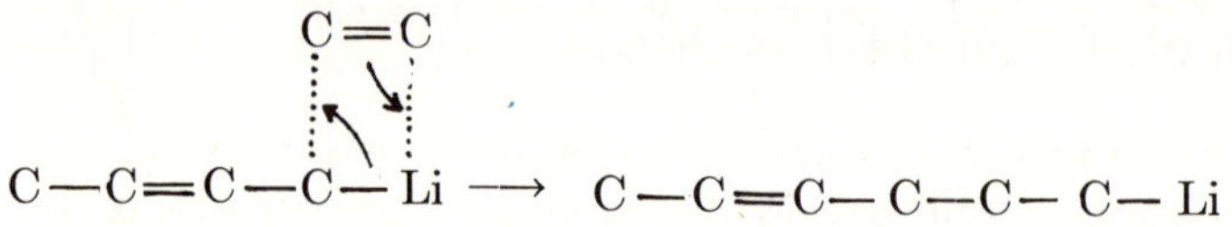

$$C-C=C-C-Li \;\; \xrightarrow{C=C} \;\; C-C=C-C-C-C-Li \tag{11a}$$

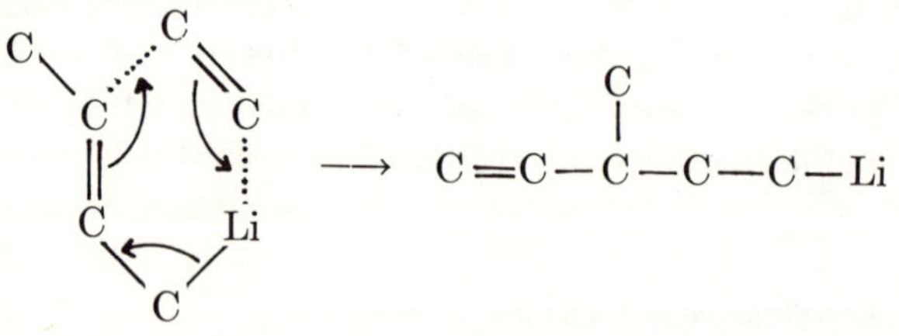

$$\xrightarrow{} \;\; C=C-\underset{\underset{C}{|}}{C}-C-C-Li \tag{11b}$$

In the absence of more acidic hydrocarbons, ethylene serves as both telogen and taxogen. The addition of ethylene to the lithium–carbon bond was previously observed by Ziegler, et al. (16). Those authors also found that ether catalyzes the addition reaction. However only growth products such as hexyl-, octyl-, and decyllithium were formed.

The product of the RLi–amine-induced telomerization of ethylene is a waxy, highly-crystalline, high-melting, low-molecular-weight polyethylene. In addition, more than one mole of polymer is formed per mole of RLi–amine complex, showing that a chain transfer mechanism is involved. In this connection the average molecular weight of the polyethylene does not depend on the ethylene pressure. Thus, ethylene must participate directly in the chain-transfer process. That process can be interpreted as an olefin displacement reaction involving a six-centered transition state analogous to that suggested above (Reaction 11b):

$$\longrightarrow \text{LiC}_2\text{H}_5 + \text{CH}_2=\text{CH}-\text{R} \qquad (12)$$

An equally plausible mechanism involves simply a transmetalation reaction between $\text{RCH}_2\text{CH}_2\text{Li}$ and ethylene—that is, $\text{RCH}_2\text{CH}_2\text{Li} + \text{CH}_2 = \text{CH}_2 \rightarrow \text{RCH}_2\text{CH}_3 + \text{CH}_2 = \text{CHLi}$. Studies involving sodium alkyls have shown that metalation can occur at vinylic or allylic positions (17).

The ethylene addition step can be interpreted as an insertion of the ethylene molecule between an amine-coordinated lithium cation and a carbanionic residue. This insertion reaction necessarily leads to the formation of a linear paraffinic carbon chain which agrees with experimental findings.

The high melting range, in spite of the low average molecular weight of the product, suggests a complete linearity of the polymer chain. Also, its high crystallinity (92% measured by x-ray diffraction) and the typical infrared absorption bands at 13.7 and 13.9 associated with the linear arrangement of solid paraffinic carbon chains (18, 19) agree with the assignment of a linear carbon chain structure. Moreover a molten film of the polyethylene wax shows an infrared absorption pattern in the 7.25 region similar to that of an authentic sample of polymethylene as reported by others (18, 19). Furthermore the infrared spectrum of a molten polyethylene sample reveals a terminal olefinic unsaturation in accordance with the proposed formation mechanism.

The number average molecular weight of the polyethylene corresponds to a β value of 0.021. Table VIII compares the β values of ethyl-

Table VIII. Values Observed with Four Telogens Under Equivalent Reaction Conditions[a]

Telogen	β
Toluene	0.75
Benzene	0.45
1-Butene	0.11
Ethylene	0.02

[a] Ethylene partial pressure, 400 psig; temperature, 110°C

ene, butene, benzene, and toluene under equivalent reaction conditions; the values are inversely proportional to the relative acidity of these hydrocarbons—that is, toluene > benzene > butene > ethylene in order of decreasing acidity. Thus the telomerization reaction provides a means of comparing the acidity of very weakly acidic unsaturated hydrocarbons. However, the measurement is kinetic in nature and does not necessarily accurately reflect the thermodynamic acidity.

Literature Cited

1. Pines, H., Schaap, L., *Advan. Catal.* (1960) **12**, 117.
2. Schaap, L., Pines, H., *J. Amer. Chem. Soc.* (1957) **79**, 4967.
3. Eberhardt, G. G., Butte, W. A., *J. Org. Chem.* (1964) **29**, 2928.
4. Langer, A. W., *Trans. N.Y. Acad. Sci.* (1965) **27**, 741.
5. Gilman, H., Morton, J. W., Jr., *Org. React.* (1954) **8**, 265.
6. Gilman, H., Gaj, B. J., *J. Org. Chem.* (1963) **28**, 1725.
7. Brown, H. C., Sujishi, S., *J. Amer. Chem. Soc.* (1948) **70**, 2878.
8. Margerison, D., Newport, J. P., *Trans. Faraday Soc.* (1963) **59**, 2058.
9. Bartlett, P. D., Tanber, S. J., Weber, W. P., *J. Amer. Chem. Soc.* (1969) **91**, 6362.
10. Ziegler, K., *Angew. Chem.* (1959) **71**, 623.
11. Broaddus, C. D., *J. Org. Chem.* (1970) **35**, 10.
12. Butte, W. A., *Hydrocarbon Process.* (1966) **45**, 277.
13. Tobin, M. C., Currano, M. J., *J. Polym. Sci.* (1957) **24**, 93.
14. Hart, H., Crocker, R. E., *J. Amer. Chem. Soc.* (1960) **82**, 418.
15. Eberhardt, G. G., Davis, W. R., *J. Polym. Sci.* (1965) **A3**, 3753.
16. Ziegler, K., Gellert, H., *Justus Liebigs Ann. Chem.* (1950) **567**, 195.
17. Broaddus, C., Logan, T., Flautt, T., *J. Org. Chem.* (1963) **28**, 1174.
18. Krimm, S., Liang, C., Sutherland, G., *J. Chem. Phys.* (1956) **25**, 549.
19. Bryant, W., Voter, R., *J. Amer. Chem. Soc.* (1953) **75**, 6113.

RECEIVED March 13, 1973.

10

Telomerization of Conjugated Diolefins with Aromatics and Olefins Using Chelated Organosodium Catalysts

WILLIAM BUNTING and ARTHUR W. LANGER, JR.

Corporate Research Laboratories, Esso Research and Engineering Co., Linden, N.J. 07060

> *Telomerization of butadiene and isoprene with aromatics and olefins proceeds rapidly at 0°–100°C using organosodium catalysts in combination with aliphatic tertiary chelating polyamines containing two to six nitrogens. The products range from monoadduct up to tacky semisolids. Chain transfer increased with increasing complexing ability of the chelating agent, increasing chelating agent concentration, increasing acidity of the telogen, increasing temperature, and decreasing monomer concentration. More than 74 mole % selectivity to pentenylbenzene was obtained from butadiene and toluene. The ratio of alpha/internal unsaturation in the monoadduct varied from 0.5 to 1.7, and it decreased at the higher temperatures because of the double bond isomerization activity of the catalyst. Catalyst efficiencies greater than 1300 grams/gram benzylsodium were obtained.*

Anionic telomerizations of conjugated diolefins with hydrocarbon acids are known but suffer from very low catalytic efficiencies. Morton et al. (1) and, later, Pappas et al. (2) used unchelated organosodium compounds to telomerize conjugated diolefins with weak hydrocarbon acids but obtained very low catalyst efficiencies (about 5 grams/gram catalyst). More recently, the anionic telomerization of butadiene and toluene by sodium on oxide supports (3) and sodium in tetrahydrofuran (4) was studied; also, a potassium amide/lithiated alumina catalyst was used to telomerize butadiene (5).

Common organosodium compounds are generally insoluble in inert solvents, and this causes considerable difficulty in their preparation and purification as well as in their use as catalysts and reagents. In earlier work with organolithium catalysts it was found that chelating tertiary diamines and higher polyamines formed soluble complexes (6) with enhanced reactivities. The prospects of obtaining similar results with organosodium compounds stimulated the research that is the subject of this paper. Substantial differences between the chelated sodium and lithium systems were expected and found. For example, chelating diamines and sodium compounds form only weak, unstable complexes although the diamines show some effect on organosodium catalysts in solution.

This report covers the use of chelated organosodium compounds as novel catalysts for telomerizing conjugated diolefins with weak hydrocarbon acids such as aromatics and olefins (7). The factors affecting the chain-transfer reaction were of particular interest because one of our objectives was to increase selectivity to low-molecular-weight species. The products are useful in synthesizing plasticizer alcohols, flame retardants, and surface coatings.

Discussion

Chelated Organosodium Compounds. The chelated organosodium compounds used in the telomerization process consist of those complexed by aliphatic, tertiary, chelating polyamines. In general, any organosodium compounds may be used that, when chelated, can initiate polymerization of the diolefin. This relates to carbanion basicity in that the initiating carbanion must have comparable or greater basicity than the allyl carbanion formed from the diolefin. Phenylsodium and benzylsodium are the most useful. Alkyl sodium compounds are too reactive to permit preforming the complex without decomposing the chelating agent. However, they can be used when the complex is formed in the presence of monomer or a suitable hydrocarbon acid to convert the alkyl anion to a carbanion of lower activity.

The preferred chelating agents are derivatives of ethylenediamine, higher polyamines, and isomers. The particular tertiary chelating polyamines discussed here together with their abbreviations are (they are all permethylated polyamines):

    TMED: tetramethylethylenediamine
    PMDT: pentamethyldiethylenetriamine
    iso-HMTT: tris-(β-dimethylaminoethyl)amine
    HMTP: heptamethyltetraethylenepentamine
    OMPH: octamethylpentaethylenehexamine

Two examples of chelated organosodium compounds are shown in Figure 1.

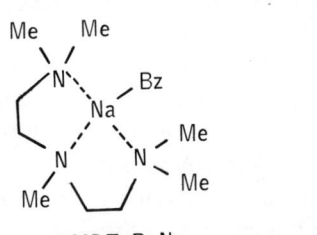
PMDT•BzNa

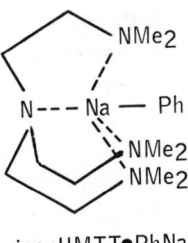
iso-HMTT•PhNa

Figure 1. Two chelated organosodium compounds

Telomerization. The telomerization is run by introducing gaseous butadiene into the atmosphere above a solution of catalyst in the hydrocarbon acid under conditions rigorously excluding air or water. The gaseous butadiene may be diluted with an inert gas and bubbled through the reactant solution. To maximize conversion to low molecular-weight products, efficient mixing of reagents is necessary. The reaction product is isolated by washing with water and distilling. The process is represented in Figure 2 by the telomerization of toluenes and butadiene.

The initial step involves attack of the chelated organosodium on the conjugated diolefin to give an allyl anion. The allyl anion can either abstract a proton from the hydrocarbon telogen (toluene) to give the monoadduct or add to another molecule of diolefin to give a new allyl

Figure 2. Telomerization of toluene and butadiene

anion. Further, proton abstraction or diolefin attack can occur at either the 1- or 3-carbon of the allyl anion. There seems to be a positive correlation between the relative reactivities of the 1- and 3-carbons in proton abstraction and in polymerization. It is possible to vary the ratio of monoadducts I:II (see Figure 2) as well as the average degree of polymerization of the reaction by varying the reaction conditions and catalyst.

Remetalation of products can occur at both benzylic and allylic positions to produce additional branched telomer structures. Although they were not all identified, the number of gas chromatographic (GC) peaks in the diadduct fraction indicated that all expected structural and double-bond isomers were obtained.

Nature of the Catalyst. The effect of varying the chelating agent in toluene–butadiene telomerization is shown in Tables I and II. The use of triglyme (a tetraether) as a chelating agent gave an inactive

Table I. Effect of Chelating Agent, 40°C[a]

Chelating Agent	Total Product (grams)	Selectivity to monoadduct, %	Alpha/Internal Unsaturation
None	6.0	34	0.5
TMED	15.8	32	0.5
PMDT	11.5	26	1.2
iso-HMTT	13.3	45	1.6
HMTP	17.5	60	1.4
OMPH	15.8	59	1.5
triglyme	no reaction		

[a] To 2.63 mmoles benzylsodium were added 2.63 mmoles chelating agent and 50 ml toluene. The reaction was heated to 40°C and maintained at 40°C while 22 cc/min gaseous butadiene were introduced into the atmosphere above the reaction for 2 hrs. At the end of that time 5 ml of water were added, and the organic layer separated and dried (K_2CO_3). Toluene was removed at water-aspirator pressure, and the residue was distilled (140°C/0.05 mm). The distillate was analyzed by GC.

Table II. Effect of Chelating Agent, 5°C[a]

Chelating Agent	Total Product (grams)	Selectivity to Monoadduct, %	Alpha/Internal Unsaturation
None	50 mg (waxy solid)	0	—
TMED	11.8	41.5	0.21
PMDT	9.5	18	0.85
iso-HMTT	8.6	6.2	1.72
HMTP	11.2	22.5	1.5

[a] Same reaction conditions as in Table I except the reaction temperature was 5°C.

catalyst. Presumably the catalyst decomposed *via* metalation of the chelating agent. In terms of selectivity to monoadduct and alpha/internal unsaturation at 40°C, the TMED–BzNa gives results similar to benzylsodium alone whereas complexes containing the higher chelating agents differ markedly from benzylsodium alone. This is a reflection of the relative stability of the complexes. At 40°C, TMED–BzNa is a relatively weak complex while PMDT–BzNa and the complexes with higher chelating agents are relatively strong complexes.

As shown in Tables I and II, unchelated benzylsodium is much less reactive than is the chelated benzylsodium. Ignoring colligative properties there are several general explanations for the nature of the catalyst. Certainly the insolubility of organosodium compounds in hydrocarbon media and the relatively high solubility of chelated organosodium compounds is part of any explanation comparing chelated with unchelated sodium compounds. Moreover, surrounding the sodium cation by a Lewis base (the chelating agent) might be expected to increase the electron density on the anion, making the anion in a chelated sodium compound a stronger base relative to the anion in an unchelated sodium compound (8, 9). The more chelating groups surrounding the cation, the greater should be the charge polarization of the carbon–sodium bond. As the charge density on the intermediate allyl anion involved in the telomerization (Figure 3) increases, the relative reactivities of the 1- and 3-carbons in proton abstraction and polymerization would be expected to change in a direction increasing the relative reactivity of the 3-carbon, which fundamentally agrees with the data given in Tables I and II.

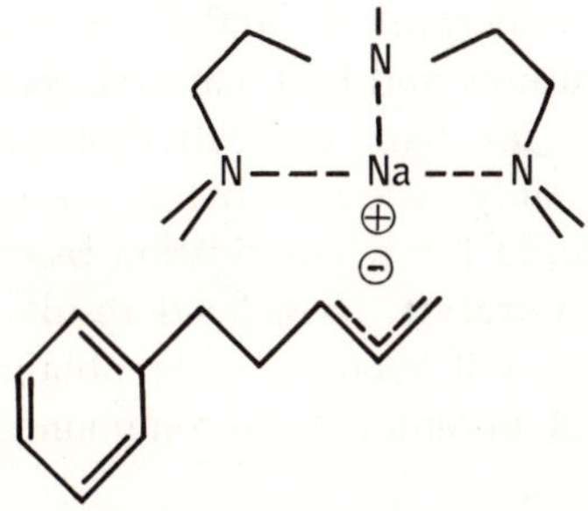

Figure 3. Allyl anion intermediate in toluene–butadiene telomerization

Undoubtedly, greater charge polarization in the metal-carbon bond occurs in the chelated compounds. However, how significant this increased charge polarization is with an already highly ionic species is a moot question. Whereas the organolithium compounds are usually considered to be fairly covalent in nature, the organosodium compounds are usually thought of as highly ionic species (10).

Another possible explanation of the results invokes steric effects in the intermediate allyl anion (Figure 3). The steric interaction between the bulky chelated cation and the benzyl group probably pushes the chelated cation toward the 1-carbon, increasing the steric shielding about the 1-carbon relative to the 3-carbon. Consequently, in going to higher chelating agents the chelated cation becomes increasingly bulky and the 3-carbon becomes more reactive relative to the 1-carbon in proton abstraction (manifested by increasing alpha/internal unsaturation in Tables I and II). Moreover, the steric requirements for reaction with butadiene (polymerization) should be greater than proton extraction from toluene (chain transfer). Consequently, selectivity to monoadduct should increase with higher chelating agents.

The data in Tables I and II are followed quite well with alpha/internal unsaturation and with selectivity to monoadduct (Table I). Conversion to monoadduct, though, is not well followed at 5°C (Table II). This may be because of a temperature-dependent change in colligative properties, lower temperature favoring higher states of aggregation.

We have no molecular-weight data for chelated organosodium compounds in solution. However, Langer has measured the molecular weights of several chelated organolithium compounds in hydrocarbon solution (11). He found TMED–n-BuLi to be monomeric at $0.1M$ and TMED–LiCHPh$_2$ to vary considerably in aggregation state with change in concentration, being monomeric only at low concentration ($0.05M$). The chelated organosodium compounds probably behave like the highly ionic chelated diphenylmethyllithium. Consequently, aggregation states greater than one are probably important in catalysis of the telomerization, particularly at low temperatures and high catalyst concentrations. A model for the catalyst structure involving colligative properties, ionic character, and steric effects is probably necessary before we can know the intimate details of the mechanism of the telomerization reaction.

Nonstoichiometric Catalyst. Beneficial results in optimizing selectivity to low-molecular-weight telomer can be obtained by deviating from the 1:1 stoichiometry of organosodium compounds to chelating agent.

Table III. Nonstoichiometric Catalyst Compositions[a]

Mmoles iso-HMTT	Mmoles BzNa	R[b]	Total Product, grams	Selectivity to Monoadduct, %	Alpha/Internal Unsaturation
2.63	2.63	1	29.3	50	1.2
5.26	2.63	2	38.8	65.8	1.52
5.26	5.26	1	39.8	73.2	1.25
10.52	2.63	4	40.3	66.8	1.41
10.52	5.26	2	40.1	74.4	1.16
2.63	10.52	0.25	39.7	63.8	0.61

[a] Butadiene 50 cc/min, was bubbled for 2 hr into 250 ml toluene containing the catalyst at 40°C. The reaction was quenched with 5 ml of water. Yields are distilled yields.
[b] Molar ratio of chelating agent to benzylsodium

Data for nonstochiometric catalyst compositions are shown in Table III. In the preceding section the large differences between TMED–BzNa at 5°C (Table II) and 40°C (Table I) are briefly discussed in terms of the strength of the complex. At 40°C TMED–BzNa behaves much like unchelated benzylsodium in the telomerization whereas at 5°C TMED–BzNa differs markedly from benzylsodium. At higher temperatures the complex between chelating agent and sodium compound is more dissociated. This equilibrium can be forced to the right by adding more chelating agent or more sodium compound:

$$\text{Chel} + \text{NaX} \underset{\Delta}{\rightleftarrows} \text{chel–NaX} \qquad K_{eq.} = \frac{[\text{chel–NaX}]}{[\text{chel}][\text{NaX}]}$$

Unfortunately, both excess sodium compound and chelated sodium compound catalyze the telomerization, even when the former is less effective, and it is difficult to separate these two reactions. With excess chelating agent, however, the equilibrium shift is clear. The data in Table III for excess chelating agent support the concept of a chelated sodium compound in equilibrium with free chelating agent and sodium compound. It is also possible, particularly below 25°C, that catalytic species of stoichiometry chel_2NaX or $\text{chel}(\text{NaX})_2$ are involved in addition to chel–NaX in the telomerization reaction.

Effect of Temperature and Butadiene Concentration. Increasing the reaction temperature increases the chain-transfer rate faster than the polymerization rate (Table IV). Activation energy for chain transfer is probably higher than that for polymerization although the lower solubility of butadiene at the higher temperatures would also contribute to the same result.

Table IV. Effect of Reaction Temperature

Temperature °C	Telomer Yield, grams	Selectivity to Monoadduct, %	Alpha/Internal Unsaturation
5	8.6	6.2	1.7
40	13.3	45	1.6
70	15.0	67	0.9

Increasing the reaction temperature not only increases conversions to monoadduct but also decreases catalyst lifetime and isomerizes the product. Under favorable conditions, more than 1300 grams product/gram BzNa were obtained. Aging the catalyst, iso-HMTT–BzNa, at 110°C for four hours results in an inactive catalyst. Presumably the catalyst slowly decomposes *via* metalation and decomposition of the chelating agent as was shown for the chelated alkyllithium catalysts. That the decrease in alpha/internal unsaturation with increasing tem-

Table V. Effect of Butadiene Addition Rate

C_4H_6 Rate (cc/min)	Time (min)	Selectivity to Monoadduct, %	Alpha/Internal Unsaturation
22	120	67	0.88
44	60	63	0.91
66	40	59	1.03
110	24	57	1.35

perature results in part from product isomerization and not only from some fundamental change in the nature of the catalyst (dissociation) is shown in Table V. Adding a constant amount of butadiene, but in shorter reaction times, increases alpha/internal unsaturation. Also, increasing the effective butadiene concentration decreases selectivity to monoadduct as expected.

Scope of the Reaction. Isoprene also has been successfully used; it gives slightly higher selectivity to monoadduct than does butadiene under identical reaction conditions, reflecting the slightly lower reactivity of isoprene in the telomerization. Benzene and octene have been used to replace toluene. They give higher molecular weight telomers, reflecting a slower chain-transfer step because of the lower acidity of allylic and aromatic protons relative to the benzylic protons on toluene (12). Also, the organosodium component of the catalyst has been replaced entirely or in part by the analogous organolithium compound. In general the presence of organolithium compounds in the telomerization process gives telomer of markedly higher molecular weight because organolithium compounds undergo chain-transfer type reactions much slower than do organosodium compounds. These reactions are summarized in Table VI.

Preparation of Oxo Alcohols. A telomer product containing about equal amounts of alpha and internal olefin monoadducts was hydroformylated using cobalt octacarbonyl at 120°C and 3000 psig, yielding a 60% conversion to aldehydes. The aldehydes were reduced to their

Table VI. Miscellaneous Telomerizations

Taxogen	isoprene	C_4H_6	C_4H_6	isoprene	C_4H_6
Telogen	toluene	1-octene	toluene	toluene	benzene
RM	BzLi	$C_5H_{11}Na$	BzLi/BzNa	BzNa	C_6H_5Na
Chelating agent	PMDT	iso-HMTT	TMED	iso-HMTT	iso-HMTT
Temperature, °C	40	40	40	70	70
Time, min	120	120	120	80	120
Selectivity to monoadduct, %	trace	trace	8.6	64	low
$\overline{M}_n$	2549	2422	—	—	998

corresponding alcohols with TMED–LiAlH$_4$ in benzene (13). Of the three alcohols produced about 60% was 6-phenyl-1-hexanol. The other two alcohols are probably 5-phenyl-2-methyl-1-pentanol and 4-phenyl-2-ethyl-1-butanol. By converting these alcohols to phthalate esters, low volatility plasticizers are obtained.

Experimental

Reactions were run under conditions rigorously excluding air or moisture. Hydrocarbon solvents were purified by passage through an alumina column and storage over sodium ribbon before use. Chelating agents were dried (sodium ribbon), fractionally distilled, and stored over calcium hydride. Organosodium compounds were obtained from Orgmet (Hampstead, N.H.) and used without further purification.

General Telomerization. Organometallic compound and solvent were added to a dried flask equipped with a stirring bar in a dry box. The reaction flask was stoppered or fitted with a thermometer and removed from the dry box. The reaction was placed under nitrogen and heated (or cooled) to the desired temperature. The conjugated diolefin was then added by introducing it into the atmosphere above the solution, bubbling through the reaction solution, or dissolving in an appropriate solvent, and adding the resultant solution dropwise. At the end of the desired reaction period, 5–10 ml of water were added with stirring. The organic layer was dried (K_2CO_3) and either fractionally distilled at water-aspirator pressure (bp of monoadduct is about 115°–120°C/15 mm, diadduct about 175°C) or bulb-to-bulb distilled (140°C/0.02 mm). Distillates were analyzed on a Carbowax-20M column at about 180°C. The monoadducts were characterized by NMR and IR.

Chelated Organolithium Catalyst. To 1.6 ml of 1.6N n-butyllithium in hexane were added 50 ml toluene and 0.6 ml PMDT, generating PMDT–BzLi *in situ*. The reaction was kept at 40°C while butadiene was introduced into the atmosphere above the reaction (22 cc/min) for 2 hrs. The reaction was then quenched with 5 ml of water. The organic layer was separated, dried (K_2CO_3), and the solvent was removed under vacuum to give 5 grams of product having $\overline{M}_n = 2549$.

Chelated Organolithium/Organosodium Catalyst. To 0.95 ml of 1.6N n-butyllithium (1.5 mmoles) were added 50 ml toluene, 0.17 grams (1.5 mmoles) benzylsodium, and 0.45 ml (∼ 3 mmoles) TMED. The reaction was heated to 40°C and kept at that temperature while gaseous butadiene was introduced (22 cc/min) into the atmosphere above the reaction for 2 hr. The reaction was then quenched with 5 ml of water, and the organic layer separated and dried (K_2CO_3). The toluene was removed at water-aspirator pressure, and the residue was distilled (140°C/0.05 mm). The distillate was analyzed by GC. Total product weight was 6.1 grams. Selectivity to monoadduct was 8.6%. The ratio of alpha-to-internal unsaturation was 1.25.

Telomerization of Toluene and Isoprene. To 0.3 gram benzylsodium were added 50 ml of toluene and 0.7 ml of iso-HMTT. A solution of 20 ml isoprene in 25 ml toluene was added dropwise to this reaction mixture at 70°C over 80 min. A work-up similar to the preceding example was used. The toluene–isoprene telomer weighed 23.8 grams. Selectivity to monadduct was 64% based on GC analysis.

Telomerization of Benzene and Butadiene. To 0.26 gram (2.63 mmoles) of phenylsodium were added 50 ml of benzene and 0.7 ml of iso-HMTT. The reaction was kept at 70°C while butadiene was introduced into the atmosphere above the reaction for 2 hrs at 22 cc/min. A work-up similar to that given under chelated organolithium catalyst gave 7 grams of benzenebutadiene telomer with $M_n = 998$.

Telomerization of 1-Octene and Butadiene. To 0.3 gram of n-amylsodium were added 50 ml of 1-octene and 0.7 ml of iso-HMTT. Butadiene gas was added to this reaction mixture at 40°C at 22 cc/min for 2 hr. A work-up similar to that given under chelated organolithium catalyst gave 4.6 grams of octene–butadiene telomer with $M_n = 2422$.

Literature Cited

1. Morton, A., Patterson, G. H., Donovan, J. P., Little, E. L., *J. Amer. Chem. Soc.* (1946) **68**, 93.
2. Pappas, J. J., Schriesheim, A., U.S. Patent **3,189,660** (1965).
3. Eberhardt, G. G., Peterson, H. J., *J. Org. Chem.* (1965) **30**, 82.
4. Kume, S., et al., *Makromol. Chem.* (1966) **98**, 109; Kume, S., *Makromol. Chem.* (1966) **98**, 120.
5. Bloch, H. S., U.S. Patent **3,355,484** (1967).
6. Langer, A. W., U.S. Patent **3,541,149** (1970); U.S. Patent **3,451,988** (1969).
7. Bunting, W., Langer, A W., U.S. Patent **3,750,200** (1972); U.S. Patent **3,742,057** (1973).
8. Originally proposed to account for the unusual reactivity of TMED–n-BuLi: Langer, A. W., *Am. Chem. Soc. Div. Polymer Chem. Preprints* (1966) 137.
9. Langer, A. W., *Trans. N.Y. Acad. Sci.* (1965) **27**, 746.
10. Cotton, F. A., Wilkinson, G., "Advanced Inorganic Chemistry," Interscience, New York, 1962.
11. Langer, A. W., unpublished results.
12. Cram, D. J., "Fundamentals of Carbanion Chemistry," p. 19, Academic, New York, 1965.
13. Langer, A. W., Whitney, T. A., *Chem. Abstr.* (1971) **75**, 38572Z; U.S. Patent **3,734,963** (1973).

RECEIVED March 13, 1973.

11

Polylithiation of Hydrocarbons

ROBERT WEST

University of Wisconsin, Madison, Wis. 53706

Certain hydrocarbons undergo polymetalation by alkyllithium compounds, often with TMEDA as a catalyst, to give polylithiated organic compounds. For acetylenes, protons both α and β to the triple bond are replaced; thus C_3Li_4 is obtained from propyne, $CH_3C_3Li_3$ from 1- or 2-butyne, and C_5Li_4 from 1,3-pentadiyne. Further reaction of these substances with alkyl sulfates or organometallic halides yields highly unsaturated organic or organometallic derivatives. Polylithiation of aromatic compounds also takes place; toluene, naphthalene, anthracene, fluorene, indene, ferrocene and 1- and 3-phenylpropyne have been converted to mixtures of polylithium compounds that have been studied by derivatization with D_2O or trimethylchlorosilane. Small amounts of perlithiated aromatic compounds appear to be formed from some compounds.

In recent years preparations of several polylithiated organic compounds have been reported. In a few cases it is even possible to make perlithio compounds, or "lithiocarbons," in which each hydrogen in a hydrocarbon molecule has been replaced by a lithium atom. Some polylithio and perlithio compounds can be made simply by the interaction of hydrocarbons or halocarbons with alkyllithium reagents, without catalysis by amines. In other cases, however, use of chelating diamines (usually TMEDA) is either clearly beneficial or absolutely essential. This paper reviews the synthesis of poly- and perlithiated compounds, with emphasis on the use of chelating diamines as aids in the metalation reactions.

Aliphatic Polylithio Compounds

Excluding the well-known lithium acetylide, C_2Li_2, the first perlithio compound to be reported was C_3Li_4, synthesized by West, Carney, and Mineo in 1965 (1) by the reaction of propyne with n-butyllithium in hexane. When the propyne (chilled) is added slowly to the hexane solution of the lithium reagent, the alkyne is polylithiated so rapidly that the hexane-insoluble salt 1-lithiopropyne never precipitates. Refluxing the mixture completes the metalation; recovery of n-butane shows that yields of C_3Li_4 higher than 80% can be obtained. The structure of C_3Li_4 is not known, but the infrared spectrum, which shows an extremely strong absorption at 1670 cm^{-1} and no band at higher frequencies, suggests that the structure is allenic, $Li_2C=C=CLi_2$ (2). C_3Li_4 often precipitates from hexane as a red powder but sometimes remains in solution, perhaps complexed with excess n-butyllithium.

Recently, Shimp and Lagow produced C_3Li_4 as the major product in the high-temperature reaction of lithium atoms with carbon vapor, along with smaller amounts of C_2Li_2 and CLi_4 (4). The same group has also shown that a mixture of polylithio compounds, including CLi_4, C_2Li_4, and C_2Li_2, is obtained in the reaction of lithium atoms with CCl_4; in the same reaction using C_2Cl_6, C_2Li_6 is the major product (5). These important studies provide an entirely new method for the synthesis of lithiocarbons, including the previously unknown perlithioalkanes.

The lithiocarbon C_3Li_4 reacts with organic and organometallic substrates to give a variety of highly unsaturated derivatives (Scheme 1). With organometallic halides such as trialkylchlorosilanes, the usual products are tetrasubstituted allenes (2). Alkyl halides often give explosions, but the alkyl sulfates produce mixtures of tetraalkyl derivatives of propyne and propadiene (3).

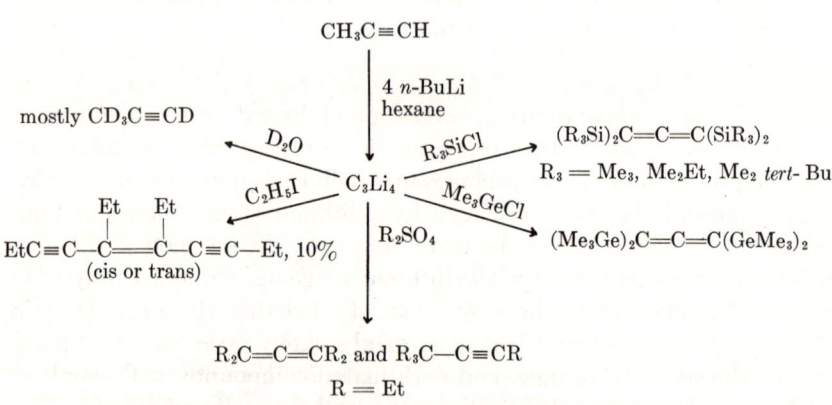

Scheme 1. *Reactions of C_3Li_4*

Polymetalation is apparently a general reaction of 1-alkynes. Lithiation of 1-butyne has been studied in detail (2). When this compound is treated with three equivalents of n-butyllithium or *tert*-butyllithium in hydrocarbon solvents, trilithio derivative, $CH_3C_3Li_3$, is formed. Nonconjugated internal acetylenes resist metalation by alkyllithium reagents

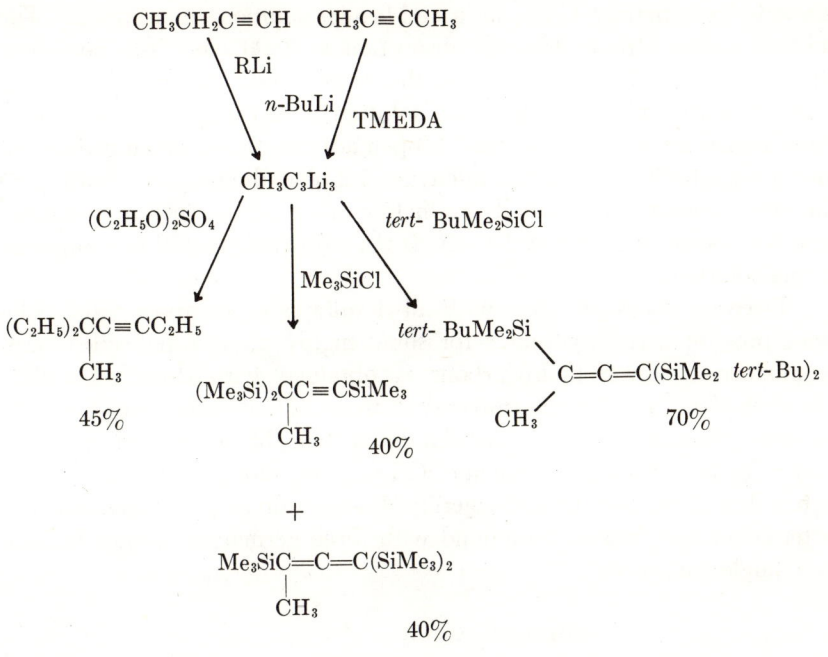

Scheme 2. *Formation and reactions of $CH_3C_3Li_3$*

under usual conditions. When TMEDA is added, however, polylithiation takes place. With TMEDA, 2-butyne forms a trilithio derivative identical to that obtained from 1-butyne (6); see Scheme 2.

Reaction of $CH_3C_3Li_3$ with diethyl sulfate gives a moderate yield of 5-ethyl-5-methyl-3-heptyne, along with smaller amounts of less highly substituted acetylenes. Derivatization of $CH_3C_3Li_3$ with trimethylchlorosilane yields about equal amounts of the tris(silyl)butyne and methylallene (Scheme 2). However the bulkier derivatizing agent *tert*-butyldimethylchlorosilane converts $CH_3C_3Li_3$ exclusively into the less-strained allenic product, 1,1,3-tris(*tert*-butyldimethylsilyl)-1,2-butadiene. This reaction exemplifies the common finding that products of silyl derivatization of lithium compounds are sterically determined. The structure of $CH_3C_3Li_3$ is uncertain, but from its low-frequency infrared absorption (1750 cm^{-1}), it probably has an allenic structure, $CH_3CLi=C=CLi_2$, similar to that also proposed for C_3Li_4 (6).

In 1971 the second perlithio compound, C_5Li_4, was obtained by lithiation of 1,3-pentadiyne (7, 8). When the latter is treated with alkyllithium compounds in the absence of TMEDA, addition of alkyllithium reagents to one of the triple bonds is the main reaction, and no highly metalated products form. However with TMEDA, polylithiation takes place to give mainly C_5Li_4 as a red-brown solution in n-butane. The solution shows strong infrared absorption at 1800 cm^{-1}; the cumulene structure, $Li_2C{=}C{=}C{=}C{=}CLi_2$, thus seems most probable for C_5Li_4.

When treated with water, the solution of C_5Li_4 gives a mixture of three hydrocarbons: the original 1,3-pentadiyne; the nonconjugated isomer 1,4-pentadiyne; and the allenyne, 1,2-pentadien-4-yne. With D_2O the corresponding deuterated products contain about 70% tetradeuterated molecules, showing that C_5Li_4 is the principal polylithio compound in the solution.

Derivatizations of C_5Li_4 with alkyl sulfates or organometallic chlorides provide useful syntheses for some highly unsaturated substances. A mixture of three hydrocarbons is obtained with dimethyl sulfate (Scheme 3). Trialkylchlorosilanes convert C_5Li_4 exclusively to the allenyne products, which are less hindered than the other isomers. However with trimethylchlorogermane, C_5Li_4 gives, along with the allenyne isomer 1,5,5,5-tetrakis(trimethylgermyl-1,3-pentadiyne) (Scheme 3). The latter is the first known compound with three germanium atoms bonded to a single carbon (8).

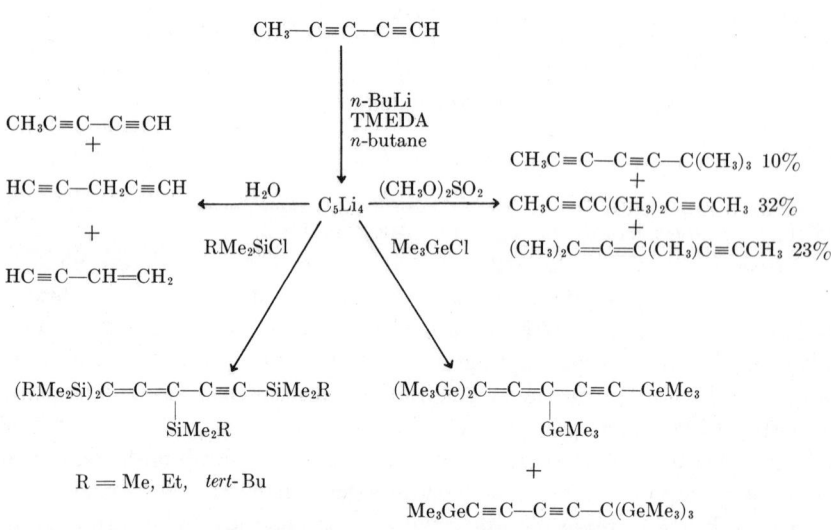

Scheme 3. Reactions of C_5Li_4

Polylithiation of Aromatic Compounds

Toluene. Lithiation of aromatic and alkylaromatic compounds is markedly catalyzed by chelating diamines. Toluene is metalated by organosodium reagents but is rather unreactive toward organolithium compounds in the absence of diamine catalysts. However in the presence

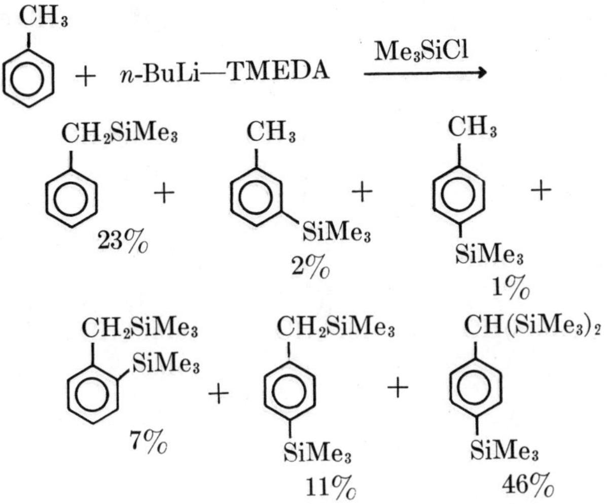

Journal of the American Chemical Society

Scheme 4. Polylithiation of toluene and derivatization with trimethylchlorosilane (10)

of TMEDA or diazabicyclo[2.2.2]octane, toluene undergoes quantitative metalation by *n*-butyllithium to form benzyllithium (*9*).

Polylithiation of toluene using excess *n*-BuLi–TMEDA produces a bright orange solution from which a red solid gradually separates. Derivatization of the reaction mixture with trimethylchlorosilane produces mono-, bis-, and tris(trimethylsilyl) isomers, as shown in Scheme 4 (*10*). Under a variety of metalation conditions of the many possible isomers only those shown in Scheme 4 were isolated. Polylithiation of benzyllithium (prepared from dibenzylmercury and lithium) with TMEDA present gave the same mixture of compounds after derivatization whereas attempts to polylithiate *o*-, *m*-, and *p*-lithiotoluenes gave only very slow reactions and low yields of polylithio compounds. This and other supporting evidence indicates that dilithiation of toluene takes place mostly at the alpha-carbon replacing two hydrogens to form $PhCHLi_2$.

The third lithiation takes place exclusively para. When ring metalation occurs at any point, the resulting aryllithium compound is unreactive toward further metalation (*see* Scheme 5).

The principal monotrimethylsilyltoluene is the meta isomer. This agrees with the results of earlier experiments which find mainly *m*-lithiotoluene in the ring-lithiated byproducts in the synthesis of benzyllithium by lithiation of toluene with *n*-butyllithium–TMEDA (*11, 12*). On the basis of the current view that metalation of aromatic compounds proceeds

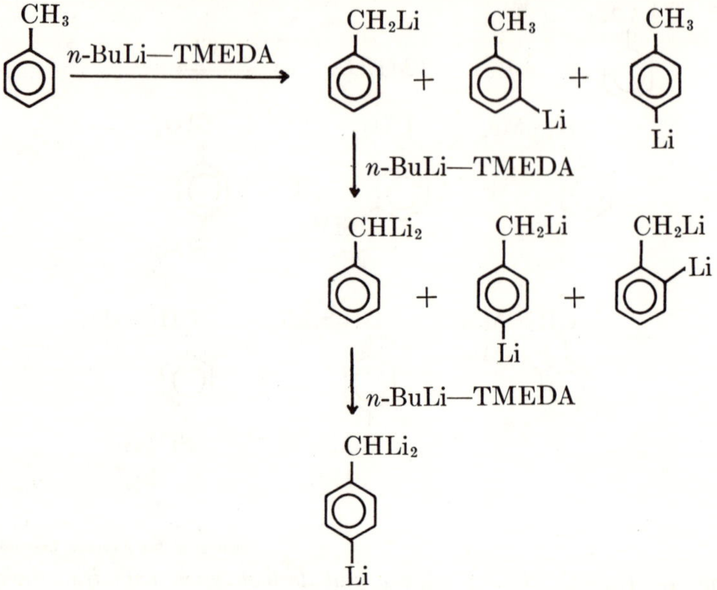

Scheme 5. Probable course of lithiation of toluene with n-BuLi–TMEDA (10)

via nucleophilic attack on hydrogen by the alkyl anion (13, 14), the meta isomer should predominate.

Further lithiation, however, seems to take place by a different mechanism. Among the disilyl compounds, only ortho and para trimethylsilyl isomers are obtained (Scheme 4). To account for this shift in isomer preference, West and Jones suggest that the high negative charge present in the benzyllithium causes a change in mechanism to one in which electrophilic attack on the carbon by the positive lithium predominates (10):

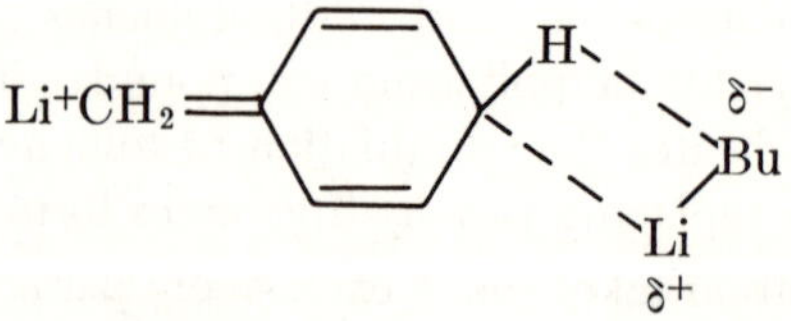

Actually, the transition state might look very similar to that for nucleophilic attack on ring hydrogen; the difference lies in the degree of participation by lithium.

Finally, the only tris(trimethylsilyl) compounds formed is the α,α,para compound (Scheme 4). Lithiation of α,α-dilithiotoluene at the ortho position, which would be electronically as favorable as para substitution, may be precluded by the hindering effect of the two bulky TMEDA molecules coordinated to the lithium atoms:

$$\begin{array}{c} \text{structure of dilithio compound with two TMEDA ligands coordinated to lithium atoms bridged by a CH unit bonded to a phenyl group} \end{array}$$

Polycyclic Aromatics. Extensive replacement of hydrogen by lithium in polycyclic aromatic hydrocarbons has been demonstrated by Halasa (15). Anthracene, biphenyl, fluorene, indene, and ferrocene (16) undergo polymetalation by n-butyllithium–TMEDA in hexane at 70°C for 24 hours. The products are insoluble mixtures of polylithio compounds containing up to 10 lithium atoms per molecule. Derivatization was accomplished using both D_2O and trimethylchlorosilane and by analyzing the mixture of deuterated or silylated products by mass spectrometry. The results for anthracene, which are typical, appear in Table I.

The product distributions for the deuterated anthracenes are the most significant numbers; these data are corrected for ^{13}C content but depend on the (untested) assumption of equal probability of loss of H and D during fragmentation. Nevertheless it is apparent that substantial amounts of anthracenes bearing up to seven lithium atoms were produced, and that traces of perlithioanthracene, $C_{14}Li_{10}$, must have been obtained. The figures for relative abundance of the trimethylsilyl derivatives are discordant and less reliable; when trimethylchlorosilane is used as a derivatizing agent, further lithiation often occurs during the derivatization. Take, for example, the lithiation-silylation of acetonitrile. Treatment of CH_3CN with three equivalents of alkyllithium followed by three equivalents of trimethylchlorosilane produces $(Me_3Si)_2C=C=NSiMe_3$, but studies of the lithiation reaction show that it does not proceed past dilithiation (to C_3HNLi_2) until Me_3SiCl is added (17).

Similar experiments demonstrate the formation of mixtures containing polylithio compounds up to $C_{13}Li_{10}$ for fluorene, up to $C_{12}H_4Li_6$ for biphenyl, and up to $C_9H_2Li_7$ for indene. With ferrocene the effect of TMEDA was studied by comparative experiments, results of which are summarized in Table II (16). The dramatic increase in polylithiation using TMEDA is clearly shown. With TMEDA the most abundant single product is the hexadeutero isomers, resulting from $C_{10}H_4Li_6$; without TMEDA the most probable product is that from $C_{10}H_6Li_4$.

Table I. Mass Spectra of Products
D₂O Derivatization

m/e	Ions	Relative Abundance	Product Distribution, %
179		9.5	—
180	$C_{14}H_8D_2$	38.0	11.33
181	$C_{14}H_7D_3$	71.4	20.14
182	$C_{14}H_6D_4$	100.00	27.75
183	$C_{14}H_5D_5$	66.6	17.02
184	$C_{14}H_4D_6$	52.2	14.73
185	$C_{14}H_3D_7$	23.8	6.55
186	$C_{14}H_2D_8$	9.5	1.92
187	$C_{14}HD_9$	4.7	0.10
188	$C_{14}D_{10}$	2.3	0.02

Polylithioferrocene was also derivatized with trimethylchlorosilane, yielding a mixture of polysilyl derivatives from which a crystalline tetrakis(trimethylsilyl)ferrocene was isolated and shown to be the 1,3,1′,3′ derivative. Polylithioferrocene and the polylithioaromatics generally catalyze the polymerization of conjugated dienes leading to star-shaped polymers.

The polylithiation of 1-phenylpropyne has been studied in somewhat greater detail. The first evidence that this compound might undergo lithiation at the aromatic nucleus as well as on the side chain was presented by Mulvaney, Folk, and Newton who found that both C_6H_5—C_3Li_3 and C_6H_4Li—C_3Li_3 were produced from $C_6H_5C{\equiv}CCH_3$ and excess n-butyllithium in refluxing hexane (18). This reaction is strongly catalyzed by TMEDA; in the presence of the chelating diamine, predominant tri- and tetralithiation occurs, even at 25°C (19).

Lithiation of 3-phenylpropyne produced similar polylithium compounds. With three equivalents of n-butyllithium in cyclohexane, 3-

Table II. Degree of Polylithiation of Ferrocene with and without TMEDA from Mass Spectra of Deuterium Derivatives (16)

Number of Li atoms	No TMEDA	with TMEDA
0	1.3	0.3
1	4.2	0.7
2	25.6	2.4
3	28.3	5.9
4	30.8	23.0
5	7.4	26.0
6	1.5	29.0
7	0.3	9.6
8	0.1	3.2

from Polylithioanthracene (15)

Me₃SiCl Derivatization

m/e	Ions	Relative Abundance
322	$C_{14}H_8(SiMe_3)_2$	80
394	$C_{14}H_7(SiMe_3)_3$	100
466	$C_{14}H_6(SiMe_3)_4$	60
538	$C_{14}H_5(SiMe_3)_5$	50
610	$C_{14}H_4(SiMe_3)_6$	20
682	$C_{14}H_3(SiMe_3)_7$	20
754	$C_{14}H_2(SiMe_3)_8$	10
826	$C_{14}H(SiMe_3)_9$	10
898	$C_{14}(SiMe_3)_{10}$	10

phenylpropyne is converted mainly to a bright-red trilithio derivative, $C_6H_5C_3Li_3$. The infrared spectrum of this substance shows only a single strong C=C band at 1780 cm⁻¹, consistent with the allenic structure PhCLi=C=CLi₂. Derivatization with trimethylchlorosilane produced the tris(trimethylsilyl) phenylallene, PhC(SiMe₃)=C=C(SiMe₃)₂ (19).

Trapping with trimethylchlorosilane served to identify the positions of nuclear lithiation of 1-phenylpropyne. The tetrasilyl products from C_6H_4Li—C_3Li_3 were identified as the ortho and para derivatives, indicating that nuclear lithiation takes place preferentially ortho or para to the side chain:

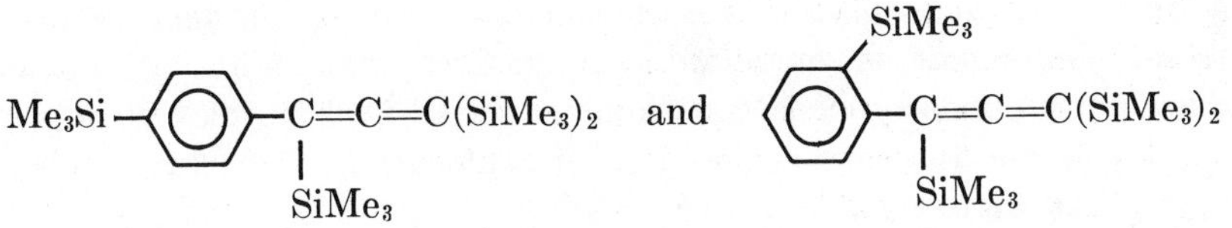

Me₃Si—⟨○⟩—C=C=C(SiMe₃)₂   and   ⟨○⟩—C=C=C(SiMe₃)₂
         | (with SiMe₃ ortho)
        SiMe₃ |
                                       SiMe₃

Under forcing conditions, 1-phenylpropyne can be much more highly lithiated. The greatest substitution was obtained by heating 1-phenylpropyne neat with a 50-fold excess of n-BuLi at 75° to 85°C for 48 hours. Quenching with D₂O produced a mixture of deuterated 1- and 3-phenylpropynes, which was studied by mass spectroscopy. The isomer distribution is shown in Table III.

The major products arise from the penta- and hexalithio compounds, but a significant amount of C_9HLi_7 and 1% of the perlithio compound C_6Li_5—C_3Li_3 apparently form. Derivatization of the polylithiated mixture using Me₃SiCl produces numerous silyl-substituted isomers, which could only be separated when the number of silicons was less than six.

The most significant product is probably the pentasilyl compound, identified as o-, p-bis(trimethylsilyl)phenyltris(trimethylsilyl)allene:

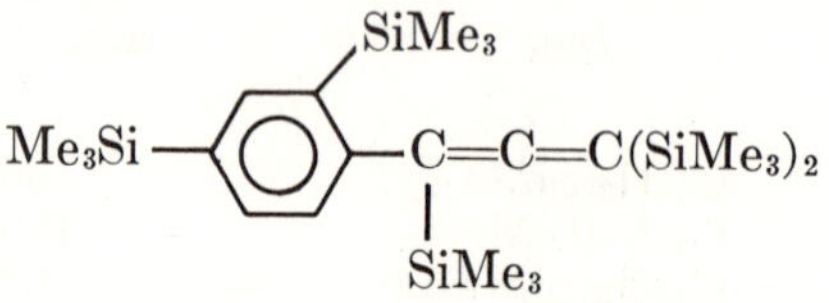

$$Me_3Si-C_6H_3(SiMe_3)(SiMe_3)-C=C=C(SiMe_3)_2$$

Table III. Products from D_2O Derivatization of Polylithiated 1-Phenylpropyne

Isomer	% Produced
d_0	0
d_1	0
d_2	0
d_3	5
d_4	12
d_5	36
d_6	37
d_7	8
d_8	1

Polylithiation of aromatic compounds offers much promise for future research. It is evident that perlithioaromatics can be prepared, albeit only mixed with less-highly lithiated species. Moreover the perlithio aromatics seem to be stable species once formed. A major problem limiting further advance in this area is the lack of a polar solvent inert to alkyllithium compounds at moderate temperatures. Activation of organolithium compounds by chelating diamines seems likely to play an important role in the further development of lithiocarbon chemistry, both in the aliphatic and aromatic series.

Literature Cited

1. West, R., Carney, P. A., Mineo, I. C., *J. Amer. Chem. Soc.* (1965) **87**, 3788.
2. West, R., Jones, P. C., *J. Amer. Chem. Soc.* (1969) **91**, 6156.
3. Priester, W., West, R., unpublished data.
4. Shimp, L. A., Lagow, R. J., *J. Amer. Chem. Soc.* (1973) **95**, 1343.
5. Chung, C., Lagow, R. J., *Chem. Comm.* (1972) 1078.
6. Chwang, T. L., Ph.D. Thesis, University of Wisconsin (1971).
7. Chwang, T. L., West, R., *Chem. Commun.* (1971) 813.
8. Chwang, T. L., West, R., *J. Amer. Chem. Soc.* (1973) **95**, 3224.
9. Eberhardt, G. G., Butte, W. A., *J. Org. Chem.* (1964) **29**, 2928.
10. West, R., Jones, P. C., *J. Amer. Chem. Soc.* (1968) **90**, 2656.
11. Chalk, A. J., Hoogeboom, T. J., *J. Organometal. Chem.* (1968) **11**, 615.
12. Broaddus, C. D., *J. Org. Chem.* (1970) **35**, 10.
13. Bryce-Smith, D., *J. Chem. Soc.* (1954) 1079.

14. Hall, G. E., Piccolini, R., Roberts, J. D., *J. Amer. Chem. Soc.* (1955) **77**, 4540.
15. Halasa, A., *J. Organometal. Chem.* (1971) **31**, 369.
16. Halasa, A., Tate, D., *J. Organometal. Chem.* (1970) **24**, 769.
17. Gornowicz, G. A., West, R., *J. Amer. Chem. Soc.* (1971) **93**, 1720.
18. Mulvaney, J. C., Folk, T. L., Newton, D. J., *J. Org. Chem.* (1967) **32**, 1674.
19. West, R., Gornowicz, G. A., *J. Amer. Chem. Soc.* (1971) **93**, 1720.

RECEIVED March 30, 1973.

12

Directed Metalation

D. W. SLOCUM and D. I. SUGARMAN

Southern Illinois University, Carbondale, Ill. 62901

> *The directed metalation reaction—lithiation with n-butyllithium of a position ortho to a substituent on an aromatic ring—is described. Aromatic systems in which the reaction has been studied are benzene, thiophene, naphthalene, and ferrocene. A systematic listing of the bond types that can be formed at the site of metalation is provided. Also of interest is the assessment of the relative directing abilities of directing substituents and comments and observations on the mechanism of the reaction. Utility of the reaction is indicated by the results from asymmetric-directed lithiation and the synthesis of heterocycles.*

It has been known for 40 years that alkyllithium compounds will react with specifically substituted aromatic compounds to effect metalation —that is, replace an aromatic proton with a metal ion. More recently the orientation in a variety of such metalations has been worked out resulting in the identification of substituents that have been demonstrated to direct metalation to an aromatic proton adjacent to said substituents. This, then, is the reaction that is now called "the directed metalation reaction." Since many of these substituents contain a directing nitrogen atom, it is appropriate that this process be reviewed here.

Within the past decade, many additional directing substituents have been discovered so that the number of synthetic derivatives available through this method is large. One of the great advantages of this reaction is its extremely high specificity. Assuming that synthesis of a specific ortho-disubstituted benzene compound were feasible *via* electrophilic substitution, difficulty in separating the ortho from the para- and even meta-substitution product might be anticipated. Use of the directed metalation reaction in this instance, assuming the same compound could be synthesized by the two methods, would yield pure ortho isomer uncontaminated by all else save starting material. In addition there are many groups that cannot be simply introduced into an aromatic system by electrophilic substitution that can be readily introduced by directed metalation. Thus electrophilic substitution and directed metalation of substituted aromatic compounds appear to complement one another nicely.

Aromatic Systems in Which Directed Metalations Have Been Effected

The great ability of n-butyllithium or n-butyllithium–TMEDA complex to effect metalation of aromatic compounds (*1*) suggests that most aromatic systems should also undergo the directed metalation reaction provided appropriate directing groups are present. One exception to this has been the [2.2]paracyclophane system where no evidence of metalation by either n-butyllithium or its TMEDA complex under a variety of conditions has been seen (*2*). To our knowledge, no other aromatic system resists metalation. Thus the number of systems in which directed metalation might be useful is potentially large although only four have been systematically studied. That only a small number of such systems have been explored up to this time is not to say that only a few such systems exist; rather it is an indication that further study is required.

Benzene. The earliest work in directed metalation was done on the benzene system in the early 1930's (*3*). That the benzene ring was the most promising system for the exploitation of this reaction appears logical since it was by far the most examined aromatic system at that time. Not only have the largest number of directing substituents been successfully demonstrated for this system, but in all probability the demand for a convenient route to a specific polysubstituted aromatic compound will be highest for this system. In all cases examined, directed metalation of a monosubstituted benzene has yielded almost exclusively ortho metalation, with only ortho-disubstituted products obtained (Reaction 1). With

$$\text{R-C}_6\text{H}_5 + \text{C}_4\text{H}_9\text{Li} \longrightarrow \text{R-C}_6\text{H}_4\text{Li} \xrightarrow{\text{Electrophile}} \text{R-C}_6\text{H}_4\text{X} \qquad (1)$$

more than one substituent on the ring the situation becomes more complex but not too much more so; in most cases, simple rules predict the position of metalation (*see* below).

Monosubstituted benzenes that undergo the directed metalation reaction are summarized in Table I. In every case a variety of ortho-disubstituted products have been prepared; these would be tedious or impossible to prepare by other routes.

Table I. Directed Metalation of Monosubstituted

Directing Substituent (R) in Equation 1	Electrophile	n-C_4H_9Li/ Substrate	Solvent
—$CH_2N(CH_3)_2$	Ph_2CO	2.0	ether/hexane
—$CH_2CH_2N(CH_3)_2$	Ph_2CO	1.2	ether/hexane
—$CHOHCH_2N(CH_3)_2$	$ClSi(CH_2)_3$	2.5	ether/hexane
—CH_2OH	CH_3I	2.6	ether/hexane
—CH_2NHCH_3	PhCHO	2.5[c]	hexane
—CH_2NHPh	Ph_2CO	1.5[c]	hexane
—OCH_3	CO_2	1.0	ether/hexane
—$CONHCH_3$	Ph_2CO	2.5	THF/hexane
—$SO_2N(CH_3)_2$	Ph_2CO	1.2	THF/hexane
—SO_2NHCH_3	Ph_2CO	2.5	THF/hexane
—CF_3	CO_2	1.5	ether
—$N(CH_3)_2$	Ph_2CO	1.0	hexane
—F	CO_2	1.0	THF

[a] Reaction proceeded to give mostly styrene *via* elimination.
[b] Also gave 8% of meta acid as product.

Ferrocene (Ruthenocene). The directed metalation reaction has proved to be of great synthetic value in the preparation of 1,2-disubstituted ferrocenes. From an organic chemist's point of view, ferrocene can be considered to have properties similar to five-membered ring heterocycles—great sensitivity to acid and oxidizing conditions. Many electrophilic reactions cannot be run on ferrocene, and metalation has thus come to be the preferred process for preparing many monosubstituted ferrocenes (4, 5). In addition, electrophilic substitution of monosubstituted ferrocenes containing conventional activating substituents gives mixtures of 1,2-, 1,3-, and 1,1'-disubstituted ferrocenes; electrophilic substitution of monosubstituted ferrocenes containing conventional deactivating substituents yields only 1,1'-disubstituted ferrocenes (7). All this reveals the inaccessibility, for the most part, of homoannularly disubstituted ferrocenes by a route involving electrophilic substitution. In a number of instances directed metalation has provided clean, concise routes to specific 1,2-disubstituted ferrocenes (Reaction 2) that were either difficult

$$\underset{\underset{|}{Fe}}{\overset{R}{\bigcirc}} + C_4H_9Li \longrightarrow \underset{\underset{|}{Fe}}{\overset{R}{\bigcirc}}{\overset{}{\text{Li}}} \xrightarrow{\text{Electrophile}} \underset{\underset{|}{Fe}}{\overset{R}{\bigcirc}}{\overset{}{\text{X}}} \qquad (2)$$

to prepare or inaccessible by conventional electrophilic methods. Synthesis of substituted ferrocenes by metalation of ferrocene itself or by the directed metalation of certain substituted ferrocenes has been reviewed (8).

Benzenes with n-Butyllithium

Metalation Period, hrs	Temp, C°	% Yield	Reference
18.00	25	84	6
11.00	25	7 [a]	66
21.00	25	61	2
18.00	25	45	2
1.50	25	64	56
4.00	25	86	56
21.00	35	65	33
0.25	65	81	46
0.25	0	82	53
0.25	0	82	52
6.00	35	48 [b]	67
2.00	68	55	68
7.00	−50	60	12

[c] TMEDA required.

Several directing groups known to be good directors in the benzene system have been found to provide directed metalation in the ferrocene system; others, however, are unique to the ferrocene system. Table II summarizes the directing substituents available for ferrocene. Of these, —$CH_2N(CH_3)_2$, —$CH_2CH_2N(CH_3)_2$, —CONHR, —CPh_2OH, —OCH_3, and —$SO_2N(CH_3)_2$ are known directors in benzene, but —CH_2OR, —Cl, and (pyridyl group) are unique to ferrocene. The —$CH_2N(CH_3)_2$ side-chain has also been found to effect directed metalation in ruthenocene (9).

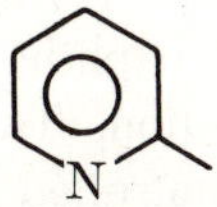

A complication that does not extend to other aromatics exists in ferrocene metalation. This is heteroannular dimetalation, which gives products that often contaminate the desired 2-metalation product. This phenomenon was investigated in one instance and from metalation of dimethylaminomethylferrocene with excess n-butyllithium, the 1,2,1'-trisubstituted product shown in Reaction 3 was isolated (10). Considerable amounts of a 1,2,1'-trisubstituted ferrocene product could also be isolated

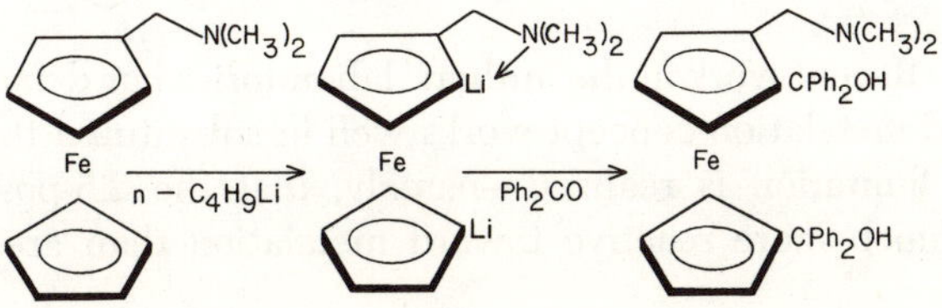

Table II. Substituents that Direct

Directing Substituent R in Equation 2	Electrophile	C_4H_9Li/ Substrate	Solvent
$-CH_2N(CH_3)_2$	Ph_2CO	2.5	ether/hexane
$-CH_2CH_2N(CH_3)_2$	Ph_2CO	1.5	ether/hexane
-2-pyridyl	Ph_2CO	25.5	ether/hexane
$-CPh_2OH$	CO_2	2.5	ether
$-CH_2OCH_3$	Ph_2CO	1.6	ether/hexane
$-OCH_3$	$(CH_2O)x$	1.6	ether/hexane
$-Cl$	CH_3I	2.0	ether/hexane
$-SO_2N(CH_3)_2$	Ph_2CO	1.2	ether/hexane
$-CH_2N\overset{H}{\underset{S}{\overset{CH_3}{\diagup}}}$	$ClSi(CH_3)_3$	2.5	ether/hexane
$-CH(CH_3)N(CH_3)_2$	$ClSi(CH_3)_3$	1.16	ether/hexane

^a Isolated as $-CH_2O-CH_3$

from the metalation and condensation of dimethylaminoethylferrocene (11). It seems quite possible that most, if not all, the monosubstituted ferrocenes that undergo directed metalation may provide 2,1'-dimetalation under certain conditions.

Naphthalene. Although not as much work has been performed on the naphthalene system as on the thiophene system (see the next section), results in these systems thus far are still interesting. Perhaps the most unusual aspect of the metalation of 1-substituted naphthalenes is that metalation takes place at either the 2- or the 8-position. To some extent the ratio of metalation at the 2-position can be controlled by judicious exercise of reaction conditions and metalating reagent. A mixture of 8- and 2-metalated intermediates has been postulated (Reaction 4). Those groups at the 1-position that have been demonstrated to provide this pattern of metalation in naphthalene are $-F$ (12), $-OCH_3$ (13, 14), and $-CH_2N(CH_3)_2$ (15). 2-Fluoronaphthalene is the only 2-substituted naphthalene compound examined, and it has been found to undergo lithiation at both the 3- and the 1-position (16).

$$\text{Naphthalene-R} + C_4H_9Li \longrightarrow \text{Naphthalene-R, Li (2-position)} + \text{Naphthalene-R, Li (8-position)} \quad (4)$$

Thiophene. Recent work in the authors' laboratories has demonstrated that the directed metalation concept works well in substituted thiophenes once a certain limitation is realized—namely, that the 2,5-positions of thiophene are much more reactive toward metalation than are the 3,4-

Metalation in the Ferrocene Series

Metalation Period, hrs	Temp, °C	% Yield	Reference
1	25	71	10
2	25	68	11
6	—	51	69
36	25	72	70
2.5	25	32.5	26
3	25	60[a]	71
3.5	25	72	71
6	25	17	72
45	25	57	37
1	—	57.7	40

positions. This gives significance to the fact that a number of 3-substituted thiophenes are metalated in the 2-position with little or no products from 5-metalation being detected (Reaction 5). A summary of the groups that provide such directed metalation in thiophenes is given in Table III.

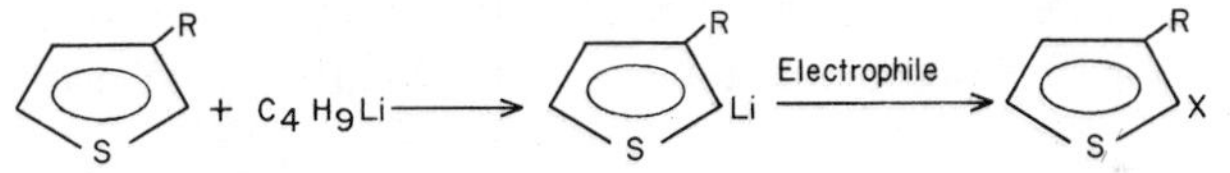

$$\text{thiophene-R} + C_4H_9Li \longrightarrow \text{thiophene-R-Li} \xrightarrow{\text{Electrophile}} \text{thiophene-R-X} \quad (5)$$

Since thiophene itself is readily metalated in the 2-position, a 3-position substituent's causing metalation to take place at the 2-position suggests that the ready metalation at a position adjacent to sulfur is further aided by the directing 3-substituent. It has also been found in a few instances that when a blocking group is placed in the 5-position of thiophene, a directing substituent in the 2-position will direct metalation to the 3-position (17). An example of this is shown in Reaction 6.

Very few examples of directed metalation in furan or pyrrole derivatives have been reported. 3-Bromothiophene has been shown to undergo metalation in the 2-position with lithium diisopropylamide (18).

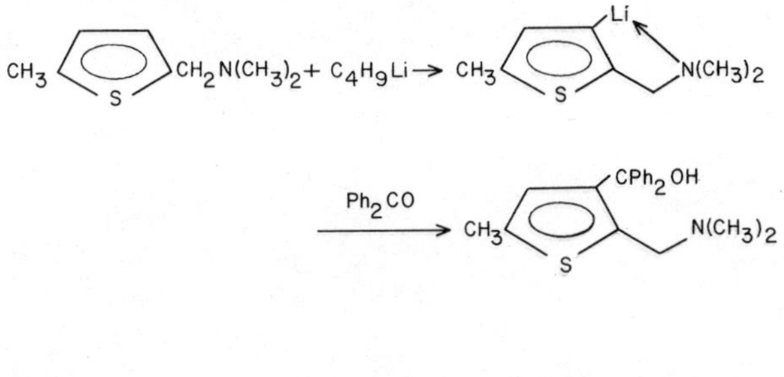

$$CH_3\text{-thiophene-}CH_2N(CH_3)_2 + C_4H_9Li \longrightarrow CH_3\text{-thiophene(Li)-}N(CH_3)_2$$

$$\xrightarrow{Ph_2CO} CH_3\text{-thiophene-}[CPh_2OH][N(CH_3)_2] \quad (6)$$

Table III. Directed Metalation of

Directing Substituent R in Equation 5	Electrophile	$C_4H_9Li/$ Substrate	Solvent
—OCH_3	CO_2	1.0	—
—$OC(CH_3)_3$	CO_2	0.98	ether
—SCH_3	CO_2	0.995	—
—CN	CO_2	1.06	ether
—Br	CO_2	0.906	ether
—$CH_2N(CH_3)_2$	$HCON(CH_3)_2$	1.2	ether/hexane
—CH_2OCH_3	$HCON(CH_3)_2$	—	—
—$CONHCH_3$	Ph_2CO	2.11	ether/hexane

Bond Types (Functional Groups) that can be Introduced at the Metalation Site

One aspect of the directed metalation reaction that makes it such a powerful synthetic tool is the large number of derivatives that can be prepared at the site of lithiation. The high concentration of negative charge on the carbon atom bonded to the lithium atom makes the former highly nucleophilic, like a Grignard reagent and about as versatile. A variety of derivatives have been prepared, and there are probably a significant number yet to come. Representative routes to most types of derivatives are recorded here with recent leading references. Most routes have been worked out with N-chelated intermediates. The following symbol is used to designate a generalized aromatic 2-lithio intermediate:

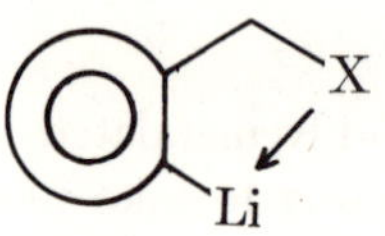

Carbon-Carbon Bonds. Synthetic methods involving the formation of carbon-carbon bonds are always of great interest. Primary, secondary, or tertiary alcohols can be prepared by the reaction of the lithio intermediate with the appropriate aldehyde or ketone (Reactions 7–9) (6,

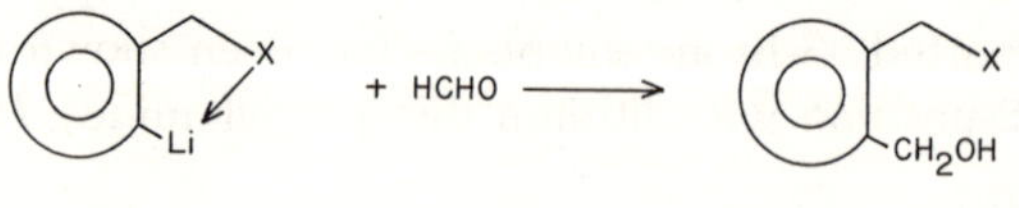

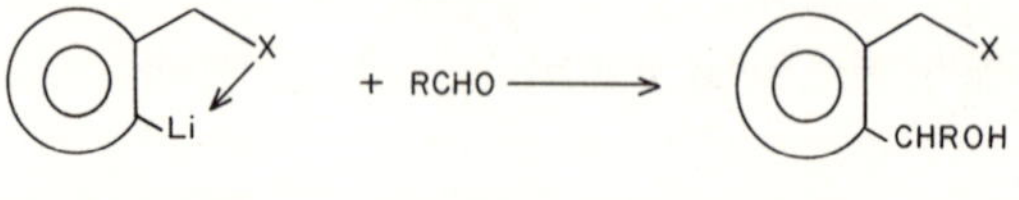

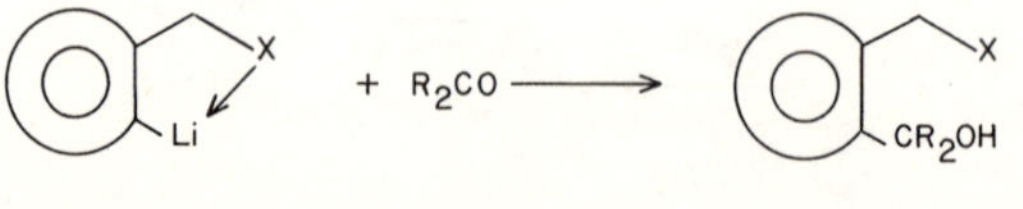

3-Substituted Thiophenes with n-Butyllithium

Metalation Period, hrs	Temp. °C	% Yield	Reference
0.5	35	86	73
0.5	34	62	74
0.5	35	70	75
1.0	−70	68	76
3.0	25	36	77
—	—	75	17
—	—	72	17
—	—	23	25

19). Ethylene oxide or other epoxides can also be used to prepare β-substituted ethyl alcohol derivatives (Reaction 10) (*6*).

Ketones may be prepared by the reaction of the lithio intermediate with a nitrile (Reaction 11) (*6*). Nitriles possessing no alpha hydrogens work best in this reaction. Formyl derivatives of aromatics may be synthesized by treating the lithiated species with dimethylformamide (Reaction 12) (*19*). Carboxylic acids are readily available by carbonation of such lithio intermediates (*14, 16*). Treatment of these lithio intermediates with aryl or alkyl isocyanates yields amides (Reaction 13) (*10*).

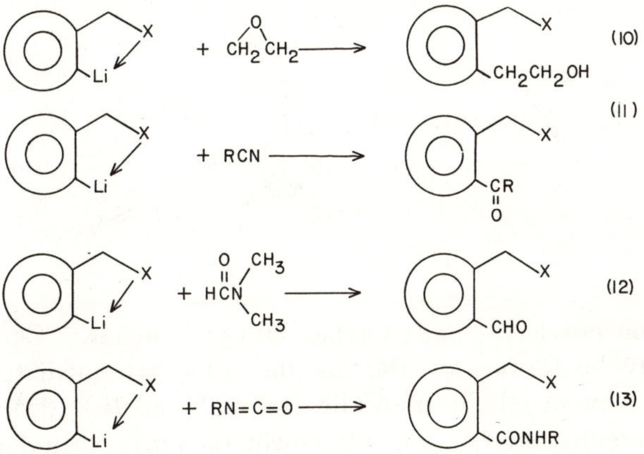

The lithium atom may be replaced with a methyl group by treating the metalated species with methyl iodide or dimethylsulfate (Reaction 14) (*20*). In this case the directing group cannot be an amine since the amine site is also alkylated in the process, complicating isolation (*21*). Other alkyl groups cannot be introduced this way because of the predilection of alkyl halides for elimination in the presence of strong base. Rather, the routes to ethyl, isopropyl, and other alkyl derivatives involve reduction of the corresponding alcohol (Reaction 15) (*22*).

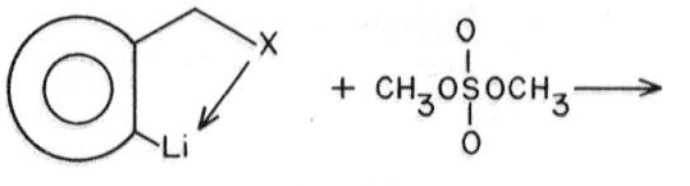

$$\text{(Reaction 14)}$$

$$\text{(Reaction 15)}$$

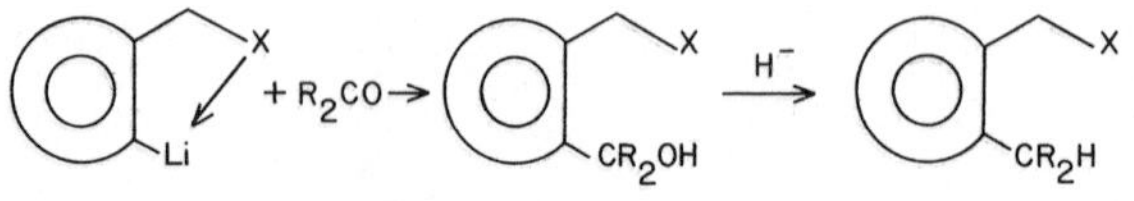

Carbon-Halogen Bonds. Carbon-halogen bonds may be prepared directly *via* the lithio intermediate or a second intermediate prepared from the lithio intermediate. For example, reaction of hexachloroethane with 2-lithiodimethylaminomethylferrocene gave the chloro derivative (Reaction 16) (*23*). However, preparation of other halogen derivatives *via* lithio intermediates has not been successful. A better and more versatile method for preparing the chloro, bromo, and iodo derivatives involves isolating a boronic acid intermediate as in Reaction 17 (*24*).

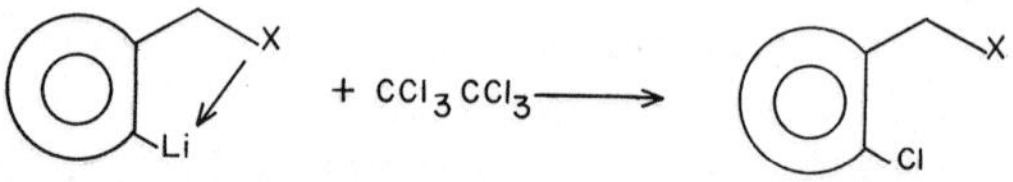

$$\text{(Reaction 16)}$$

$$\text{(Reaction 17)}$$

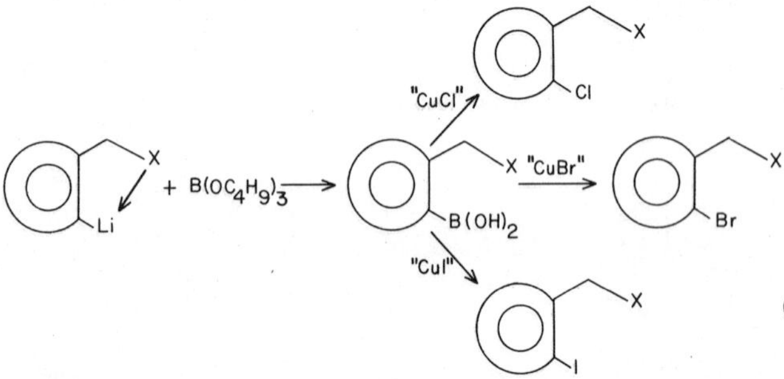

Carbon-Nitrogen and Carbon-Oxygen Bonds. Carbon-nitrogen bonds may be formed by treating the lithio intermediate with either methoxylamine or ethyl nitrate (Reactions 18 and 19) (*25*). These reactions apparently involve a displacement on nitrogen and result in the preparation of the amino and nitro derivative, respectively. Carbon-oxygen bonds may be prepared *via* a boronic acid intermediate and reaction with cuprous acetate (Reaction 20) (*71*), a reaction completely analogous to the synthesis of halogen derivatives described in Reaction 17.

Other Carbon-Heteroatom Bonds. Carbon-mercury bonds are readily formed by treating the metalated species with mercuric chloride (Reaction 21) (*27*). The resulting chloromercury derivative in one case is a useful intermediate in the preparation of 2-iododimethylaminomethylferrocene (*27*).

Carbon-silicon bonds can be formed by treating the lithio intermediate with halogen-containing silanes (Reaction 22) (28). Carbon-phosphorus bonds can be prepared similarly (Reaction 23) (29).

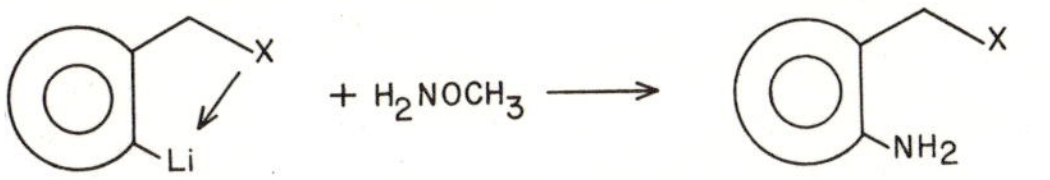

$$\text{Ar(Li)-X} + H_2NOCH_3 \longrightarrow \text{Ar(NH}_2\text{)-X} \quad (18)$$

$$\text{Ar(Li)-X} + C_2H_5ONO_2 \longrightarrow \text{Ar(NO}_2\text{)-X} \quad (19)$$

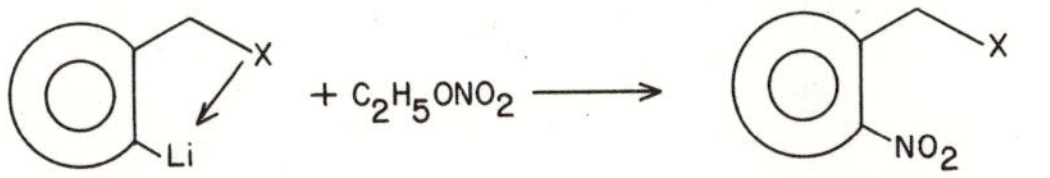

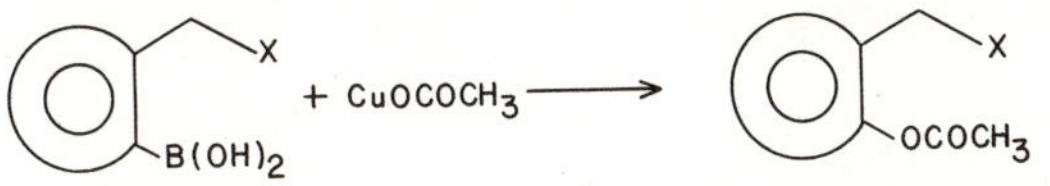

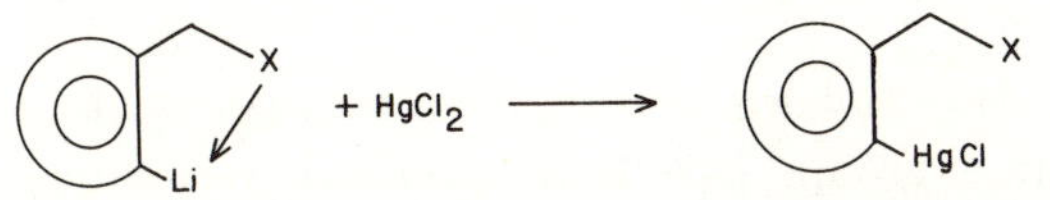

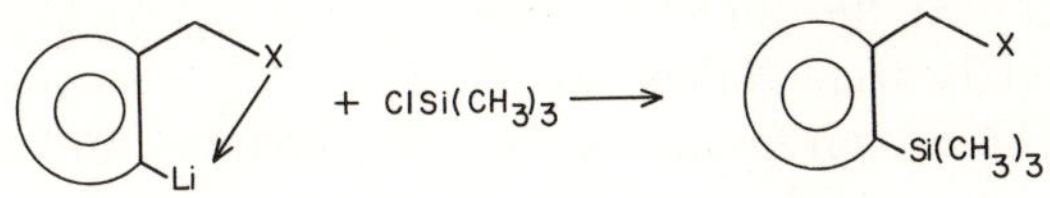

$$\text{Ar(B(OH)}_2\text{)-X} + CuOCOCH_3 \longrightarrow \text{Ar(OCOCH}_3\text{)-X} \quad (20)$$

$$\text{Ar(Li)-X} + HgCl_2 \longrightarrow \text{Ar(HgCl)-X} \quad (21)$$

$$\text{Ar(Li)-X} + ClSi(CH_3)_3 \longrightarrow \text{Ar(Si(CH}_3\text{)}_3\text{)-X} \quad (22)$$

$$\text{Ar(Li)-X} + ClPPh_2 \longrightarrow \text{Ar(PPh}_2\text{)-X} \quad (23)$$

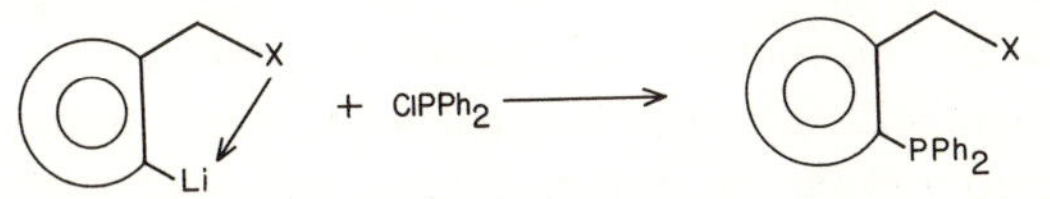

Relative Directing Abilities of Substituents

When more than one directing group is present in an aromatic molecule it is important to know which directing group will exert the principal effect—that is, which is the stronger director. Competitive metalation of nine of the ortho-directing substituents for the benzene ring has recently been examined in our laboratories. These are: —$CH_2N(CH_3)_2$, —$CH_2CH_2N(CH_3)_2$, —CONHR, —OCH_3, —$N(CH_3)_2$, —CF_3, —F, —$SO_2N(CH_3)_2$, and —SO_2NHCH_3. Ratings were based on the competitive lithiation of the appropriate para-disubstituted benzenes. Data are now available for the competitive metalation of the methoxy group vs. the eight other directing groups (30). These results dictate that —$CH_2N(CH_3)_2$, —CONHR, —$SO_2N(CH_3)_2$, and —SO_2NHCH_3 are stronger directors than —OCH_3, and —$CH_2CH_2N(CH_3)_2$, —$N(CH_3)_2$, —F, and —CF_3 are weaker directors (Reactions 24 and 25). From some additional data now available, it is apparent that —$CONHCH_3$ is a stronger director than —$CH_2N(CH_3)_2$ (31). It is also likely that the

sulfonamides are the strongest directors known so that a ranking of
$-SO_2NHCH_3$, $-SO_2N(CH_3)_2$ > $-CONHCH_3$ > $-CH_2N(CH_3)_2$
> $-OCH_3$ > $-CH_2CH_2N(CH_3)_2$, $-N(CH_3)_2$, $-F$, $-CF_3$ might be
inferred for the benzene system, at least.

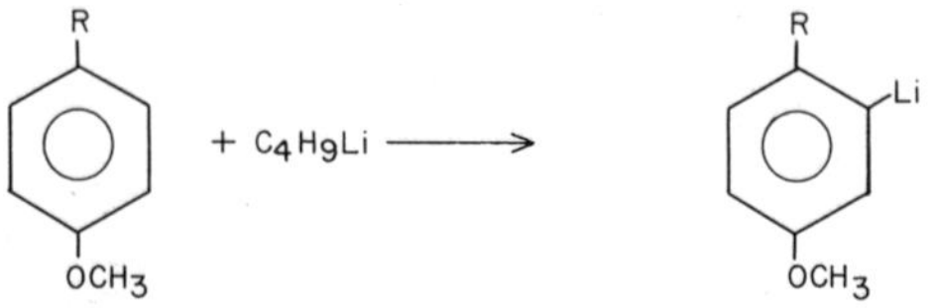

R–C₆H₄–OCH₃ + C₄H₉Li ⟶ R–C₆H₃(Li)–OCH₃ (24)

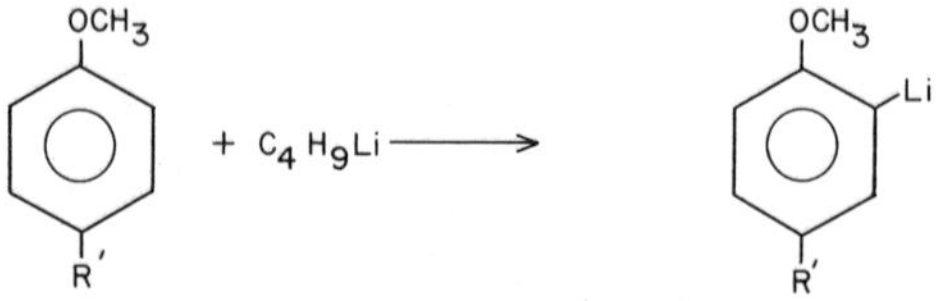

CH₃O–C₆H₄–R' + C₄H₉Li ⟶ CH₃O–C₆H₃(Li)–R' (25)

In all cases where a meta-disubstituted benzene containing two ortho-directing groups has been examined, metalation has taken place ortho to each of the directing groups—that is, the 2-position of a 1,3-disubstituted benzene (Reaction 26) (*30, 32*). For ortho-disubstituted benzenes where both groups were ortho directors, the stronger directing group (as determined above) was found to control the metalation site. Modification of the metalation pattern by steric effects of some of the more bulky substituents was not realized. An extreme example of this has now been examined. *o-tert*-Butylanisole has been metalated with *n*-butyllithium (Reaction 27) (*33*). As anticipated, the yield of metalation product was significantly diminished compared with the metalation of anisole. However, even the 5% yield of product realized was that from metalation ortho to the methoxy group—that is, the site of metalation had not changed. Metalation of this compound with *n*-butyllithium–TMEDA complex brought a 30% yield of the same ortho metalation product previously described (Reaction 27). Thus the steric effect originally noted could be overcome by a stronger metalating reagent.

Extension of this study of the efficiency of directing groups to other aromatic systems should provide further insight into the reliability of the above ranking.

The Directing Mechanism

In all directed metalations studied, the lithium atom is directed to a proton adjacent to the directing substituent. No single explanation can be proposed now to account for all the known examples of the directed metalation reaction. Rather, a combination of varying degrees of a

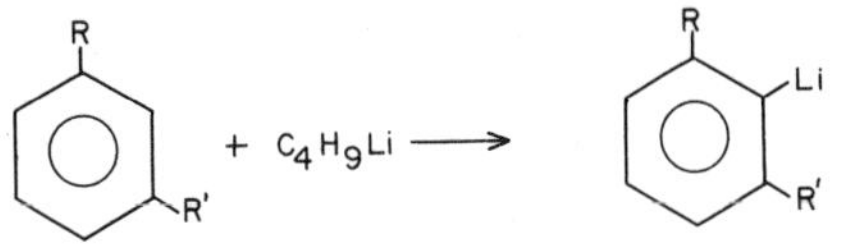

[Reaction (26): ArH with R, R' substituents + C₄H₉Li → ortho-lithiated product]

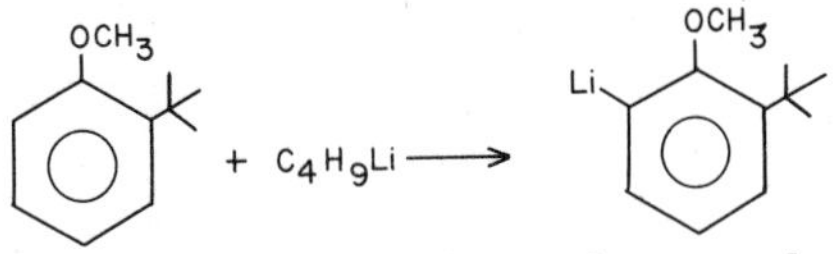

[Reaction (27): anisole derivative with OCH₃ and t-Bu + C₄H₉Li → ortho-lithiated product]

coordination mechanism coupled with an inductive effect seems most appropriate.

A good example of the intervention of a coordination mechanism is that in the ortho metalation of dimethylbenzylamine (6). The methylene group essentially insulates the ring from any inductive influence of the nitrogen atom. The fact that this molecule can be ortho metalated strongly indicates that some other effect is operating. Such an effect involves the coordinated lithio intermediate depicted in Reaction 28. A coordination mechanism would also seem to be the most likely directive effect with —$CH_2CH_2N(CH_3)_2$, —CONHR, and —CH_2NHR side chains. A most intriguing demonstration of the coordinating effect of nitrogen in dimethylbenzylamine is provided by a study of ring vs. side-chain metalation with alkyl sodio reagents (34). The benzylamine was initially metalated at the ortho position, but after 20 hours, rearrangement to the more stable alpha position was complete (Reaction 29). Moreover, the rearrangement could be reversed by adding lithium bromide to the solution containing the alpha-metalated species (Reaction 30). These results can be interpreted to signify that the alpha-metalated species was

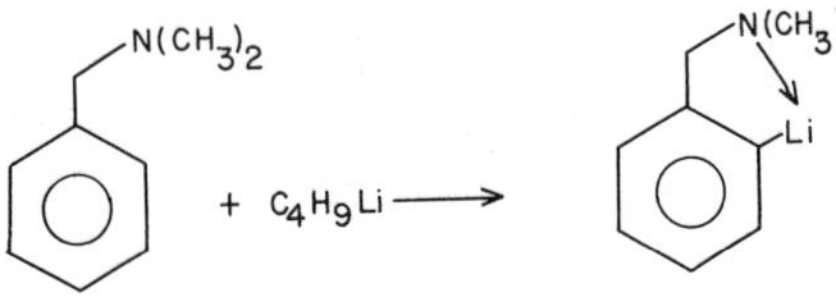

[Reaction (28): C₆H₅CH₂N(CH₃)₂ + C₄H₉Li → ortho-lithiated intermediate with Li coordinated to N(CH₃)₂]

[Reaction (29): C₆H₅CH₂N(CH₃)₂ + C₄H₉Na/Hexane → ortho-Na intermediate → (Ph₂CO) ortho adduct; after 20 hr → α-Na species (CHN(CH₃)₂) → (Ph₂CO) alpha adduct]

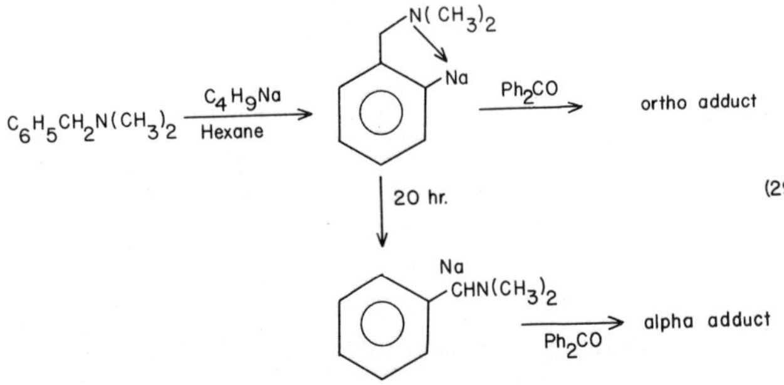

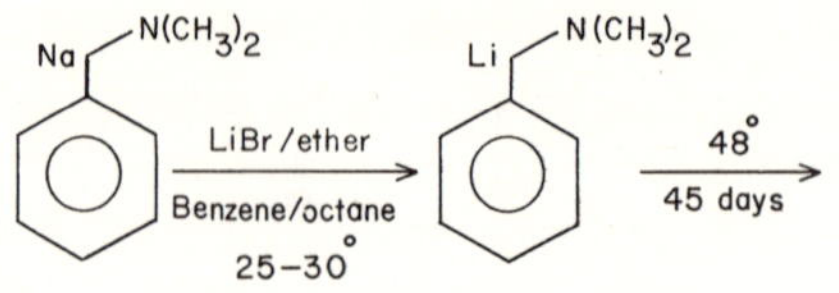

more carbanionic in the case of the sodio derivative. Hence, it was more conducive to resonance stabilization at the benzyl position, and the ortho position metalation site is greatly stabilized by coordination in the case of the lithio intermediate. Thermodynamic and kinetic roles have been reversed in these two instances. No meta or para product was detected in either sequence.

Substituents such as —SO$_2$NR$_2$, —CF$_3$, —Cl, —F, —OCH$_3$, and —N(CH$_3$)$_2$ significantly polarize the aromatic ring and might be said to operate by some combination of inductive and field effects. Certainly the sulfonamides and —CF$_3$ possess significant field effect contribution while the remaining four substituents—each with an electronegative atom bond to the ring—must have significant inductive contribution. Some coordination may also contribute to transition states involving —Cl, —F, —OCH$_3$, and —N(CH$_3$)$_2$, but drawing coordinate structures such as that for the 2-lithiation of dimethylbenzylamine (Reaction 28) for these substituents involves postulation of a four-membered ring. This can be

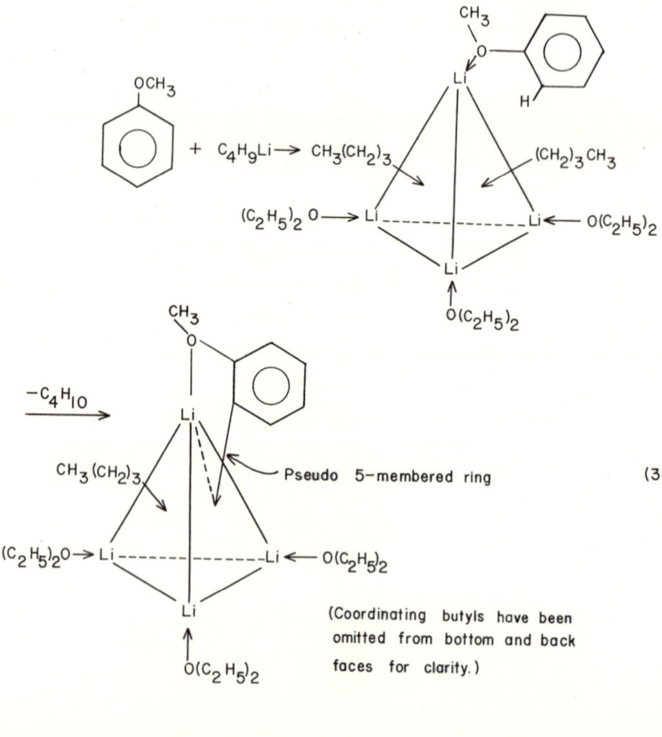

(Coordinating butyls have been omitted from bottom and back faces for clarity.)

avoided when the tetrameric structure of n-butyllithium (30) is invoked, because a pseudo five-membered ring can then be drawn (Reaction 31).

The interplay of inductive effect and coordination is brought out in the competitive metalation of p-fluoroanisole and p-dimethylaminoanisole. In each case the methoxy group controls the site of metalation (30). Coordination effects fall in the order —N(CH$_3$)$_2$ > —OCH$_3$ > —F, while the inductive order would be just the reverse of this. Since neither order was observed, a combination of effects is presumed to be operating.

Asymmetric Directed Lithiation

The resolution of racemic mixtures is certainly the most widely used method of preparing optically active compounds. An alternate method of preparing certain optically active compounds, usually quicker and often more practical, is asymmetric induction. Possibly the newest example of such induction is asymmetric lithiation.

The principle of asymmetric lithiation involves both the fact that formation of diastereomeric intermediates should involve different energies of activation, and the idea that, in directed lithiations, a lithium atom is coordinated with nitrogen or some other heteroatom (see above). When the coordinating nitrogen atom resides in a chiral environment, one of the two possible diastereomeric lithio intermediates is energetically favored for steric or other reasons. Thus one of the two possible intermediates should be formed preferentially, with the resulting condensation products reflecting the stereoselectivity of the lithiation.

One of the earliest descriptions of an asymmetric lithiating reagent was reported by Nozaki and co-workers in 1968 (35). (—)-Sparteine was used to coordinate n-butyllithium, and this complex stereoselectively added to several carbonyl compounds (Reaction 32). Moreover, the Skattebol-Moore method (which consists of dehalogenating gem-dihalocyclopropanes with an alkyllithium complex) by Nozaki to synthesize allenes gave optically active products when the n-butyllithium/(—)-sparteine complex was used (36).

Once it was demonstrated that it adds stereoselectively, the n-butyllithium/(—)-sparteine complex was used to prepare a series of optically active ferrocenes (36). Treatment of isopropyl ferrocene with a 2.5-molar excess of the lithiating complex followed by reaction with an electrophile

sparteine-N→Li-C$_4$H$_9$ + PhCHO ⟶ Ph—CHOH | C$_4$H$_9$ (32)

6% optical yield

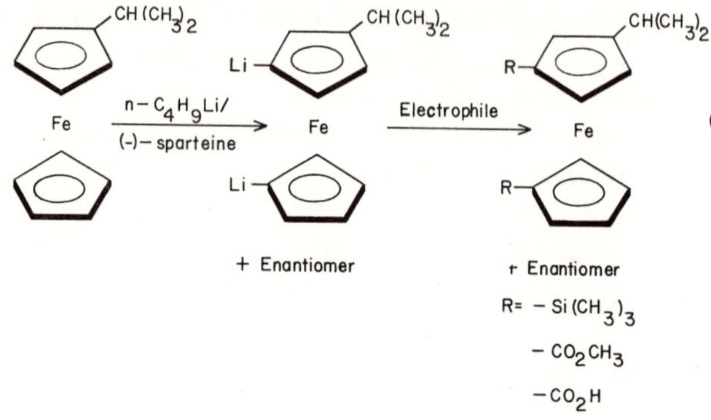

POLYAMINE-CHELATED ALKALI METAL COMPOUNDS

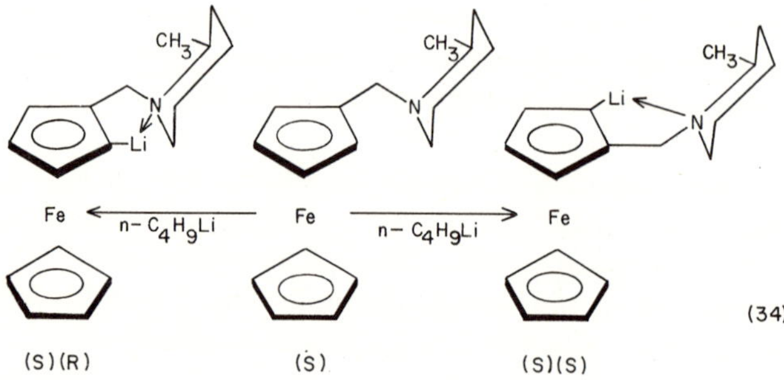

(33)

R= $-Si(CH_3)_3$
$-CO_2CH_3$
$-CO_2H$

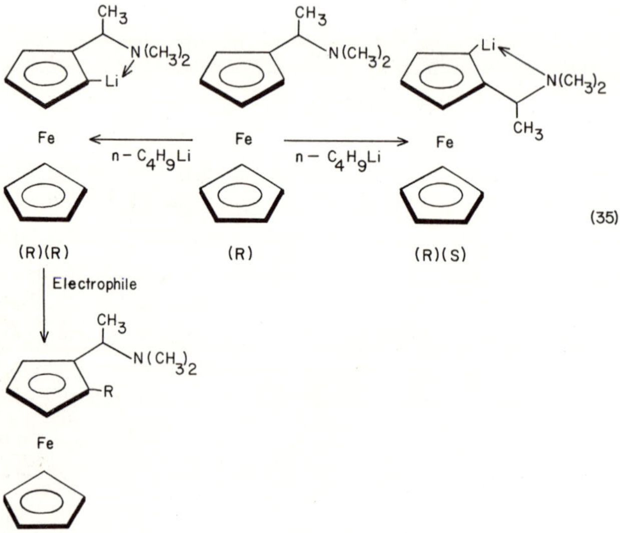

(34)

(35)

yielded 3,1′-disubstituted isopropyl ferrocenes in 3% optical yield (Reaction 33). We term this an asymmetric metalation procedure.

Carrying the concept of asymmetric lithiation one step further, Nozaki and co-workers incorporated the asymmetry-inducing complexing reagent with the metalated molecule (ferrocene) itself (37, 38). 1-Ferrocenylmethyl-2-methylpiperidine was resolved and treated with n-butyllithium to give a mixture of diastereomeric lithio intermediates by directed metalation (Reaction 34). An optical yield of 93% was initially claimed for this reaction, but subsequent work by Ugi and co-workers (39) resulted in the suggestion that only a 67% optical yield was obtained.

Further syntheses involving asymmetric lithiation have been reported by Ugi (40). Optically active 1-ferrocenylethyldimethylamine was used to obtain stereoselective syntheses in 96% optical yield (Reaction 35). Knowledge of the configuration of the starting amine allowed the absolute configuration of the principal lithio intermediate to be inferred as the (R)(R) diastereomer. Additional support for this assignment has been published (41).

A very interesting compound may be prepared *via* this method and used in the stereoselective syntheses of peptides. Ugi has found that asymmetrically induced four-component syntheses will form optically active peptides (Reaction 36) (42, 43). Compounds of the type R^*—NH_2, must, to be useful in this synthesis, meet these criteria:

1) Condensation of the amine with the other components, if necessary *via* the Schiff base of the amine and aldehyde, should take place rapidly and in high yield.

2) As a component of the condensation, the amine must also possess the effect of an asymmetrically inducing steric matrix and provide for a highly steroselective synthesis of the newly formed amino acid unit in the desired configuration.

$$-NH-\underset{\underset{R_1}{|}}{CH}-CO_2H \ + \ \underset{\underset{R^*}{|}}{NH_2} \ + \ \underset{\underset{R_2}{|}}{CHO} \ + \ CN-\underset{\underset{R_3}{|}}{CH}-CO^-$$

$$\downarrow$$

$$-NH-\underset{\underset{R^*}{|}}{\overset{\overset{R_1}{|}}{CH}}-CO-N-\underset{*}{\overset{\overset{R_2}{|}}{CH}}-CO-NH-\overset{\overset{R_3}{|*}}{CH}-CO- \qquad (36)$$

$$\downarrow$$

$$-NH-\overset{\overset{R_1}{|*}}{CH}-CO-NH-\overset{\overset{R_2}{|*}}{CH}-CO-NH-\overset{\overset{R_3}{|*}}{CH}-CO- \ + \overset{*}{R}-X$$

* Denotes optically active

3) The residual R* must be readily cleaved from the intermediate polymer under mild conditions—for example, in cold formic or trifluoroacetic acid—preferably in such a way that the amine can be regenerated.

The only compounds meeting all these criteria were ferrocene compounds prepared by Ugi *via* the asymmetric directed metalation method, shown in Reaction 35.

Goldberg and Bailey (*44*) have used the asymmetric-directed metalation procedure as a route to compounds demonstrating (for the first time) pseudoasymmetry in ferrocenes. A pseudoasymmetric 1,2-disubstituted ferrocene was prepared by procedures such as those illustrated in Figure 1.

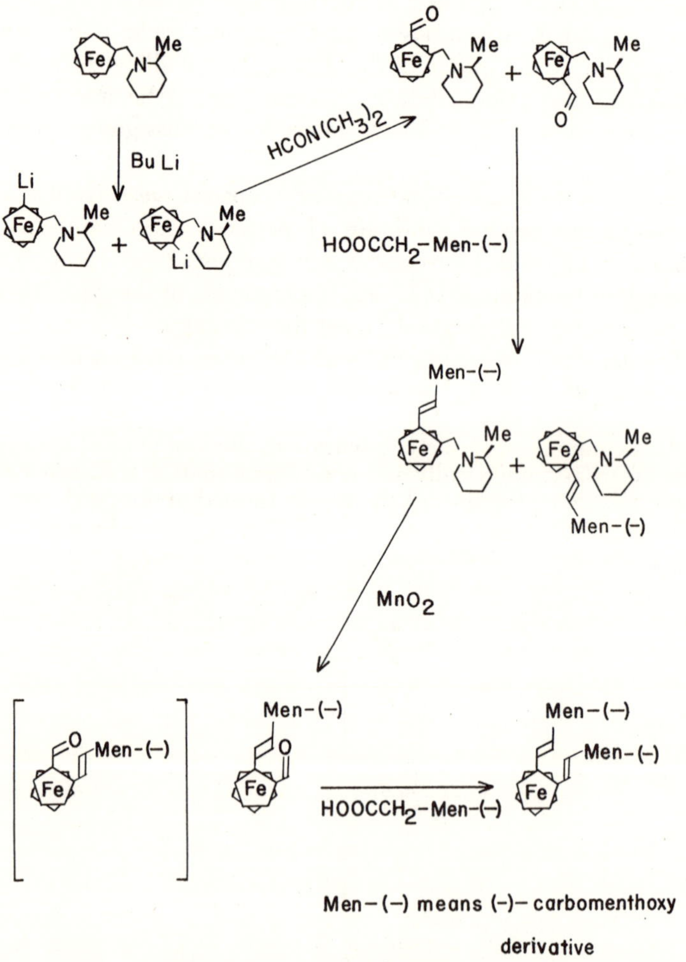

Figure 1. Preparation of a pseudoasymmetric 1,2-disubstituted ferrocene

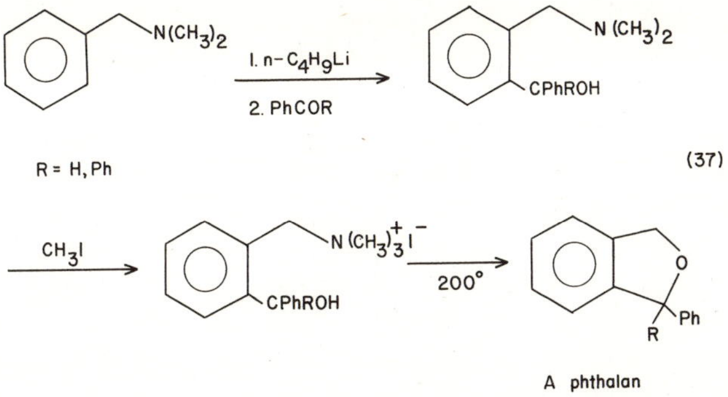

[Reaction scheme 37: ortho-N(CH₃)₂ benzylamine + 1. n-C₄H₉Li, 2. PhCOR → ortho-substituted with CPhROH (R = H, Ph); then CH₃I → methiodide N(CH₃)₃⁺I⁻ with CPhROH; then 200° → A phthalan (benzofused ring with O, CR-Ph)]

Our conclusion is that the asymmetric lithiation procedure and the asymmetric directed metalation reaction are of great potential value for synthesizing a variety of chiral compounds as well as being elegant and profound exercises in stereochemistry.

Heterocyclic Synthesis via Directed Metalation

One of the most useful synthetic applications of the directed metalation reactions is in preparing heterocyclic systems. Of the available directing groups, those involving N-chelated intermediates have been by far the most useful. In several instances the route provided by ortho lithiation constitutes the only available method for preparing certain heterocycles. In other cases such syntheses, although not the only routes available, represent a considerable improvement over more conventional methods, especially considering the number of steps in the overall synthesis and yields. Furthermore many of the heterocyclic compounds produced *via* directed metalation procedures are of extreme interest in that they are natural products or derivatives thereof.

Initially used to prove the 1,2-disposition of the condensation products of the respective lithio intermediates, cyclization to heterocyclic compounds rapidly developed into a relatively valuable synthetic tool. The bulk of the initial work in this area was performed by Hauser and co-workers; later, significant contributions (especially in natural product heterocycle synthesis) were made in India by Narasimhan and associates.

One of the first uses of directed metalation as a route to heterocycles was the synthesis of phthalans (2,3-benzo-1,4-dihydrofurans) by the thermally induced cyclization of the methiodides of ortho-substituted dimethylbenzylamines (Reaction 37) (45). The amine was lithiated in the ortho position by n-butyllithium and condensed with benzaldehyde and benzophenone. The corresponding alcohols obtained upon aqueous work-up were converted to their respective methiodides. Heating the methiodides to 200°C for one hour under nitrogen gave the phthalans

(38)

A phthalimidine

shown in Reaction 37. A quite analogous procedure with dimethylaminomethylferrocene as the starting material gave the ferrocene analog of the phthalan derived from the benzophenone condensation (*10*).

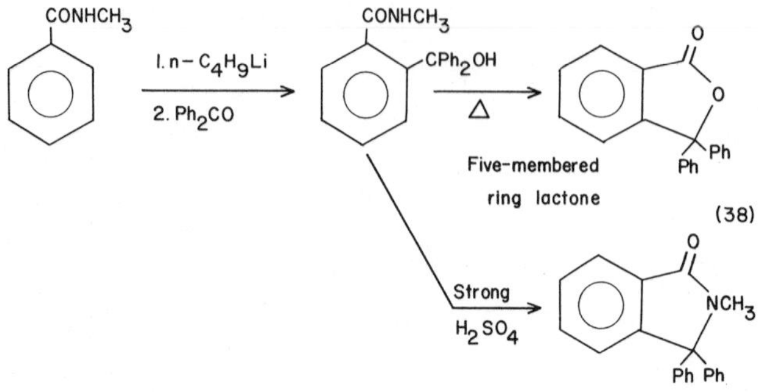

Discovery that the *N*-substituted carboxamide group could direct metalation(*46*) led to the eventual establishment of synthetic routes to a number of heterocycles, including substituted lactones, phthalimidines, and isocarbostyrils (*47*). Cyclization of the products of condensation of the lithio intermediates of *N*-methylbenzamide yielded five-membered heterocycles (Reaction 38) (*48*). *o*-Methyl-*N*-methylbenzamide, in which the methyl group was lithiated, gave six-membered ring heterocycles (Reaction 39) (*49*). The mechanism of the cyclization step in the latter procedure has been dealt with in some depth in the literature (*50*); an independent reinvestigation, however, has cast doubt on the validity of the dihydroisocarbostyril structures proposed (*51*).

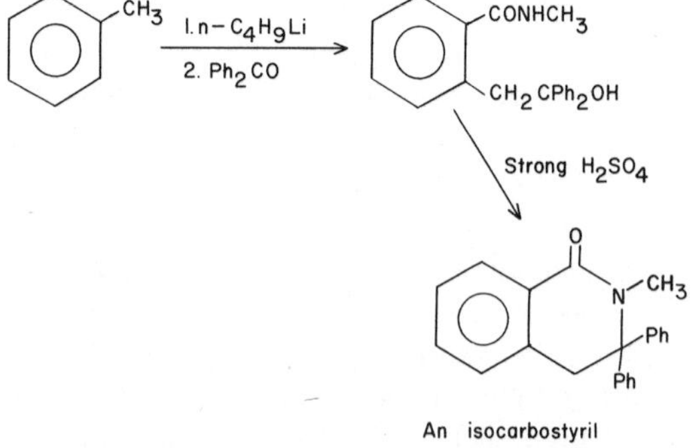

Sulfonamides as ortho-directing substituents for metalating aromatic systems opened the door to synthetic routes to cyclic sulfonic esters (sultones) and amides (sultams). The first step in this procedure for preparing sultams involved the 2-metalation of *N*-alkylbenzenesulfonamides (*52*) and condensation with a variety of ketones. The tertiary alco-

(39)

An isocarbostyril

hols thus produced were thermally dehydrated to form the corresponding sultams (Reaction 40) (52). For the preparation of sultones, N,N-dimethylbenzenesulfonamide was likewise 2-metalated (53) and condensed with benzophenone (Reaction 41). This product undergoes two reactions (54). Upon treatment with cold, concentrated sulfuric acid/methanol, the tertiary alcohol was cyclized to the sultone. At −78°C, the same reagents gave the methyl ether. A similar effect—namely changing reaction conditions to obtain different products—was observed with tertiary alcohols produced by condensation of carbonyl compounds with 2-lithiobenzamide (48). Heat or weak acid converted 2-diphenylhydroxymethyl-N-methylbenzamide to the five-membered lactone ring (Reaction 38), whereas treatment with strong acid gave the corresponding lactam (Reaction 38). Mild acid in the case of a cyclic ether in the ferrocene series opened the ring (Reaction 42) (55).

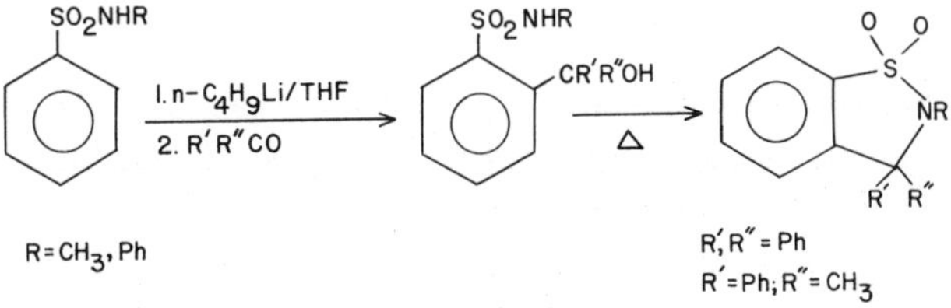

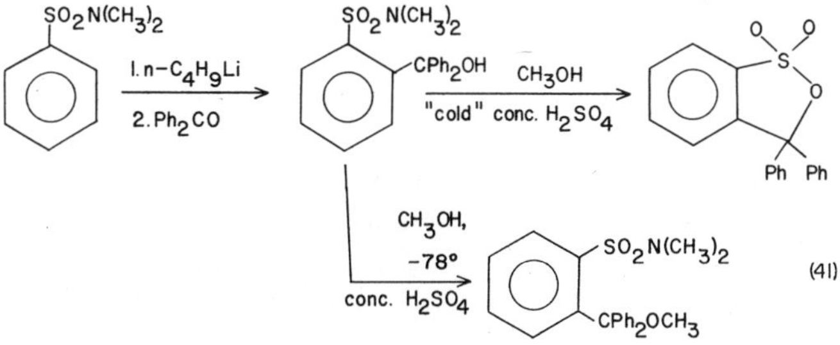

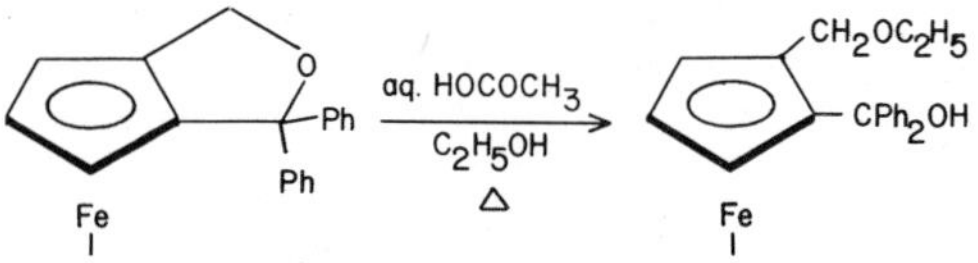

Hauser et al. (56) have reported the use of N-methylaminomethyl as an ortho-directing substituent. Treatment with acid of the secondary and tertiary alcohols produced by condensation of the 2-lithio intermediate—in this case with benzaldehyde and acetophenone, respectively—produced isoindolines (Reaction 43).

Narasimhan and Ranade (57) have incorporated the 2-metalation procedure into the preparation of isoquinoline. Furthermore their observation that metalation occurs in between two meta-oriented ortho directing groups (Reaction 26) allowed them to synthesize 5-methoxyisoquinolines *via* the directed metalation process (*see* below). Both condensation of 2-lithio-N,N-dimethylbenzylamine with ethylene oxide followed by cyclization and subsequent dehydrogenation (Reaction 44), and condensation of 2-lithio N,N-dimethyl-β-phenethylamine with paraformaldehyde followed by cyclization and subsequent aromatization (Reaction 44) yielded isoquinoline.

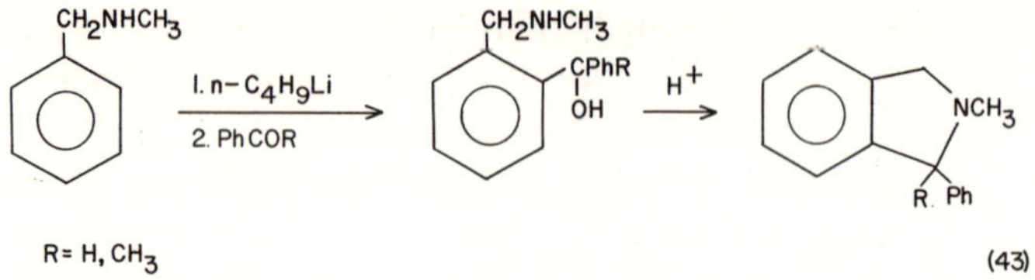

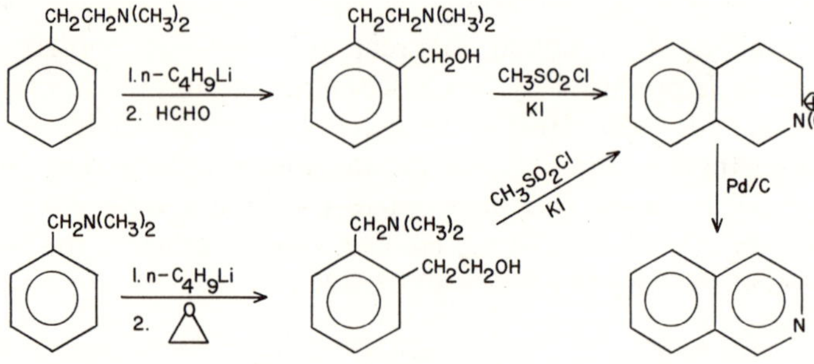

Methoxy- and ethoxy-substituted quinolines were metalated and condensed with a variety of electrophiles to yield, after other steps, a variety of natural products and derivatives (*58*). Among the condensing agents were ethylene oxide and allyl bromide, and the heterocyclic

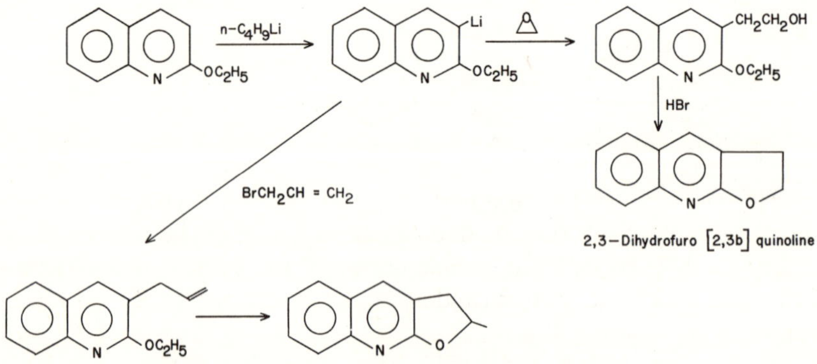

Figure 2. Synthesis of furoquinolines

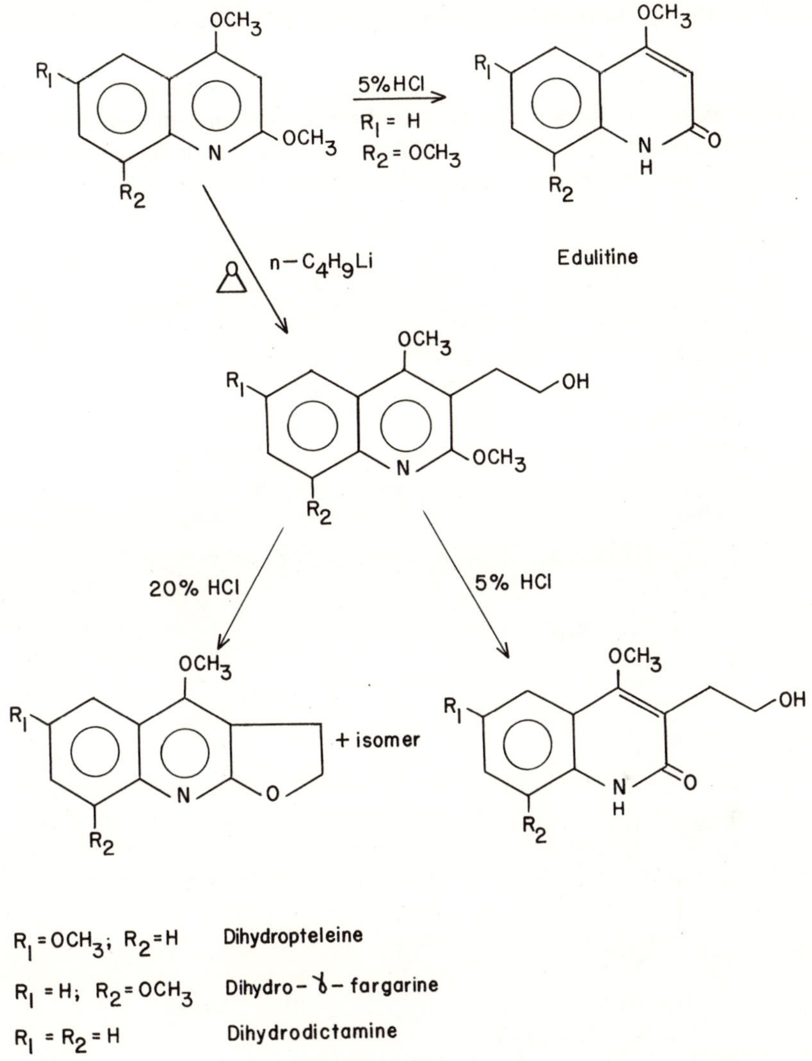

Figure 3. Synthesis of edulitine, dihydropteleine, dihydro-γ-fargarine, and dictamine

$R_1 = OCH_3$; $R_2 = H$ Dihydropteleine

$R_1 = H$; $R_2 = OCH_3$ Dihydro-γ-fargarine

$R_1 = R_2 = H$ Dihydrodictamine

natural products synthesized included furoquinolines (Figure 2) and edulitine, dihydropteleine, dihydro-γ-fargarine, and dictamine (Figure 3).

Narasimhan and Bhide (59) have also devised an elegant route for transforming laudanosine to tetrahydropalmitine *via* a directed metalation procedure (Reaction 45). This reaction sequence is important because there was previously no synthetic route for this transformation.

Narasimhan and Bhide (60) also found that the dimethylaminomethyl and N-methyl carboxamide functional groups are stronger ortho directors than the methoxy group. Applying this knowledge, they were able to synthesize methoxy-substituted isoquinolines and isocoumarins. Among the derivatives of isocoumarin prepared were mellein and 8-methoxyisocoumarin (Figure 4) (61, 62). Related directed metalations have been performed on some perhydroindian derivatives (Reaction 46) (63, 64).

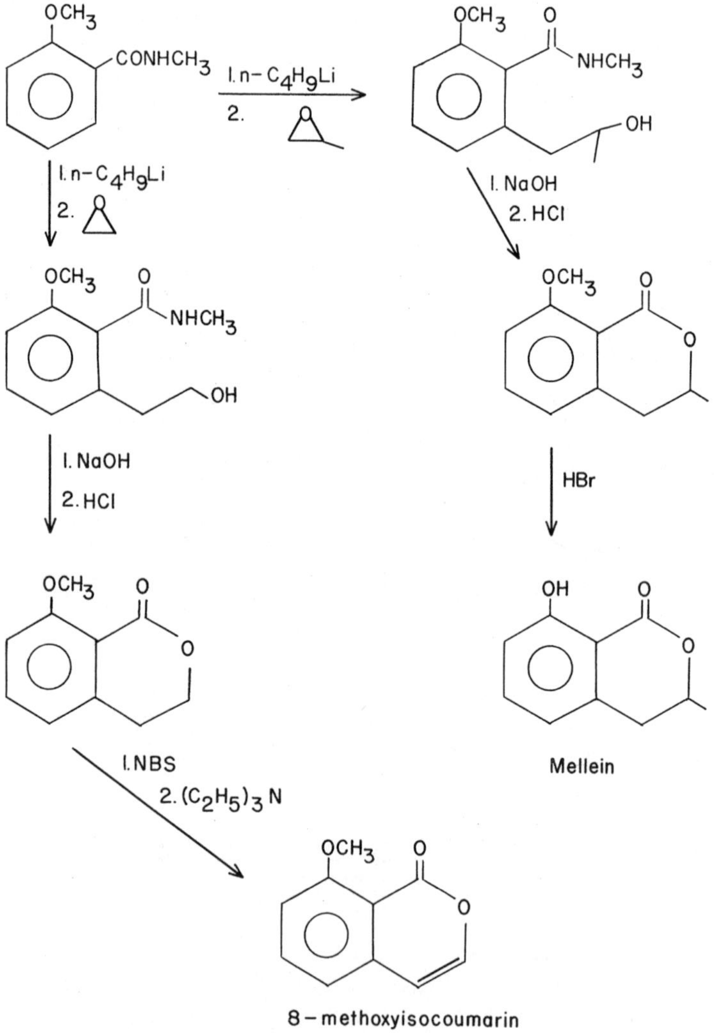

Figure 4. Preparation of isocoumarin derivatives mellein and 8-methoxyisocoumarin

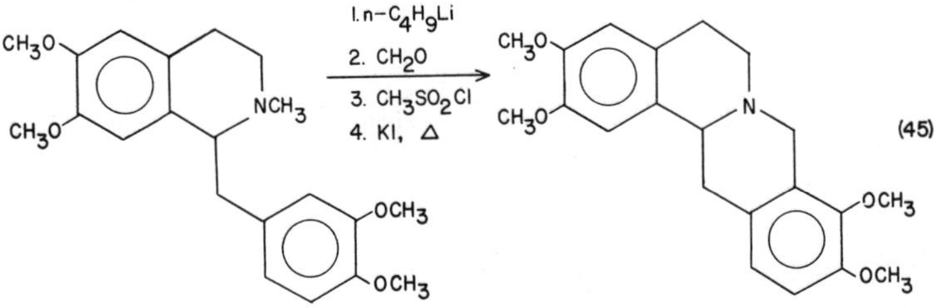

Increasing numbers of researchers are contributing to heterocyclic syntheses using directed metalation reactions although the appearance of this work in the literature is just beginning. Recently Lombardino (65) reported the synthesis of a complex heterocyclic system containing carbon, nitrogen, and sulfur (Reaction 47). Synthetic routes to these compounds are compared—namely, that *via* directed metalation and that involving electrophilic substitution. Yields from the directed metalation sequence were higher and the number of steps lower than those from electrophilic substitution, thus demonstrating the potential value of directed metalation in preparing compounds whose synthesis has already been established by an alternate route.

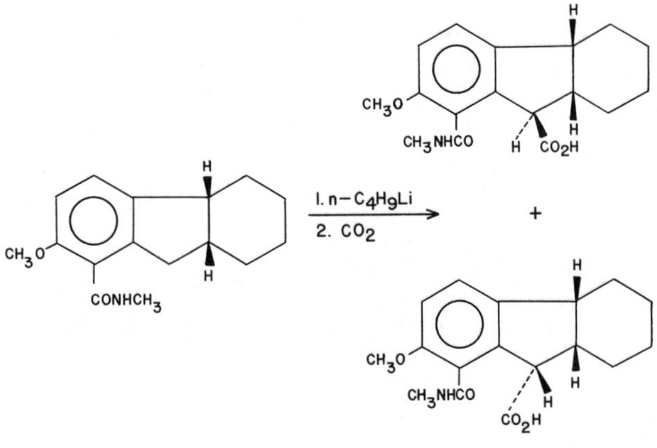

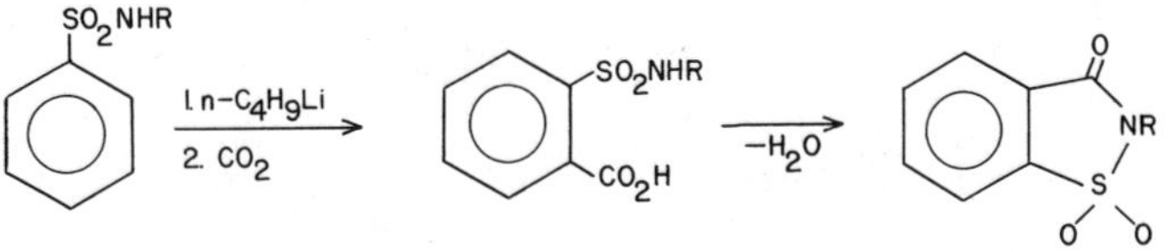

Acknowledgment

The authors are grateful for proofreading assistance by W. Achermann, R. Marchal, and R. Fellows. Special acknowledgment is made to M. Van Ness for typing the manuscript and to B. Slocum for preparing the illustrations.

Literature Cited

1. Mallan, J. M., Bebb, R. L., *Chem. Rev.* (1969) **69**, 693.
2. Achermann, W., Slocum, D. W., unpublished results.
3. Gilman, H., Morton, J. W., *Org. React.* (1954) **8**, 258.
4. Plesske, K., *Angew. Chem., Internat. Ed., Engl.* (1962) **1**, 312, 394.
5. Bublitz, D. E., Rinehart Jr., K. L., *Org. React.* (1969) **17**, 1.
6. Jones, F. N., Vaulx, R. L., Hauser, C. R., *J. Org. Chem.* (1963) **28**, 3461.
7. Rosenblum, M., "Chemistry of the Iron Group Metallocenes," Wiley, New York, 1965.
8. Slocum, D. W., Engelmann, T. R., Ernst, C., Jennings, C. A., Jones, W., Koonsvitsky, B. P., Lewis, J., Shenkin, P., *J. Chem. Educ.* (1969) **46**, 144.
9. Hoofer, O., Schlögl, K., *J. Organomet. Chem.* (1968) **13**, 443.
10. Slocum, D. W., Rockett, B. W., Hauser, C. R., *J. Amer. Chem. Soc.* (1965) **87**, 1241.
11. Slocum, D. W., Jennings, C. A., Engelmann, T. R., Rockett, B. W., Hauser, C. R., *J. Org. Chem.* (1971) **36**, 377.
12. Gilman, H., Soddy, T. S., *J. Org. Chem.* (1957) **22**, 1715.
13. Graybill, B. M., Shirley, D. A., *J. Organomet. Chem.* (1968) **31**, 443.
14. Shirley, D. A., Cheng, C. F., *J. Organomet. Chem.* (1970) **20,**, 251.
15. Gay, R. L., Hauser, C. R., *J. Amer. Chem. Soc.* (1967) **89**, 2297.
16. Kinstle, T. H., Bechner, J. P., *J. Organomet. Chem.* (1970) **22**, 497.
17. Slocum, D. W., Gierer, P. L., *Chem. Comm.* (1971) 305.
18. Davies, G. M., Davies, P. S., *Tetrahedron Lett.* (1972) 3507.
19. Marr, G., Rockett, B. W., Rushworth, A., *J. Organomet. Chem.* (1969) **16**, 141.
20. Slocum, D. W., Stonemark, F. S., *J. Org. Chem.* (1973) **38**, 1677.
21. Slocum, D. W., Jones, W. E., Crimmins, T. F., Hauser, C. R., *J. Org. Chem.* (1969) **34**, 1973.
22. Stonemark, F. S., Ph.D. Thesis, Southern Illinois University, 1971.
23. Gay, R. L., Crimmins, T. F., Hauser, C. R., *Chem. Ind. (London)* (1966) 1635.
24. Marr, G., Moore, R. E., Rockett, B. W., *J. Chem. Soc. C* (1968) 24.
25. Gierer, P. L., Ph.D. Thesis, Southern Illinois University, 1972.
26. Slocum, D. W., Koonsvitsky, B. P., *Chem. Commun.* (1969) 846.
27. Slocum, D. W., Engelmann, T. R., *J. Organomet. Chem.* (1970) **24**, 753.
28. Marr, G., *J. Organomet. Chem.* (1967) **9**, 147.
29. Marr, G., Hunt, T., *J. Chem. Soc. C* (1969) 1970.
30. Slocum, D. W., Jennings, C. A., "Abstracts of Papers," 161st National Meeting, ACS, Los Angeles, March 1971 ORGN 186.
31. Sugarman, D. I., Slocum, D. W., unpublished results.
32. Grocock, D. E., Jones, T. K., Hallas, G., Hepworth, J. D., *J. Chem. Soc. C* (1971) 3305.
33. Slocum, D. W., Koonsvitsky, B. P., *J. Org. Chem.* (1973) **38**, 1675.
34. Puterbaugh, W. H., Hauser, C. R., *J. Amer. Chem. Soc.* (1963) **85**, 2467.
35. Nozaki, H., Aratani, T., Toraya, T., *Tetrahedron Lett.* (1968) 4097.
36. Nozaki, H., Aratani, T., Toraya, T., Noyori, R., *Tetrahedron* (1971) 905.

37. Aratani, T., Gonda, T., Nozaki, H., *Tetrahedron* (1970) **26**, 5453.
38. Aratani, T., Gonda, T., Nozaki, H., *Tetrahedron Lett.* (1969) 2265.
39. Gokel, G., Hoffmann, P., Kleinamm, H., Klusacek, H., Marquarding, D., Ugi, I., *Tetrahedron Lett.* (1970) 1771.
40. Marquarding, D., Klusacek, H., Gokel, G., Hoffmann, P., Ugi, I., *J. Amer. Chem. Soc.* (1970) **92**, 5389.
41. Battelle, L. F., Bau, R., Gokel, G. W., Oyakawa, R. T., Ugi, I. K., *J. Amer. Chem. Soc.* (1973) **95**, 482.
42. Ugi, I., *Rec. Chem. Prog.* (1969) **30**, 289.
43. Marquarding, D., Klusacek, H., Gokel, G., Hoffmann, P., Ugi, I., *Angew. Chem. Internat. Edit.* (1970) **9**, 371.
44. Goldberg, S. I., Bailey, W. D., *Tetrahedron Lett.* (1971) 4087.
45. Vaulx, R. L., Jones, F. N., Hauser, C. R., *J. Org. Chem.* (1964) **29**, 505.
46. Puterbaugh, W. H., Hauser, C. R., *J. Org. Chem.* (1964) **29**, 853.
47. Barnish, I. T., Mao, C. L., Gay, R. L., Hauser, C. R., *Chem. Comm.* (1968) 564.
48. Mao, C. L., Barnish, I. T., Hauser, C. R., *J. Heterocycl. Chem.* (1969) **6**, 475.
49. Mao, C. L., Barnish, I. T., Hauser, C. R., *J. Heterocycl. Chem.* (1969) **6**, 83.
50. Mao, C. L., Henoch, F. E., Hauser, C. R., *Chem. Comm.* (1968) 1595.
51. Bailey, D. M., DeGrazia, C. G., *Tetrahedron Lett.* (1970) 633.
52. Watanabe, H., Gay, R. L., Hauser, C. R., *J. Org. Chem.* (1968) **33**, 900.
53. Watanabe, H., Schwarz, R. A., Hauser, C. R., Lewis, J., Slocum, D. W., *Can. J. Chem.* (1969) **47**, 1543.
54. Watanabe, H., Schwarz, R. A., Hauser, C. R., *Chem. Comm.* (1968) 287.
55. Slocum, D. W., Silverman, B., Rockett, B. W., Hauser, C. R., *J. Org. Chem.* (1967) **32**, 464.
56. Ludt, R. E., Hauser, C. R., *J. Org. Chem.* (1971) **36**, 1607.
57. Narasimhan, N. S., Ranade, A. C., *Chem. Ind. (London)* (1967) 120.
58. Narasimhan, N. S., Paradkar, M. V., Alurkar, R. H., *Tetrahedron* (1971) 1351.
59. Narasimhan, N. S., Bhide, B. H., *Chem. Ind. (London)* (1969) 621.
60. Narasimhan, N. S., Bhide, B. H., *Tetrahedron Lett.* (1968) 4159.
61. Narasimhan, N. S., Bhide, B. H., *Chem. Comm.* (1970) 1552.
62. Narasimhan, N. S., Bhide, B. H., *Tetrahedron* (1971) 6171.
63. House, H. O., Hanners, W. E., Racah, E. J., *J. Org. Chem.* (1972) **37**, 985.
64. House, H. O., Hudson, C B., Racah, E. J., *J. Org. Chem.* (1972) **37**, 989.
65. Lombardino, J. G., *J. Org. Chem.* (1971) **36**, 1843.
66. Slocum, D. W., Engelmann, T. R., Jennings, C. A., *Aust. J. Chem.* (1968) **21**, 2319.
67. Roberts, J. D., Curtin, D. Y., *J. Amer. Chem. Soc.* (1946) **68**, 1658.
68. Slocum, D. W., Book, G., Jennings, C. A., *Tetrahedron Lett.* (1970) 3443.
69. Booth, D. J., Rockett, B. W., *J. Chem. Soc. C* (1968) 656.
70. Benkeser, R. A., Fitzgerald, W. P., Melzer, M. S., *J. Org. Chem.* (1961) **26**, 2596.
71. Slocum, D. W., Koonsvitsky, B. P., Ernst, C. R., *J. Organometal. Chem.* (1972) **38**, 125.
72. Slocum, D. W., Achermann, W., Teymouri, E., unpublished results.
73. Gronowitz, S., *Ark. Kemi.* (1958) **12**, 239.
74. Gronowitz, S., *Ark. Kemi.* (1960) **16**, 363.
75. Gronowitz, S., *Ark. Kemi.* (1958) **13**, 269.
76. Gronowitz, S., Eriksson, B., *Ark. Kemi.* (1963) **21**, 335.
77. Gronowitz, S., *Ark. Kemi.* (1954) **7**, 361.

RECEIVED March 13, 1973.

13

Synthetic Applications of N-Chelated Organolithium Compounds

M. D. RAUSCH and A. J. SARNELLI

University of Massachusetts, Amherst, Mass. 01002

> *The discovery that organolithium compounds are made considerably more reactive by coordination with chelating tertiary diamines has greatly stimulated studies involving synthetic applications of these reagents. Most of the studies by far are concerned with the metalation reaction—hydrogen-metal interconversion—since the chelated organolithium intermediates are powerful metalating agents. Metalation studies involving benzene, toluene and other alkylbenzenes, thioanisole, anisole, benzylamines, methyl sulfides, methylphosphines, methylsilanes, ferrocene, bis(benzene)chromium, and many other compounds have been described during the past several years. Chelated organolithium reagents also exhibit enhanced reactivity in addition to double and triple bonds and in other reactions. All these studies suggest that chelated organolithium reagents will play an important role in the future development of organic and organometallic syntheses.*

The history of organolithium chemistry dates essentially from about 1930, when Karl Ziegler first prepared organolithium reagents from organic halides and lithium metal. In his pioneering paper that year (*1*), Ziegler predicted the great potential utility of organolithium reagents in both organic and organometallic syntheses. As the utility of these reagents in syntheses developed, it became clear that certain solvent systems were "better" than others. In other words, ethers are better than hexane and other alkanes as solvents because the organolithium compounds are appreciably more reactive in ethers. The relative reactivity of the organolithium reagent varies with the nature of the ether as well. Gilman *et al.* (*2, 3*), for example, found that ethyl ether and tetrahydrofuran together make a much more effective solvent for metalation reactions than does ethyl ether alone. On the other hand, some organolithium compounds attack ethers; therefore, many organolithium reagents (such as n-butyllithium) are most often prepared and sold in hydrocarbon solvents.

Also, addition of small quantities of Lewis bases such as amines to alkyllithium reagents in hydrocarbons markedly affects reactivity, especially in connection with various anionic polymerization reactions. Findings such as these prompted a number of research groups in the early 1950's to study in detail the role of Lewis bases in the structures of organolithium compounds (4, 5). In each case it was concluded that coordination complexes form when amines are added to organolithium reagents in hydrocarbons.

During 1964 and 1965 three groups of investigators—A. W. Langer, Jr., of Esso Research and Engineering; G. G. Eberhardt of Sun Oil; J. F. Eastham of the University of Tennessee—reported independently that reactivities of organolithium compounds are particularly enhanced by chelating ditertiary aliphatic amines, such as N,N,N',N'-tetramethylethylenediamine (TMEDA) or sparteine. A new era in the history and development of organolithium chemistry thus began.

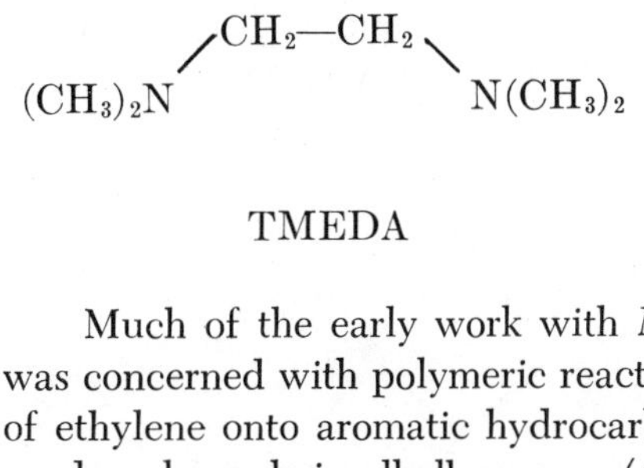

TMEDA 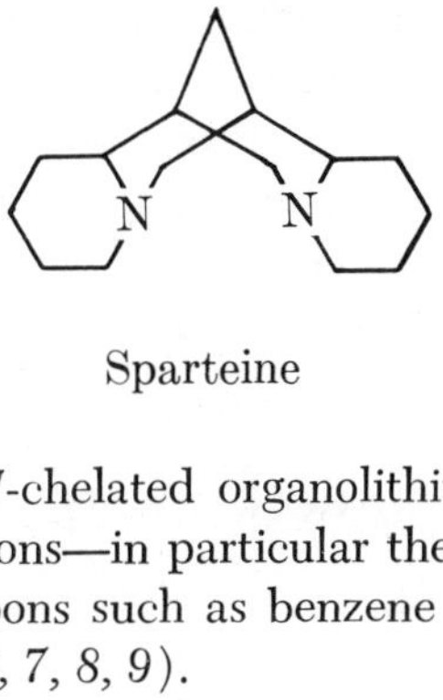   Sparteine 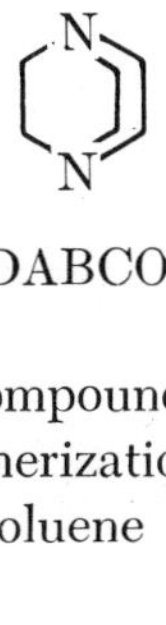   DABCO

Much of the early work with N-chelated organolithium compounds was concerned with polymeric reactions—in particular the telomerization of ethylene onto aromatic hydrocarbons such as benzene and toluene to produce long-chain alkylbenzenes (6, 7, 8, 9).

$$\text{Ar—H} + n\text{-CH}_2\text{=CH}_2 \xrightarrow{\text{BuLi + diamine}} \text{Ar—(CH}_2\text{—CH}_2)_n\text{—H} \quad \text{(telomerization)}$$

It was obvious from the outset, however, that the remarkable carbanionic reactivity of these new organometallic compounds made them valuable intermediates in organic and organometallic syntheses. Langer (6, 7) found, for example, that when the BuLi–TMEDA complex was prepared in benzene at 50°C and phosphorus trichloride was added after one hour, triphenylphosphine could be isolated in 92% yield. By contrast, BuLi alone does not react with benzene up to 100°C. Langer also observed that BuLi–TMEDA metalates toluene quantitatively within minutes at 25°C to yield benzyllithium–TMEDA and that this method represents by far the best synthesis of benzyllithium.

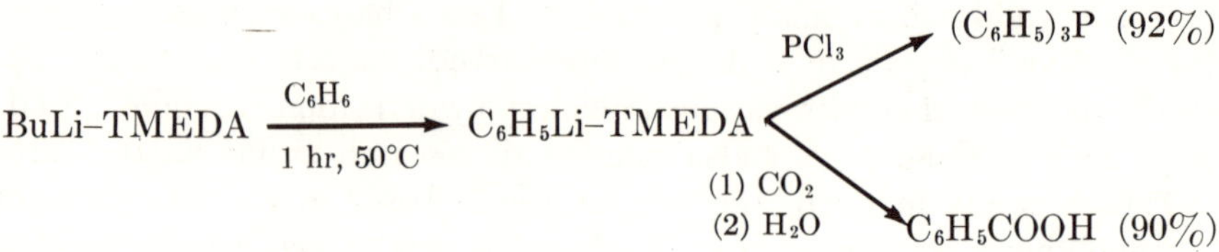

$$\text{BuLi–TMEDA} \xrightarrow[\text{1 hr, 50°C}]{C_6H_6} C_6H_5Li\text{–TMEDA} \begin{array}{c} \xrightarrow{PCl_3} (C_6H_5)_3P \ (92\%) \\ \xrightarrow[(2)\ H_2O]{(1)\ CO_2} C_6H_5COOH \ (90\%) \end{array}$$

$$\text{BuLi–TMEDA} \xrightarrow[25°]{C_6H_5CH_3} C_6H_5CH_2Li\text{–TMEDA} \quad \text{(quantitative)}$$

Some of these chelated organolithium complexes have been isolated and characterized by Langer (6, 7) and by Eastham and co-workers (10). The greatly enhanced reactivity of the chelated organolithium complexes is most likely the result of an increased ionic character of the carbon–lithium bond, which is caused by strong complexation between the chelating agent and the lithium atom. Eastham (10) has also suggested that low steric requirements of the tertiary amine are an important factor in promoting chelation of an organolithium compound and enhancing its reactivity since 1,4-diazabicyclo[2.2.2]octane (DABCO) is especially effective in this regard.

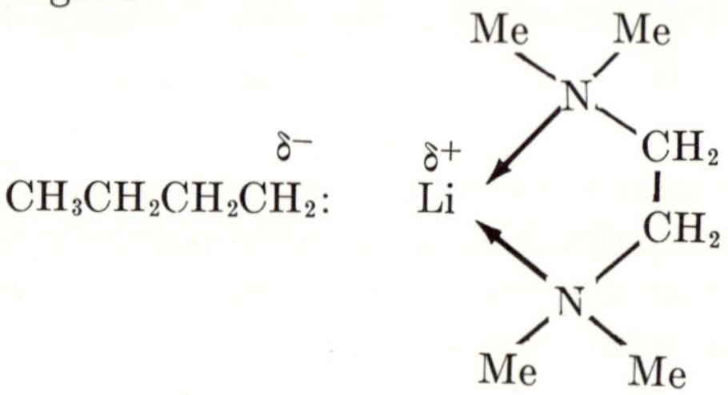

$$CH_3CH_2CH_2CH_2\!: \overset{\delta-}{} \quad \overset{\delta+}{Li} \quad \begin{array}{c} Me \quad Me \\ \diagdown N \diagup \\ \diagup \quad \diagdown CH_2 \\ \mid \\ \diagdown \quad \diagup CH_2 \\ \diagup N \diagdown \\ Me \quad Me \end{array}$$

Metalations of Aromatic Hydrocarbons

Our interest in the synthetic utility of N-chelated organolithium compounds was prompted by this early work. Since detailed experimental procedures were not generally available at that time, we initiated a study of the effects of time, temperature, stoichiometry, etc., on the reactions of BuLi–TMEDA with benzene. We found that optimum metalation of benzene occurs when the preformed BuLi–TEMEDA complex, prepared from equimolar amounts of the organolithium reagent and diamine, is allowed to react with benzene at room temperature for three hours (*11*). Carbonation and hydrolysis of the reaction mixture result in benzoic acid in average yields of 92%. When the reaction mixture is refluxed for three hours, the yield of benzoic acid (and presumably phenyllithium) decreases to 49% perhaps because of subsequent attack of either BuLi or phenyllithium on the diamine.

Using the optimum conditions for benzene metalation as indicated by the carbonation studies, we found that other reactions typical of phenyllithium proceed in very high yield when phenyllithium–TMEDA is prepared this way. Some typical reactions are outlined below.

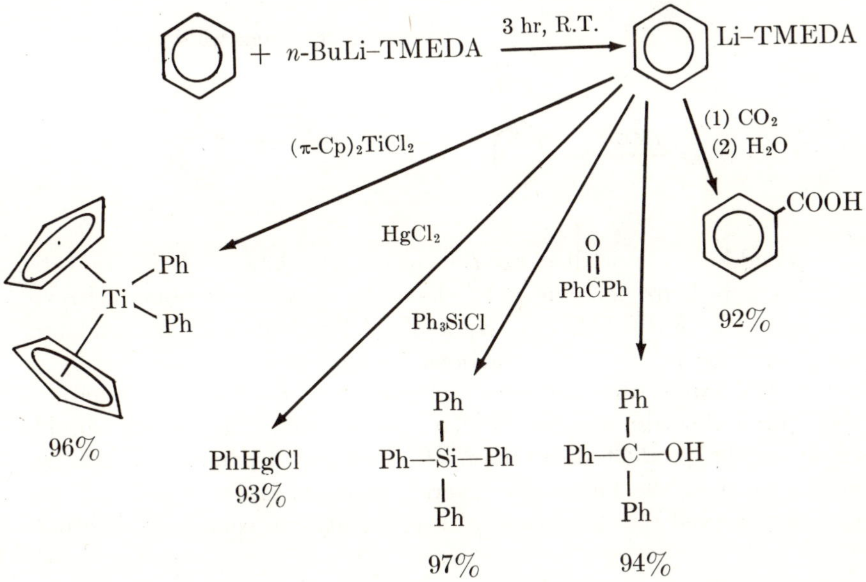

[Reaction scheme: Benzene + n-BuLi–TMEDA, 3 hr, R.T. → Ph–Li–TMEDA, which reacts with:
- $(\pi\text{-Cp})_2\text{TiCl}_2$ → Cp$_2$Ti(Ph)$_2$, 96%
- HgCl$_2$ → PhHgCl, 93%
- Ph$_3$SiCl → Ph–Si(Ph)$_2$–Ph, 97%
- PhCPh (O) → Ph–C(Ph)–OH, 94%
- (1) CO$_2$ (2) H$_2$O → PhCOOH, 92%]

Metalation of toluene by BuLi–TMEDA has been studied in detail by Chalk and Hoogeboom (*12*). A 10:1 ratio of toluene to metalating agent was used; the reactions were followed by quenching with either dimethyl sulfate or trimethylchlorosilane. Products were examined by gas-liquid chromatography. Metalation occurs at all four nonequivalent sites on toluene, and the isomer distributions within a sample occur in constant proportions, independent of the extent of reaction and the nature of the quenching reagent. About 90% of the metalation takes place at the benzylic position, about 5% at the meta position on the ring, and the rest about equally at the ortho and para positions. The authors were able to show that rapid isomerization of the various organolithium reagents does not occur and that the constant isomer distribution depends on kinetic factors only.

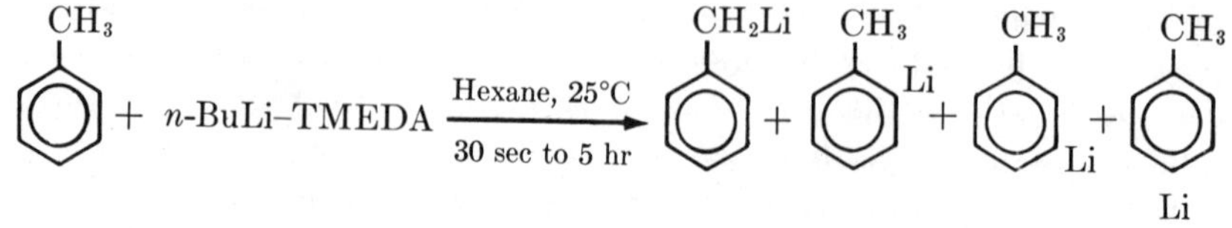

CH₃-C₆H₅ + n-BuLi–TMEDA →(Hexane, 25°C, 30 sec to 5 hr)→ [CH₂Li-C₆H₅] + [ortho-Li-C₆H₄-CH₃] + [meta-Li-C₆H₄-CH₃] + [para-Li-C₆H₄-CH₃]

	CH₂Li product	ortho	meta	para
Derivatized by Me₃SiCl:	90.4%	2.0%	5.6%	2.0%
Derivatized by Me₂SO₄:	89.8	2.1	5.5	2.6

(isomer distributions within a sample)

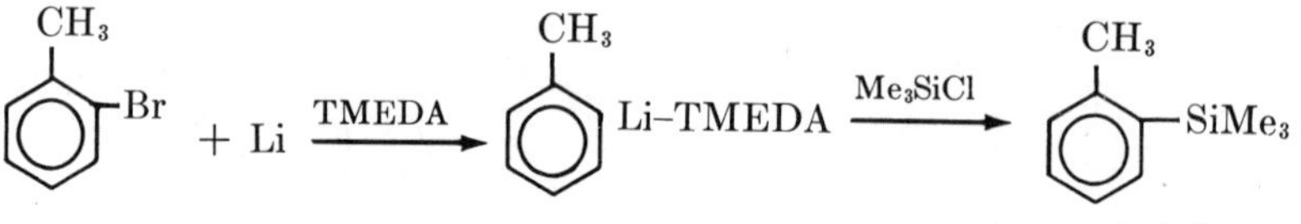

o-Br-C₆H₄-CH₃ + Li —TMEDA→ o-Li(TMEDA)-C₆H₄-CH₃ —Me₃SiCl→ o-SiMe₃-C₆H₄-CH₃ (only)

Broaddus (13) studied extensively the metalations of the aromatic hydrocarbons benzene, toluene, ethylbenzene, cumene, *tert*-butylbenzene, and anisole by BuLi–TMEDA. He used a 4:1 ratio of arene to metalating agent, and he followed the reactions by gas-liquid chromatography of the methyl esters resulting from carbonation and subsequent esterification. Broaddus also concluded that reaction time has no measurable effect on the product distribution of metalated products and that rearrangement of the kinetically favored products is not significant. He further observed that the extent of ring metalation, compared with ben-

Substrate	Time, hr	Benzyl, %	Ortho, %	Meta, %	Para, %
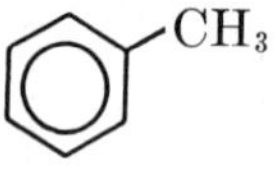 C₆H₅–CH₃	0.25	89	3	9 (ortho and meta combined)	
	0.5	90	2	8	,,
	1.0	88	3	9	,,
	2.0	92	2	6	,,
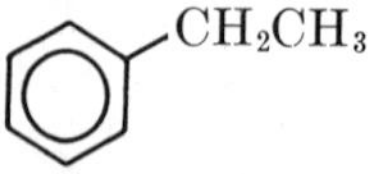 C₆H₅–CH₂CH₃	0.5	38	9	36	17
	1.0	37	9	36	17
	6.5	38	9	36	17
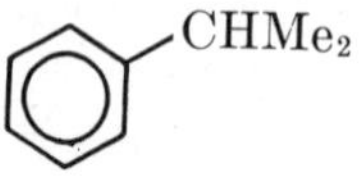 C₆H₅–CHMe₂	2.0	3	10	57	30
	24.0	3	8	59	30
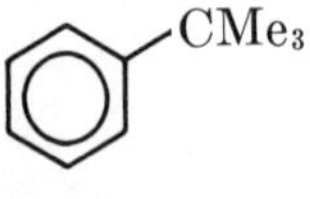 C₆H₅–CMe₃	4.0			68	32

zylic metalation, follows the order cumene >> ethylbenzene > toluene. According to his explanation of this trend, the transition states leading to benzyl carbanionic species are destabilized by an increased number of methyl groups relative to those involved in metalation at ring sites. Broaddus also found that in a competitive metalation of a benzene-anisole mixture, the ortho position of anisole undergoes metalation about 100 times faster than does benzene.

When a metalating agent such as BuLi–TMEDA is present in a much greater molar excess than the aromatic substrate, polymetalation frequently occurs. Polymetalation reactions of toluene, anthracene, biphenyl, fluorene, and indene have been extensively studied by West *et al.* (*14*) and Halasa (*15*). These results, as well as those cited above, illustrate that while metalation can be done conveniently with BuLi–TMEDA, complex metalated intermediates are often obtained, depending frequently on the reaction stoichiometry. This lack of selectivity can obviously limit the synthetic utility of a given reaction system.

Hauser and co-workers (*16, 17*) studied the metalation of N-alkyl-aromatic amines using BuLi–TMEDA. N-Methylbenzylamine undergoes dimetalation mainly at the nitrogen atom and the o-benzyl positions, as evidenced by deuteration studies. N,N-Dimethyl-o-toluidine undergoes metalation primarily in the 2-methyl position while N,N-dimethyl-p-toluidine undergoes only ortho ring metalation. This study showed that the BuLi–TMEDA complex is an appreciably better metalating agent in

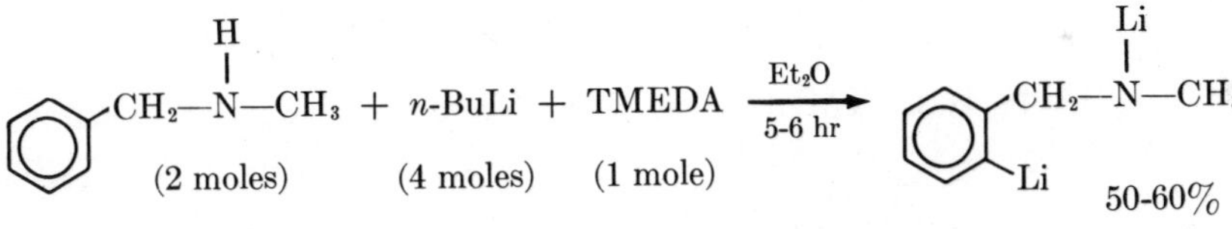

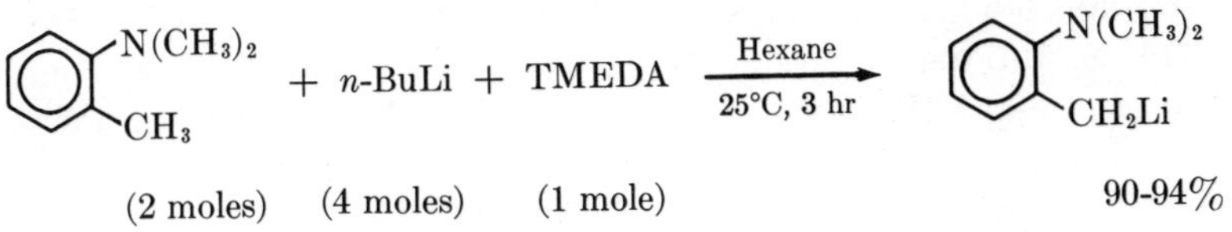

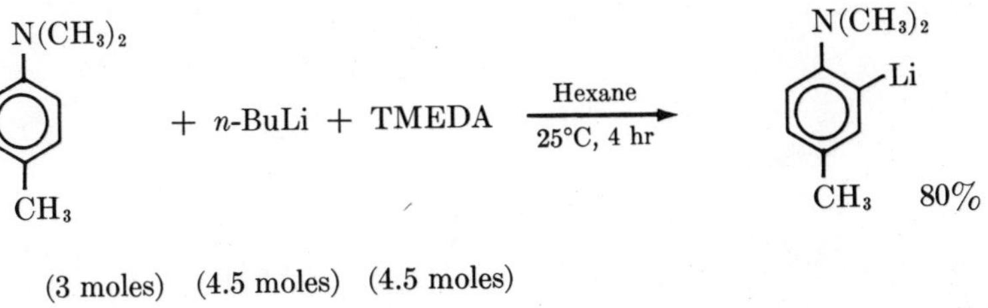

PhCH$_2$–NH–CH$_3$ + n-BuLi + TMEDA $\xrightarrow[\text{5-6 hr}]{\text{Et}_2\text{O}}$ o-Li-C$_6$H$_4$-CH$_2$–N(Li)–CH$_3$ 50-60%

(2 moles)   (4 moles)   (1 mole)

o-(CH$_3$)C$_6$H$_4$-N(CH$_3$)$_2$ + n-BuLi + TMEDA $\xrightarrow[\text{25°C, 3 hr}]{\text{Hexane}}$ o-(LiCH$_2$)C$_6$H$_4$-N(CH$_3$)$_2$ 90-94%

(2 moles)   (4 moles)   (1 mole)

p-(CH$_3$)C$_6$H$_4$-N(CH$_3$)$_2$ + n-BuLi + TMEDA $\xrightarrow[\text{25°C, 4 hr}]{\text{Hexane}}$ 2-Li-4-CH$_3$-C$_6$H$_3$-N(CH$_3$)$_2$ 80%

(3 moles)   (4.5 moles)   (4.5 moles)

reactions with these amines than is BuLi alone. The complex is not only an effectively stronger base, as suggested by better yields and shorter metalation periods, but also a more selective one.

Slocum and co-workers (18) recently examined the effect of TMEDA on directed metalation reactions. N,N-Dimethylbenzylamine, β-phenylethyldimethylamine, N,N-dimethylaniline, and anisole, all known to undergo ortho metalation with BuLi but at a relatively slow rate, were each lithiated under conditions similar to those previously reported, except that one equivalent of TMEDA was added to each metalation reaction. In three cases, the metalation rate was increased significantly with only a slight loss in overall yield. These workers also found that the site of ring metalation with p-methoxy-N,N-dimethylbenzylamine can be reversed by using TMEDA.

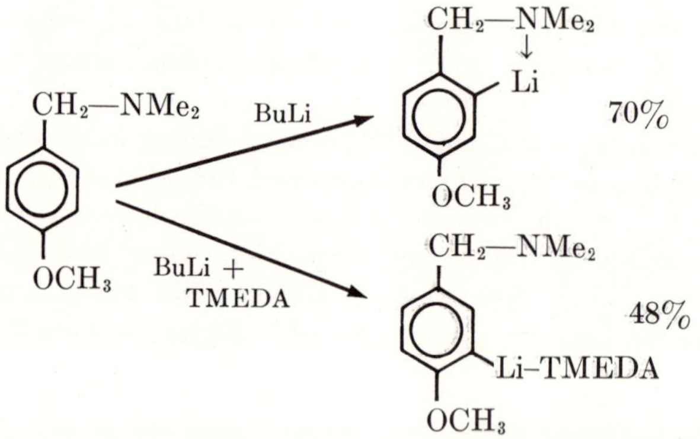

Thus, using BuLi–TMEDA vs. BuLi alone offers synthetic utility in substantially increasing metalation rates of certain monosubstituted benzenes as well as controlling the metalation sites in compounds containing more than one ortho-directing substituent.

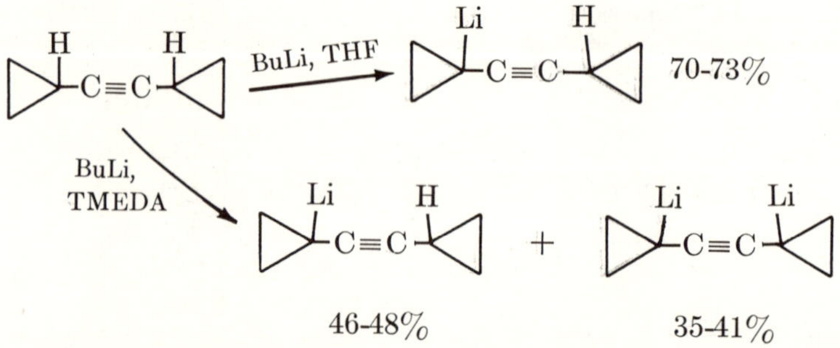

Several additional recent reports have also demonstrated the effects of TMEDA on product yields and selectivity in metalations involving BuLi. For example, Köbrich and Merkel (19) found that dicyclopropylacetylene is converted solely to a monolithio derivative by BuLi at room temperature. By contrast, BuLi–TMEDA produced, in addition to the monolithio derivative, a dilithio derivative in about the same yield.

Shirley and Cheng (*20*) reported in 1969 that metalation of 1-methoxynaphthalene by BuLi produces a mixture of the 2-lithio and the 8-lithio derivatives in an overall yield of 28% as shown by carbonation. When the same reaction conditions are used except that one equivalent of TMEDA is added, a shift in product composition to > 99.3% 2-metalation and < 0.3% 8-metalation occurs, and the yield of 1-methoxy-2-naphthalenecarboxylic acid increases to 60%. On the other hand, metalation of 1-methoxynaphthalene with *tert*-BuLi in pentane–cyclohexane solvent gives a product representing 97% 8-metalation and 3% 2-metalation. These selective metalations therefore constitute a useful synthetic route to 1,2- and 1,8-disubstituted naphthalene derivatives.

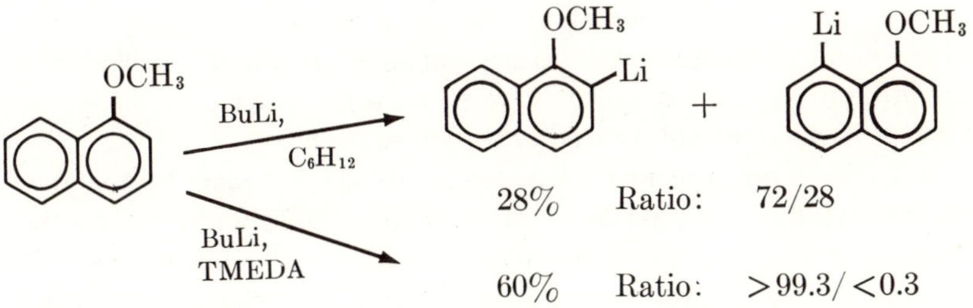

While studying the reactions of perylene, Ziegler (*21*) observed that treatment of perylene with methyllithium in boiling benzene gives only a 0.1–2.0% yield of the alkylation product, 1-methylperylene; 90 to 95% of the starting hydrocarbon is invariably recovered.

Perylene (90-95% recovery)

1-Methylperylene (0.1-2.0% yield)

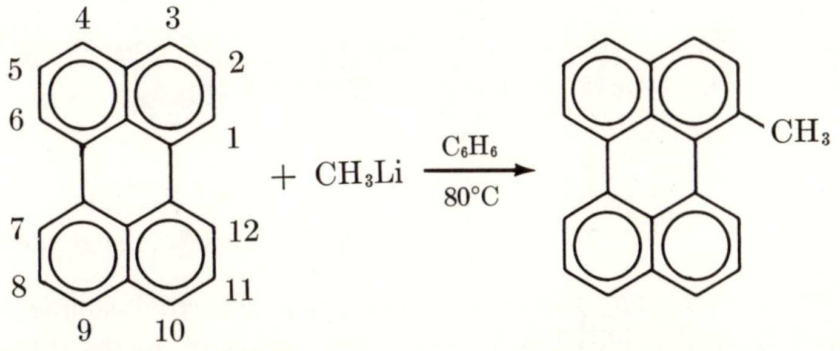

Adding TMEDA to the reaction mixture increases the yield of alkylation product to 10%. Subsequent work by Ziegler and Laski (*22*)

showed that perylene methylation using MeLi/TMEDA in benzene at 80°C for a shorter reaction time produces a 33% yield of crude crystalline methyldihydroperylenes from which 1-methylperylene can be isolated by dehydrogenation using Pc/C. These alkylation reactions presumably proceed *via* addition of the organolithium reagent to form a lithium alkyldihydroaromatic intermediate; lithium hydride is then eliminated, and the alkylated aromatic hydrocarbon forms.

Metalations of Organic Molecules Containing Hetero Atoms

The reactions discussed so far deal with metalations of aromatic hydrocarbons. Another group of metalation reactions of N-chelated organolithium complexes involve organic molecules containing hetero atoms. Second-row elements such as S, P, Si, etc., possess d orbitals that seem capable of stabilizing the partial negative charge on attached CH_2Li groups by dative π bonding—that is, d-orbital resonance stabilization can occur. Once again the greatly enhanced reactivity of N-chelated organolithium reagents has proved very useful synthetically in forming a series of carbanions substituted with hetero atoms.

As an example, the methyl group in thioanisole can be metalated using n-butyllithium to produce phenylthiomethyllithium; by contrast, anisole under the same conditions is metalated at an ortho position. This reaction has been of little practical use in synthesis since the maximum yield of the lithium intermediate is only about 35%. However, Corey and Seebach (23) recently developed an excellent procedure for generating phenylthiomethyllithium in essentially quantitative yield by reaction between equimolar amounts of thioanisole, n-butyllithium, and DABCO in tetrahydrofuran at 0°C.

$$\text{PhS-CH}_3 + \text{BuLi} + \text{DABCO} \xrightarrow[\text{45 min}]{\text{THF. 0°C}} \text{PhS-CH}_2\text{Li}$$

(1 mole)   (1 mole)   (1 mole)   97% (D_2O)
                                                            93% (Benzophenone)

Peterson (24) has also shown that even dimethyl sulfide can be metalated by BuLi–TMEDA to give high yields of methylthiomethyllithium. The reaction is rapid at room temperature and essentially complete within four hours in hexane as the solvent. Methylthiomethyllithium is quite valuable as an intermediate in synthesizing carbon functionally substituted organosulfur compounds since it has the unique advantage of giving derivatives in the sulfide oxidation state. These derivatives can be subsequently converted readily into their corresponding sulfoxides, sulfones, and sulfonium compounds by known procedures.

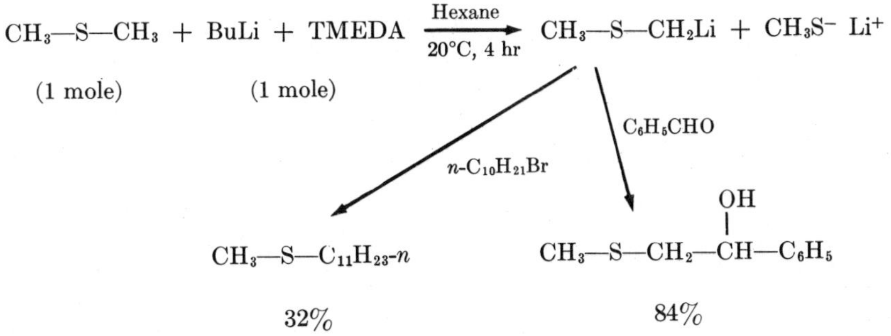

$$CH_3-S-CH_3 + BuLi + TMEDA \xrightarrow[20°C, 4\ hr]{Hexane} CH_3-S-CH_2Li + CH_3S^-\ Li^+$$

(1 mole)   (1 mole)

with $n\text{-}C_{10}H_{21}Br$ giving $CH_3-S-C_{11}H_{23}\text{-}n$ (32%)

with C_6H_5CHO giving $CH_3-S-CH_2-\underset{\underset{OH}{|}}{CH}-C_6H_5$ (84%)

Several groups have studied the feasibility of metalating the weakly acidic methylsilanes to form the corresponding silylmethyllithium compounds, a process of considerable theoretical and synthetic interest. Peterson (25) studied a three-day reaction between BuLi–TMEDA and tetramethylsilane at room temperature. Derivatization of the reaction mixture with trimethylchlorosilane gives a 36% yield of bis(trimethylsilyl)methane and an 18% yield of a product resulting from partial

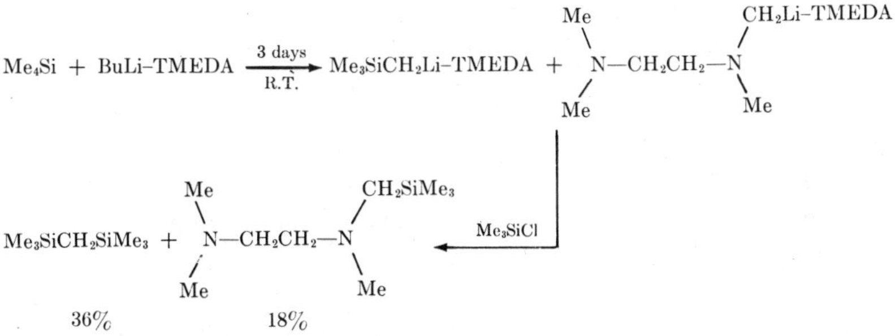

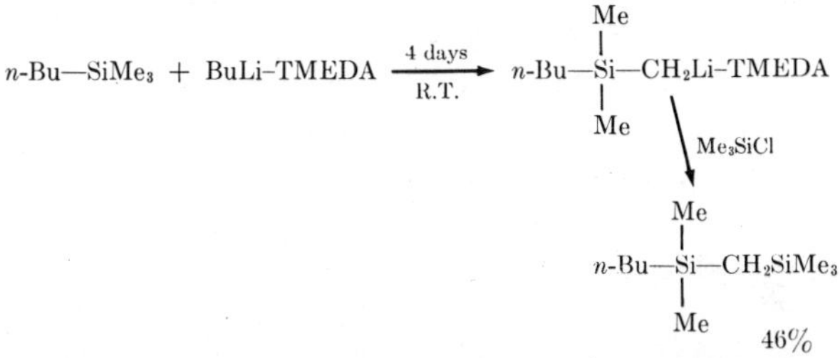

$$Me_4Si + BuLi-TMEDA \xrightarrow[R.T.]{3\ days} Me_3SiCH_2Li-TMEDA + \underset{Me}{\overset{Me}{\diagdown}}N-CH_2CH_2-N\underset{Me}{\overset{CH_2Li-TMEDA}{\diagup}}$$

$$\downarrow Me_3SiCl$$

$$Me_3SiCH_2SiMe_3 + \underset{Me}{\overset{Me}{\diagdown}}N-CH_2CH_2-N\underset{Me}{\overset{CH_2SiMe_3}{\diagup}}$$

36%   18%

$$n\text{-}Bu-SiMe_3 + BuLi-TMEDA \xrightarrow[R.T.]{4\ days} n\text{-}Bu-\underset{\underset{Me}{|}}{\overset{\overset{Me}{|}}{Si}}-CH_2Li-TMEDA$$

$$\downarrow Me_3SiCl$$

$$n\text{-}Bu-\underset{\underset{Me}{|}}{\overset{\overset{Me}{|}}{Si}}-CH_2SiMe_3$$

46%

metalation of the TMEDA ligand. Metalation of n-butyltrimethylsilane with BuLi–TMEDA for four days at room temperature also proceeds readily at one of the methyl groups.

Gornowicz and West (26) found that tetramethylsilane can also be metalated in ca. 40% yield by tert-butyllithium in the presence of TMEDA for four days at room temperature. In addition, trimethylchlorosilane also reacts with this N-chelated organolithium reagent to produce not only the coupling product, *tert*-butyltrimethylsilane, but also the organolithium intermediate resulting from metalation of one of the methyl groups. Under the same conditions, trimethylchlorosilane can be metalated much more rapidly than tetramethylsilane, indicating that the protons of trimethylchlorosilane are significantly more acidic than those of tetramethylsilane. Steric hindrance to coupling seems to be essential for metalating trimethylchlorosilane since with *n*-butyllithium (either with or without TMEDA) trimethylchlorosilane gives exclusively the coupling product, *n*-butyltrimethylsilane.

$$Me_3SiCl + \textit{tert}\text{-BuLi} + TMEDA \xrightarrow[< 1 \text{ min}]{\text{Pentane, } 15°\text{-}30°C} Me_3Si\text{—}\textit{tert}\text{-Bu} + LiCH_2\text{—}\underset{\underset{Me}{|}}{\overset{\overset{Me}{|}}{Si}}Cl$$

(2 moles)   (0.5 mole)   (0.12 mole)   20%

$$Me_3SiCH_2\underset{\underset{Me}{|}}{\overset{\overset{Me}{|}}{Si}}\text{—}\textit{tert}\text{-Bu} \xleftarrow{\textit{tert}\text{-BuLi}} Me_3SiCH_2\underset{\underset{Me}{|}}{\overset{\overset{Me}{|}}{Si}}Cl \xleftarrow{Me_3SiCl}$$

40%   20%

$$Me_4Si + \textit{tert}\text{-BuLi} + TMEDA \xrightarrow[4 \text{ days}]{\text{Pentane, } 25°C} Me_3SiCH_2Li \xrightarrow{Me_3SiCl \text{ added}} Me_3SiCH_2SiMe_3$$

(1.2 mole)   (1 mole)   (0.25 mole)   40%

Metalations of various methylphosphines by BuLi–TMEDA were studied by Peterson (27) and by Rausch and Ciappenelli (11). Whereas methyldiphenylphosphine, for example, is inert to BuLi in hydrocarbon solvents, addition of an equivalent of TMEDA gives (diphenylphosphino)-methyllithium in 40-75% yield. Dimethylphenylphosphine and dimethyl-*n*-octadecylphosphine can also be converted into α-phosphinoalkyllithium compounds in similar yields with BuLi–TMEDA. The utility of these reagents as intermediates in synthesizing carbon functionally substituted phosphines was demonstrated by reactions with diphenylchlorophosphine, benzophenone, carbon dioxide, etc.

Metalations of Organometallic π Complexes

Metalations of various organometallic π complexes with N-chelated organolithium complexes have also received considerable attention in recent years. Perhaps the best known organometallic π complex is ferrocene. Although it was discovered very early in the development of fer-

$$\text{Ph}_2\text{P–CH}_3 + \text{BuLi–TMEDA} \xrightarrow[\text{then THF added}]{\text{Hexane, 2 hr, R.T.,}} \text{Ph}_2\text{P–CH}_2\text{Li–TMEDA}$$

(1 mole)   (1 mole)   40–75%

(Characterized by Ph_2PCl and benzophenone)

Using BuLi alone and hexane, no reaction.

Using BuLi alone and Et_2O as solvent, 21% metalation in 48 hours.

$$\text{Ph(CH}_3\text{)P–CH}_3 + \text{BuLi–TMEDA} \xrightarrow{\text{Hexane, 1 hr, R.T.}} \text{Ph(CH}_3\text{)P–CH}_2\text{Li–TMEDA} \quad 65\%$$

$$(n\text{-}C_{12}H_{25})(CH_3)\text{P–CH}_3 + \text{BuLi–TMEDA} \xrightarrow{\text{Hexane, 6 hr, R.T.}} (n\text{-}C_{12}H_{25})(CH_3)\text{P–CH}_2\text{Li–TMEDA} \quad 44\%$$

rocene chemistry that this remarkable compound can be metalated with organolithium reagents, such reactions invariably lead to mixtures of mono- and dilithio intermediates. Rausch and Ciappenelli (11) subsequently found that conversion of ferrocene to 1,1'-dilithioferrocene can be done virtually quantitatively by treating ferrocene with slightly more than two equivalents of BuLi–TMEDA in hexane solution for 6 hrs. The 1,1'-dilithioferrocene prepared this way has proved to be an important intermediate in forming heteroannularly disubstituted ferrocene compounds. Thus, carbonation and hydrolysis of this dilithium reagent gives 1,1'-ferrocenedicarboxylic acid in 98% yield while reactions with either benzophenone or pyridine give 1,1'-bis(diphenylhydroxymethyl)-ferrocene and 1,1'-di(2-pyridyl)ferrocene in yields of 80% and 30%, respectively.

Subsequent studies (28) have led to the development of one-step, high-yield syntheses of 1,1'-diiodo-, 1,1'-dibromo-, and 1,1'-dichloroferrocene from reactions involving 1,1'-dilithioferrocene prepared with either the halogens, p-toluenesulfonyl halides, or polyhalogenated alkanes at

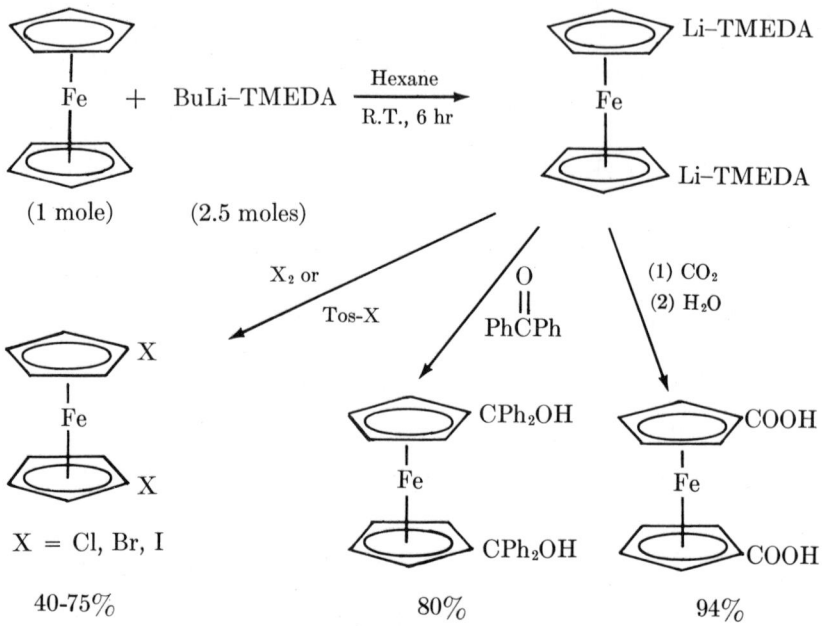

low temperatures. The ready availability of these 1,1'-dihalogenated ferrocenes in high purity has made possible a series of oligomeric 1,1'-polyferrocenes whose mixed-valence derivatives are being evaluated as organometallic semiconductors (29, 30, 31).

Rausch, Moser, and Meade recently isolated and characterized a series of N-chelated lithioferrocenes. The organolithium reagent obtained from the dimetalation of ferrocene with BuLi–TMEDA contains two molecules of chelating agent; the reagent is a very air-sensitive, pyrophoric solid. It can be stored in the solid state under nitrogen for long periods, however, and is a useful solid intermediate (32, 33). Ferrocenyllithium, isolated from a reaction between bromoferrocene and BuLi, surprisingly exhibits appreciable air stability although its TMEDA derivative is very air-sensitive.

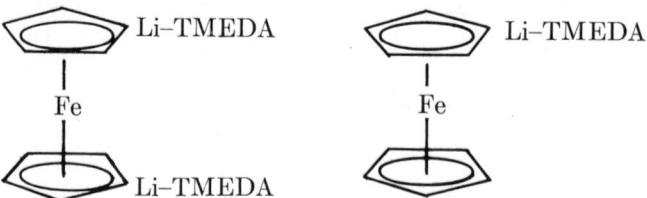

Studies of ferrocene polymetalation as well as the metalation of substituted ferrocenes using BuLi–TMEDA have also been reported recently. Halasa and Tate (*34*) metalated ferrocene using a 10-fold excess of BuLi–TMEDA in hexane at 70°C. They found that up to seven atoms of lithium can be introduced after five hours, as evidenced by deuteration and trimethylsilation.

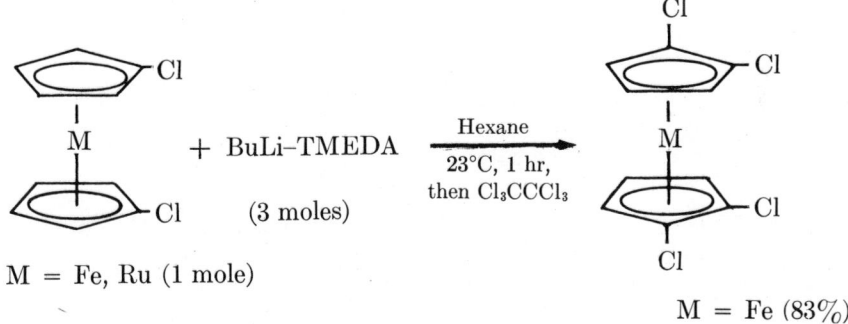

Hedberg and Rosenberg (*35*, *36*) have shown that 1,1'-dichloroferrocene and ruthenocene undergo dimetalation readily with BuLi–TMEDA. Treatment of the dimetalated intermediates with hexachloroethane gives the corresponding 1,1',2,2'-tetrachlorometallocenes in high yields, and these tetrahalogenated derivatives have served as intermediates to the very thermally and chemically stable perchlorometallocenes.

Huffman and Cope (*37*) recently used BuLi–TMEDA to metalate 2-chloromethylferrocene. The eventual products of this reaction are about equal amounts of 2-methyl- and 3-methylbutylferrocene, suggesting the possible intermediacy of methylferrocyne.

Elschenbroich (*38*) studied the metalation of di(benzene)chromium, using a 5:1 molar ratio of BuLi–TMEDA to this π-arene complex. The extent and orientation of metalation were studied by mass spectrometry of the products after quenching with D_2O. A lithium substituent on dibenzenechromium strongly activates the molecule toward further meta-

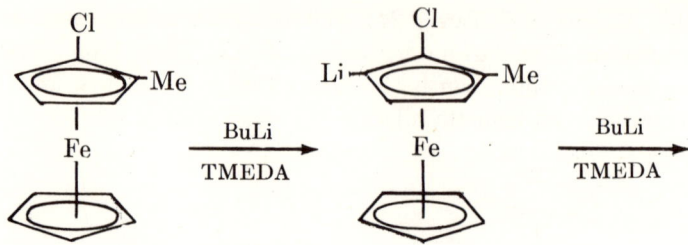

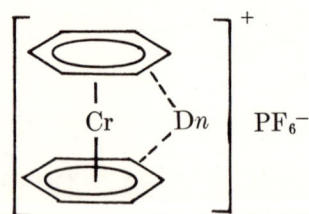 + BuLi–TMEDA $\xrightarrow{\text{(1) Cyclohexane, 70°C, 1 hr}}$ $\xrightarrow{\text{(2) D}_2\text{O, (3) O}_2; \text{NH}_4\text{PF}_6}$

(1 mole)   (5 moles)

$$\left[\begin{array}{c}\text{benzene–Cr–benzene} \cdots Dn\end{array}\right]^+ PF_6^-$$

Mass spectral analysis:

$C_{12}H_{12}Cr$	mono-D	di-D	tri-D	tetra-D
33%	11%	52%	2%	2%

lation, and heteroannular dimetalation is a major result even at early stages of the reaction. Competitive metalation studies of dibenzenechromium–benzene mixtures also demonstrated the enhanced metalation rate and therefore the increased kinetic C—H acidity of the π-complexed benzene ring. By contrast, BuLi itself does not react with dibenzenechromium, even after a prolonged interaction in various solvents (39).

Addition of a chelating diamine such as TMEDA to a metalation reaction using an organolithium reagent does not always lead to enhanced yields of the desired metalation product. Thus, Nesmeyanov et al. (40) reported in 1968 that benzene–tricarbonylchromium would undergo metalation with BuLi in tetrahydrofuran solution at $-40°C$; carbonation and hydrolysis give benzoic acid–tricarbonylchromium in 19% yield. Moser (41, 42) found that addition of an equivalent of TMEDA to this reaction mixture fails to produce any benzoic acid–tricarbonylchromium upon carbonation and hydrolysis, presumably because of prior decomposition of the organolithium intermediate that forms first under these conditions. When the metalation of benzene–tricarbonylchromium is carried out using BuLi–TMEDA in hexane, however, 17–20% yields of the complexed acid can be obtained.

Currently, the best route to phenyllithium–tricarbonylchromium is one in which the reaction between benzene–tricarbonylchromium and BuLi is conducted in 1:1 ethyl ether–tetrahydrofuran as the solvent system at $-40°C$ for one hour. Under these conditions, carbonation gives benzoic acid–tricarbonylchromium in 61% yield, and the intermediate organolithium reagent has been used to form other functionally substituted derivatives of benzene–tricarbonylchromium in good yield (41, 42).

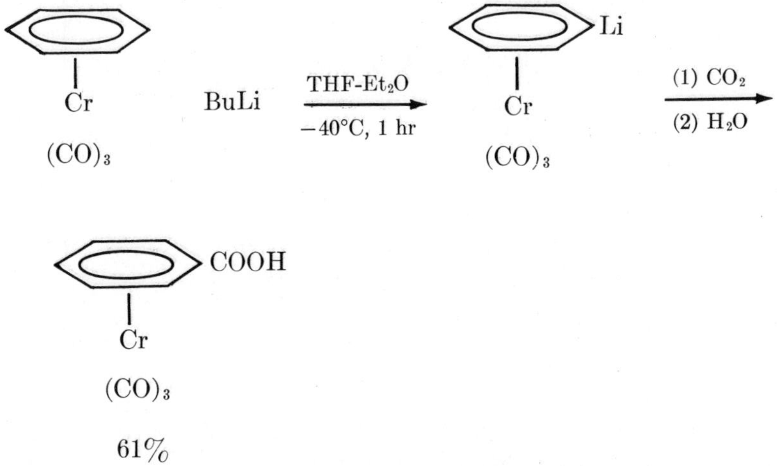

Nucleophilic Addition to Double and Triple Bonds

All the above reactions involve hydrogen–lithium interconversion (metalation). There are a limited amount of data available indicating that N-chelated organolithium intermediates also undergo nucleophilic addition to carbon–carbon double and triple bonds much more readily than does the organolithium reagent alone. Indeed, this enhanced reactivity toward addition reactions is a key factor in the telomerization of ethylene onto aromatic hydrocarbons (6, 7, 8, 9).

In 1966 Mulvaney et al. (43, 44) reported that diphenylacetylene reacted with two equivalents of BuLi in ethyl ether to produce a dilithium intermediate which, on hydrolysis, gives trans-α-n-butylstilbene in 39% yield. These workers also noted that a 45% recovery of diphenylacetylene can be realized.

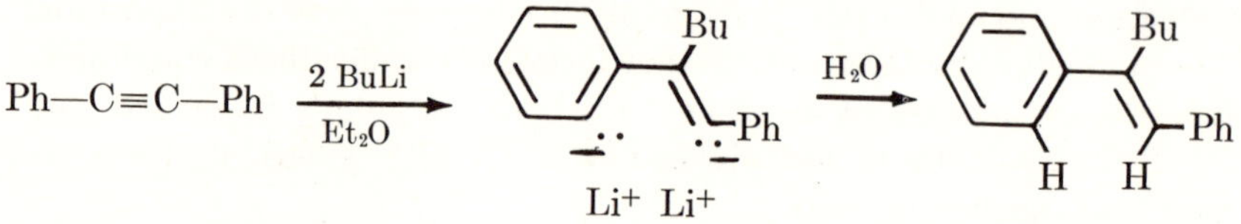

$$Ph-C \equiv C-Ph \xrightarrow{\text{2 BuLi}}_{\text{Et}_2\text{O}} \text{[dilithium intermediate]} \xrightarrow{\text{H}_2\text{O}} \text{[product]}$$

trans—α—n—butylstilbene
39% (+45% PhC≡CPh)

Klemann and Rausch (45) sought to improve the yields of the dilithium reagent since it appeared to be potentially useful for synthesizing metallocyclic compounds. Although BuLi and diphenylacetylene do not react over many hours in hydrocarbon solvents, a similar reaction in the

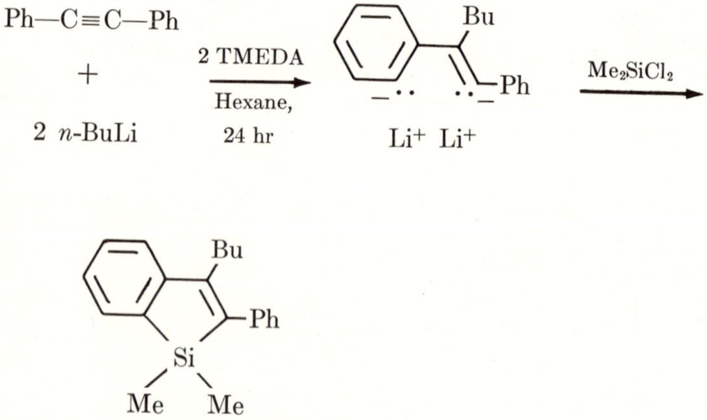

$$Ph-C \equiv C-Ph + 2\,n\text{-BuLi} \xrightarrow[\text{Hexane, 24 hr}]{\text{2 TMEDA}} \text{[dilithium]} \xrightarrow{\text{Me}_2\text{SiCl}_2}$$

92%

presence of two equivalents of TMEDA readily produces the desired dilithium reagent resulting from addition and ortho metalation. Subsequent reaction with dimethyldichlorosilane gives the metallocyclic product, 3-n-butyl-1,1,2-triphenyl-1-silaindene, in 92% yield. Similar reactions with titanocene dichloride, phenyldichlorophosphine, dimethyldichlorotin, MCl_4 (M = Si, Ge, Sn), etc., have also produced new metalloindene and spirocyclic derivatives (45, 46, 47).

Similar findings were later reported by Mulvaney and Newton (48), who noted that treating the dilithium reagent resulting from the addition-metalation reaction with deuterium oxide produces a 69% yield of trans-α-n-butylstilbene which contains 1.91 deuterium atoms per molecule. Further, treatment of diphenylacetylene with 2.5 moles of PhLi–TMEDA in hexane for 6 hrs at reflux followed by deuterolysis gives an 80% yield of triphenylethylene containing 1.62 deuterium atoms per molecule. By contrast, phenyllithium alone reacts very slowly (24 hrs) with diphenylacetylene in ethyl ether to give, after carbonation, only an 11% yield of triphenylacrylic acid.

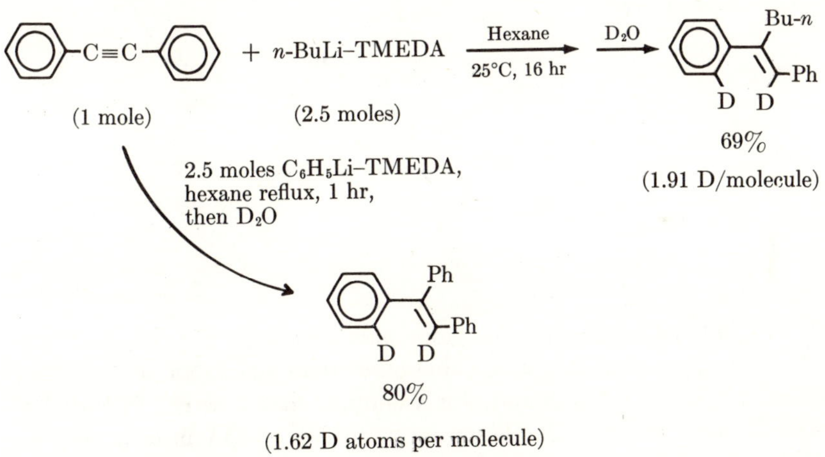

Another example of the greatly enhanced reactivity of BuLi–TMEDA compared with BuLi alone in addition reactions is exemplified by the relative reactivities of these reagents toward triphenylcyclopropene (49, 50). BuLi itself is unreactive toward this cyclic alkene; however, BuLi–TMEDA reacts readily under similar conditions. It was hoped that the powerful metalating capacity of the N-chelated organolithium reagent would produce triphenylcyclopropenyllithium, an "antiaromatic" anion. After treatment with deuterium oxide, however, the deep-red reaction mixture gave instead a 58% yield of an oil that was shown by NMR to contain four products (two sets of stereoisomers) resulting from addition of BuLi to the double bond of the cyclopropene ring followed by ring opening. Related metalation studies on *trans*-1,2,3-triphenylcyclopropane also have been carried out recently. Again, ring opening occurs to produce a mixture of *cis*- and *trans*-stilbenes.

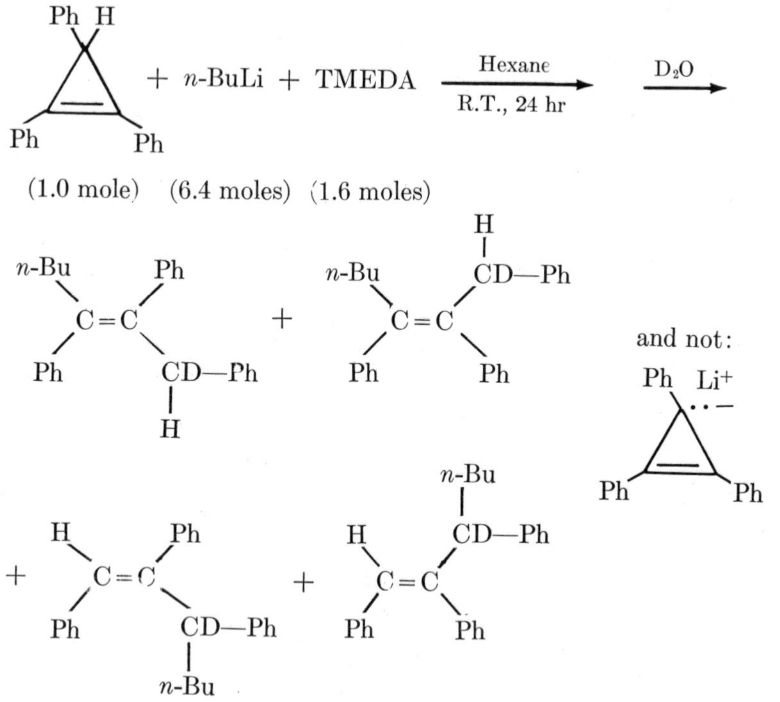

[Reaction scheme: Ph,H-substituted triphenylcyclopropene (1.0 mole) + n-BuLi (6.4 moles) + TMEDA (1.6 moles) → Hexane, R.T., 24 hr → D₂O →

Products:
- n-Bu/Ph C=C Ph/CD—Ph(H)
- n-Bu/Ph C=C Ph/Ph (with H—CD—Ph substituent)
- H/Ph C=C Ph/CD—Ph(n-Bu)
- H/Ph C=C Ph/Ph (with n-Bu—CD—Ph substituent)

and not: Ph,Ph-cyclopropene with Ph Li⁺]

N-Chelated organolithium compounds also undergo reactions having potentially important synthetic value other than metalation and addition. Peterson (*51, 52*) has found, for example, that whereas N,N-dialkylmethylamines undergo metalation with BuLi–TMEDA in only very low yields, the parent nitrogen-substituted organolithium compound N,N-dimethylaminomethyllithium can be readily formed in high yield by a transmetalation reaction between BuLi–TMEDA and (N,N-dimethylaminomethyl)tributyltin. This metal–metal interchange occurs readily at room temperature, and treatment of the resulting N,N-dimethylaminomethyllithium–TMEDA complex with benzaldehyde gives the corresponding carbinol in 86% yield. The method can also be extended to the synthesis of other N-substituted methyllithium reagents.

One report has also demonstrated the enhanced reactivity of BuLi–TMEDA compared with BuLi itself in the halogen–metal exchange reaction. Hallas and Waring (*53*) have reported that consistently good yields

$$n\text{-Bu}_3\text{Sn—CH}_2\text{—NMe}_2 + \text{BuLi–TMEDA} \xrightarrow[0°C, 1\ hr]{\text{Hexane}}$$
(1 mole)   (1 mole)

$(n\text{-Bu})_4\text{Sn} + \text{Me}_2\text{N—CH}_2\text{Li–TMEDA}$
Derivatized with C_6H_5CHO; 86%

of p-dimethylaminophenyllithium can be obtained by reaction between p-bromo-N,N-dimethylaniline and BuLi–TMEDA. Treatment of the intermediate organolithium complex with carbon dioxide gives a 70% yield of p-dimethylaminobenzoic acid while treatment with water gives N,N-dimethylaniline in 65% yield. Also, formation of p-dimethylaminophenyllithium from p-bromo-N,N-dimethylaniline and BuLi alone takes place in only moderate yield. Once again the enhanced reactivity of N-chelated organolithium reagents relative to unchelated RLi intermediates is demonstrated.

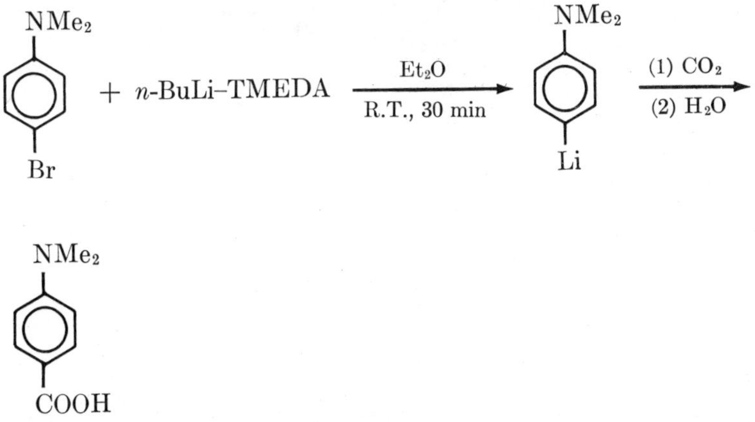

p-Br-C₆H₄-NMe₂ + n-BuLi–TMEDA $\xrightarrow[\text{R.T., 30 min}]{\text{Et}_2\text{O}}$ p-Li-C₆H₄-NMe₂ $\xrightarrow[(2)\ H_2O]{(1)\ CO_2}$

p-HOOC-C₆H₄-NMe₂

70%

Organosodium Complexes

Finally, a recent report demonstrates the very powerful metalating abilities of N-chelated organosodium complexes. Trimitsis et al. (54) found that an equimolar amount of TMEDA and n-amylsodium form a bright-blue suspension capable of converting dimethylarenes quantitatively into their α,α'-dianions at room temperature within two hours. Both 1,3-dimethylnaphthalene and m-xylene reacted with two equivalents of n-AmNa–TMEDA to form the corresponding disodium reagents in over 90% yield. This area also seems attractive for further study.

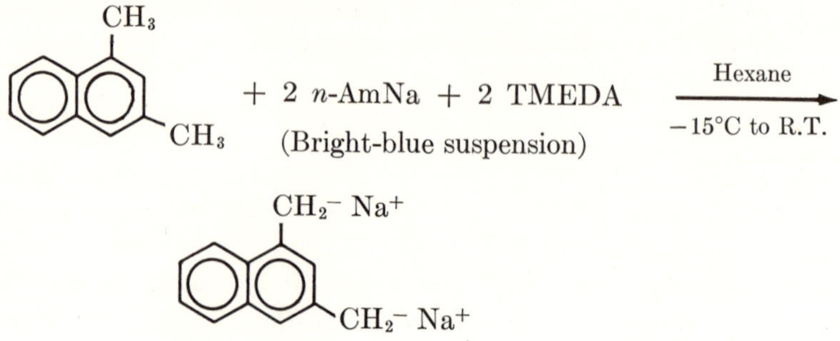

[Reaction scheme: 1,2-dimethylnaphthalene + 2 n-AmNa + 2 TMEDA (Bright-blue suspension), Hexane, −15°C to R.T. → 1,2-bis(CH$_2^-$Na$^+$)naphthalene]

CH$_3$I: 100%
CO$_2$: 74%
Benzophenone: 50%

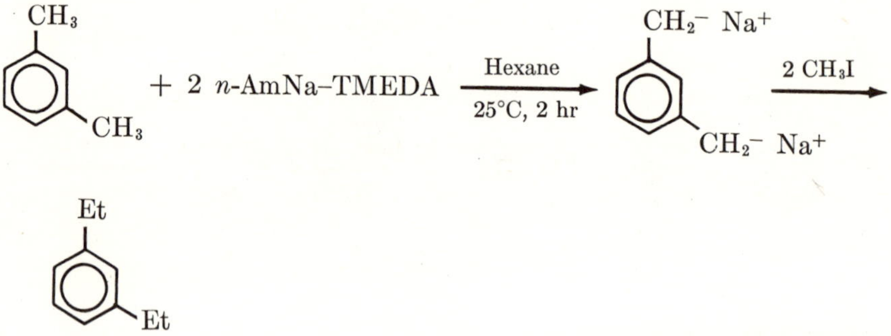

[Reaction scheme: m-xylene + 2 n-AmNa–TMEDA, Hexane, 25°C, 2 hr → 1,3-bis(CH$_2^-$Na$^+$)benzene, 2 CH$_3$I → 1,3-diethylbenzene]

90%

Acknowledgment

The authors thank A. W. Langer, Jr. for helpful discussions concerning this work and the National Science Foundation for financial support.

Literature Cited

1. Ziegler, K., Colonius, H., *Ann.* (1930) **479**, 135.
2. Gilman, H., Gorsich, R. D., *J. Org. Chem.* (1957) **22**, 687.
3. Gilman, H., Gray, S., *J. Org. Chem.* (1958) **23**, 1476.
4. Eastham, J. F., Gibson, G. W., *J. Amer. Chem. Soc.* (1963) **85**, 2171.
5. Brown, T. L., Gerteis, R. L., Bafus, D. A., Ladd, J. A., *J. Amer. Chem. Soc.* (1964) **86**, 2135.
6. Langer, Jr., A. W., *Trans. N.Y. Acad. Sci.* (1965) **27**, 741.
7. Langer, Jr., A. W., *Am. Chem. Soc., Div. Polymer Chem., Polymer Preprints* (1966) **7(1)**, 132.
8. Eberhardt, G. G., Butte, W. A., *J. Org. Chem.* (1964) **29**, 2928.
9. Eberhardt, G. G., Davis, W. R., *J. Polym. Sci., Part A* (1965) **3**, 3753.

10. Screttas, C. G., Eastham, J. F., *J. Amer. Chem. Soc.* (1965) **87**, 3276.
11. Rausch, M. D., Ciappenelli, D. J., *J. Organometal. Chem.* (1967) **10**, 127.
12. Chalk, A. J., Hoogeboom, T. J., *J. Organometal. Chem.* (1968) **11**, 615
13. Broaddus, C. D., *J. Org. Chem.* (1970) **35**, 10.
14. West, R., Jones, P. C., *J. Amer. Chem. Soc.* (1968) **90**, 2656.
15. Halasa, A. F., *J. Organometal. Chem.* (1971) **31**, 369.
16. Ludt, R. E., Crowther, G. P., Hauser, C. R., *J. Org. Chem.* (1970) **35**, 1288.
17. Ludt, R. E., Hauser, C. R., *J. Org. Chem.* (1971) **36**, 1607.
18. Slocum, D. W., Book, G., Jennings, C. A., *Tetrahedron Lett.* (1970) 3443.
19. Köbrich, G., Merkel, D., *Chem. Commun.* (1970) 1452.
20. Shirley, D. A., Cheng, C. F., *J. Organometal. Chem.* (1969) **20**, 251.
21. Ziegler, H. E., *J. Org. Chem.* (1966) **31**, 2977.
22. Ziegler, H. E., Laski, E. M., *Tetrahedron Lett.* (1966) 3801.
23. Corey, E. J., Seebach, D., *J. Org. Chem.* (1966) **31**, 4097.
24. Peterson, D. J., *J. Org. Chem.* (1967) **32**, 1717.
25. Peterson, D. J., *J. Organometal. Chem.* (1967) **9**, 373.
26. Gornowicz, G. A., West, R., *J. Amer. Chem. Soc.* (1968) **90**, 4478.
27. Peterson, D. J., *J. Organometal. Chem.* (1967) **8**, 199.
28. Kovar, R. F., Rausch, M. D., Rosenberg, H., *Organometal. Chem. Syn.* (1970/1971) **1**, 173.
29. Rausch, M. D., Roling, P. V., Siegel, A., *Chem. Commun.* (1970) 502.
30. Rausch, M. D., Roling, P. V., *J. Org. Chem.* (1972) **37**, 729.
31. Cowan, D. O., LeVanda, C., Pank, J., Kaufman, F., *Accounts Chem. Res.* (1973) **6**, 1.
32. Rausch, M. D., Moser, G. A., Meade, C. F., *J. Organometal. Chem.* (1973) **51**, 1.
33. Bishop, T. J., Davison, A., Katcher, M. L., Lichtenberg, D. W., Merrill, R. E., Smart, J. C., *J. Organometal. Chem.* (1971) **27**, 241.
34. Halasa, A. F., Tate, D. P., *J. Organometal. Chem.* (1970) **24**, 769.
35. Hedberg, F. L., Rosenberg, H., *J. Amer. Chem. Soc.* (1970) **92**, 3239.
36. Hedberg, F. L., Rosenberg, H., *J. Amer. Chem. Soc.* (1973) **95**, 870
37. Huffman, J. W., Cope, J. F., *J. Org. Chem.* (1971) **36**, 4068.
38. Elschenbroich, C., *J. Organometal. Chem.* (1968) **14**, 157.
39. Fritz, H. P., Fischer, E. O., *Z. Naturforsch.* (1957) **12b**, 67.
40. Nesmeyanov, A. K., Kolobova, N. E., Anisimov, K. N., Makarov, Yu. V., *Izv. Akad. Nauk SSSR, Ser. Khim.* (1968) **11**, 2665.
41. Moser, G. A., Ph.D. Thesis, University of Massachusetts, 1972.
42. Moser, G. A., Rausch, M. D., unpublished data.
43. Mulvaney, J. E., Gardlund, Z. G., Gardlund, S. L., *J. Amer. Chem. Soc.* (1963) **85**, 3897.
44. Mulvaney, J. E., Gardlund, Z. G., Gardlund, S. L., Newton, D. J., *J. Amer. Chem. Soc.* (1966) **88**, 476.
45. Rausch, M. D., Klemann, L. P., *J. Amer. Chem. Soc.* (1967) **89**, 5732.
46. Klemann, L. P., Ph.D. Thesis, University of Massachusetts, 1969.
47. Klemann, L. P., Rausch, M. D., unpublished data.
48. Mulvaney, J. E., Newton, D. J., *J. Org. Chem.* (1969) **34**, 1936.
49. Mulvaney, J. E., Savage, D., *J. Org. Chem.* (1971) **36**, 2592.
50. Jablonski, C., Rausch, M. D., unpublished work.
51. Peterson, D. J., *J. Organometal. Chem.* (1970) **21**, 63.
52. Peterson, D. J., *J. Amer. Chem. Soc.* (1971) **93**, 4027.
53. Hallas, G., Waring, D. R., *Chem. Ind. (London)* (1969) 620.
54. Trimitsis, G. B., Tuncay, A., Beyer, R. D., *J. Amer. Chem. Soc.* (1972) **94**, 2152.

RECEIVED March 26, 1973.

14

Asymmetric Synthesis *via* Lithium Chelates

THOMAS A. WHITNEY and ARTHUR W. LANGER, JR.

Corporate Research Laboratories, Esso Research and Engineering Co., Linden, N.J. 07036

> *Reactions of a variety of prochiral carbonyl substrates with Chel* · LiR, where Chel* = trans-N,N,N',N'-tetramethyl-1,2-cyclohexanediamine (TMCHD), were studied. Optically active carbinols were obtained and had an enantiomeric excess of up to 30% without sacrificing one asymmetric center to create a new one. Either of the absolute configurations of the product can be readily obtained by changing the absolute configuration of the chelating agent or by interchanging the R groups in the reaction of Chel* · LiR + R'COR''. For example:*

$$(-)\text{-TMCHD} \cdot \text{LiC}_4\text{H}_9 + \text{C}_6\text{H}_5\text{CHO} \rightarrow (-)\text{-C}_6\text{H}_5\text{CH(OH)C}_4\text{H}_9$$

$$(-)\text{-TMCHD} \cdot \text{LiC}_6\text{H}_5 + \text{C}_4\text{H}_9\text{CHO} \rightarrow (+)\text{-C}_6\text{H}_5\text{CH(OH)C}_4\text{H}_9$$

> *trans-1,2-Diaminocyclohexane (DACH) is a particularly attractive entry into optically active chelating agents for lithium reagents. Both enantiomers were obtained readily by an improved resolution procedure.*

Asymmetric synthesis has been investigated since Emil Fischer's classic publication on sugar chemistry in 1894 (*1*) and has since been the subject of numerous studies (*2, 3*). Marckwald (*4*) defined asymmetric synthesis as "those reactions which produce optically active substances from symmetrically constituted compounds with the intermediate use of optically active materials but with the exclusion of all other analytical processes." A broader definition of asymmetric synthesis is "a process which converts a prochiral unit into a chiral unit so that unequal amounts of stereoisomeric products result" (*see* Ref. 3, p. 5).

Of the various schemes for achieving asymmetric syntheses in reactions other than polymerizations, two have received considerable attention: enzymatic reactions (5) and reactions involving hydride transfer from the alpha or beta position of an optically active organometallic reagent:

• Biochemical methods

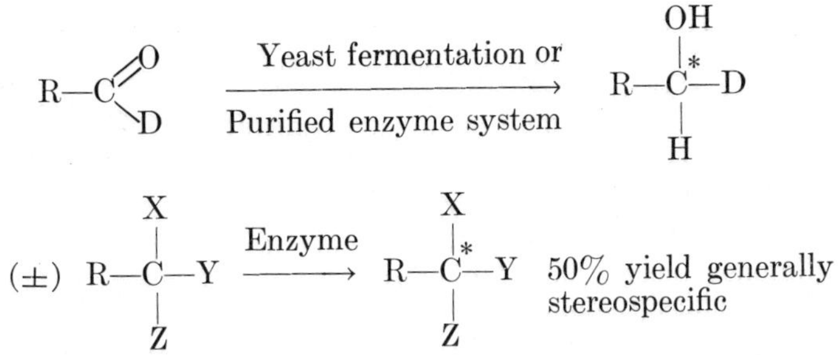

$$R-C{\overset{O}{\underset{D}{\lessgtr}}} \xrightarrow{\text{Yeast fermentation or Purified enzyme system}} R-\underset{H}{\overset{OH}{\underset{|}{C^*}}}-D$$

$$(\pm)\ R-\underset{Z}{\overset{X}{\underset{|}{C}}}-Y \xrightarrow{\text{Enzyme}} R-\underset{Z}{\overset{X}{\underset{|}{C^*}}}-Y \quad 50\%\ \text{yield generally stereospecific}$$

• Hydride transfer

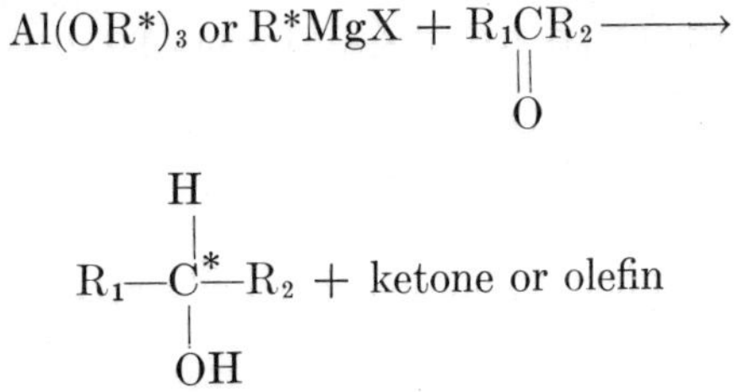

$$Al(OR^*)_3 \text{ or } R^*MgX + R_1\underset{O}{\overset{||}{C}}R_2 \longrightarrow$$

$$R_1-\underset{OH}{\overset{H}{\underset{|}{C^*}}}-R_2 + \text{ketone or olefin}$$

Both methods, however, have disadvantages. Biochemical transformations can have limited application, and there is always the problem of finding the proper bacteria, animal preparation, or enzyme and culture medium to effect a new synthesis. In addition, product isolation—such as in the production of an optically active α-deuteroalcohol, where a small amount of product must be isolated from a large quantity of spent fermentation liquor—can present formidable separation problems. Product isolation from enzyme systems, especially immobilized enzymes, could be much simpler, however.

Hydride-transfer reactions suffer from the several shortcomings. First, a conventional optical resolution must usually be performed to obtain an optically active carbinol, which is then converted to the halide when the Grignard method is to be used. The actual reduction is generally not the only reaction pathway; hence carbinol by-product is produced. More undesirable, however, is the fact that the asymmetric center of the organometallic reagent is sacrificed when the new chiral center is created. Unless the reaction is stereospecific, which is rarely the case, a net overall decrease in chirality results.

While this work was in progress an alternate method of asymmetric synthesis *via* hydride transfer was reported, in which the asymmetric center of the chiral moiety is not sacrificed (*3*, p. 204). This method uses the reaction product of $LiAlH_4$ and varying amounts of optically active amino carbinols, such as (−)-quinine, (+)-cinchonidine, and (−)-ephedrine, to reduce prochiral substrates. In this system the hydride anion species is sigma bonded to the optically active residue, and a maximum of three hydrides are available for further reaction. The aminocarbinols could sometimes be recovered for reuse. In the instant system the chiral chelating agent forms coordinate bonds to the lithium *cation,* and four hydrides are available for subsequent reaction.

Conceptually, an optically active, asymmetric lithium compound, Chel* · LiR (where Chel* denotes the optically active, chelating agent) should induce stereoselective reactions at the Li-R bond. This should occur since reaction can proceed *via* two diastereomeric transition states of unequal energy. If an energy difference of about 2 kcal/mole could be achieved, 100% optical bias could be realized. Nevertheless, the optically active chelates could thus be used to prepare optically active products in electrophilic reactions without destroying one asymmetric center to create a new one as the chelating agent could be recovered unchanged and recycled:

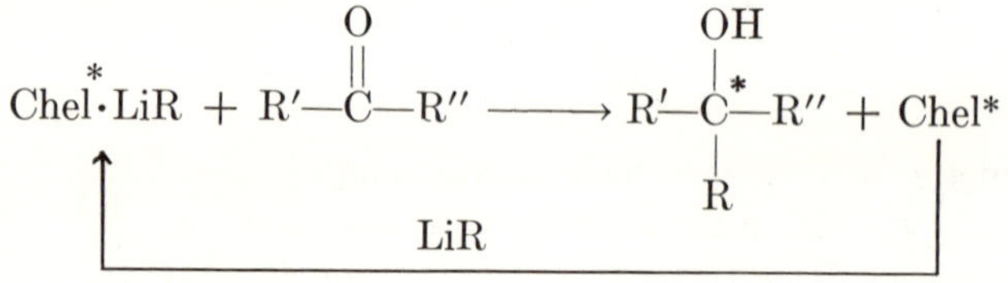

$$\text{Chel}^*\cdot\text{LiR} + R'-\underset{\|}{\overset{O}{C}}-R'' \longrightarrow R'-\underset{R}{\overset{OH}{\underset{|}{\overset{|}{C^*}}}}-R'' + \text{Chel}^*$$

(with LiR recycling)

Chel* Precursor

Before attempting asymmetric syntheses *via* the above scheme, careful thought was given to the choice of the Chel* precursor. It was deemed that (a) the compound should be a racemic mixture (*6*); (b) resolution should be easy—that is, very high optical purity should be obtained from only one crystallization of an appropriate salt; (c) inexpensive resolving agents (*e.g.,* tartaric acid) should be used; (d) the Chel* precursor should be easily resolvable even when grossly chemically impure; (e) both enantiomers should be obtainable in very high optical purity; and (f) absolute configuration of the compound should be known.

These considerations led to the choice of *trans*-1,2-diaminocyclohexane (DACH) as the optimum initial Chel* precursor since both (R,R)-(−)-DACH and (S,S)-(+)-DACH may be obtained from the racemic mixture *via* the (+)-tartrate and (+)-bitartrate salts, respectively (*7*,

8, 9). The literature procedures were followed initially to separate *cis*- and *trans*-DACH (*10*) and to resolve the latter. Variations of the published procedure (*8*) were studied to determine the effect on optical yield of the (−)-antipode. The best results were obtained when the reaction was run with no special precautions. Purification of the DACH was found to be unnecessary.

(S,S)-(+)-DACH is less readily available than (−)-DACH. The former is initially obtained from the mother liquor as an optically impure (+)-bitartrate, which is converted to the dihydrochloride; the latter salt is fractionally crystallized repeatedly from water and, finally the (+)-DACH–2HCl salt is mechanically separated from the featherlike aggregates of the racemic salt (*8*). This cumbersome procedure was found to be unnecessary to secure (+)-DACH of high optical purity. By taking advantage of the racemic mixture property of DACH, less than 50% chemically and optically pure (+)-DACH is readily upgraded by fractional crystallization from the melt or hydrocarbon solution. Furthermore, (+)-DACH of very high optical purity could be obtained by a single crystallization of the neutral salt of unnatural (−)-tartaric acid. Thus facile procedures were developed for preparing both DACH antipodes inexpensively and in quantity.

Eschweiler-Clarke (*11*) methylation of (+)- and (−)-DACH gave (R,R)-(−)- and (S,S)-(+)-*N,N,N',N'*-tetramethylcyclohexanediamine [(+)- and (−)-TMCHD] in high yield.

Results and Discussion

Previous investigations have shown that chelated organolithium reagents are highly reactive and synthetically versatile (*12, 13, 14*). In addition, the chemistry of chelated complex metal hydrides has been investigated, including their use for reducing carbonyl compounds (*15*). The results of this investigation of the reaction of optically active chelated lithium compounds and prochiral carbonyl substrates are summarized in Table I.

The results summarized in the table were obtained without our trying to optimize reaction conditions for maximum stereospecificity. Generally, the reactions were begun at −75° to −80°C. After all reactants were combined, the reaction mixture was held at that temperature for 30 minutes, then allowed to warm to room temperature. This procedure was followed mainly to study the effect of ketone structure on the optical yield of the carbinol product. Although the effect of temperature on reaction stereospecificity was not studied in detail, comparison of the results of runs 6 and 9 suggest that lower temperatures should give

Table I. Summary of Reactions of Optically Active

Run	Chelate	Substrate
1	(−)-TMCHD·LiC$_4$H$_9$	C$_6$H$_5$CHO
2	(−)-TMCHD·LiC$_6$H$_5$	C$_4$H$_9$CHO
3	(−)-TMCHD·LiAlH$_4$	C$_6$H$_{13}$COCH$_3$[b]
4	(−)-TMCHD·LiAlH$_4$	C$_6$H$_{13}$COCH$_3$[c]
5	(−)-TMCHD·LiAlH$_4$	C$_6$H$_5$COC$_4$H$_9$
6	(+)-TMCHD·LiAlH$_4$[d]	C$_6$H$_{13}$COCH$_3$[c]
7	(−)-TMCHD·LiAlH$_4$	α-Tetralone[c]
8	(−)-TMCHD·LiAlH$_4$	β-Tetralone
9	(−)-TMCHD·LiAlH$_4$	C$_6$H$_{13}$COCH$_3$[g]
10	(−)-TMCHD·LiAlD$_4$	C$_6$H$_5$CHO[b]

Chelated Lithium Compounds and Prochiral Substrates

Product	$[\alpha]^{25}_{589}$	Optical Purity %
OH | C_6H_5C-C_4H_9 | H	$-2.68°$ (C, 14.3, B^a)	8.65
OH | $C_6H_5CC_4H_9$ | H	$+2.98°$ (C, 13.3, B^a)	9.5
OH | $C_6H_{13}CCH_3$ | H	$-1.07°$ (C, 13.5, B^a)	10.7
OH | $C_6H_{13}CCH_3$ | H	$-1.17°$ (C, 13.5, B^a)	11.7
OH | $C_6H_5CC_4H_9$ | H	$+1.75°$ (C, 13.7, B^a)	5.6
OH | $C_6H_{13}CCH_3$ | H	$+1.06°$ (C, 14.4, B^a)	10.6
α-Tetralole	$-0.97°^e$ (C, 2.50, C^b)	3.9
β-Tetralole	$-2.32°$ (C, 7.8, C^f)	8.2
OH | $C_6H_{13}CCH_3$ | H	$-0.40°$ (C, 13.3, B^a)	4.0
OH | C_6H_5C-D | H	$-0.16°$ (Neat)	10.3

Table I.

Run	Chelate	Substrate
11	(−)-TMCHD·LiAlH$_4$	C$_6$H$_5$COCHO^i
12	(−)-TMCHD·LiAlH$_4$	HOCH$_2$CH$_2$COCH$_3$^{b}
13	(−)-TMCHD·LiAlH$_4$	HO(CH$_2$)$_3$COCH$_3$^{b}
14	(−)-TMCHD·LiAlH$_4$	C$_6$H$_5$COCH$_3$^{j}

^{a} B = benzene.
^{b} Molar ratio of chelate to subtrate = 1:2.
^{c} Molar ratio of chelate to substrate = 1:4.
^{d} The (+) −TMCHD had [α]$^{25}_{589}$ + 51.4° (C, 5.35, 95% EtOH) or 97% optical purity.
^{e} Rotation taken at 17°C.

higher optical yields. Particular attention was paid to complete removal of the residual optically active chelating agent from the product.

Comparison of the results from runs 1, 2, and 5 shows that the absolute configuration of the product can be varied without changing the absolute configuration of the chelating agent. The same result is achieved with the latter change (*cf.* run 3 with 6). The use of optically active chelated LiAlD$_4$ constitutes a very facile route to optically active α-deuteroalcohols (run 10). α-Deuteroalcohols have previously been prepared by reduction of deuteroaldehydes in actively fermenting media (*16*), with isolated enzyme systems, and by asymmetric reductions of aldehydes by chiral Grignard reagents *via* hydride transfer (*17*). Both methods suffer from the disadvantages discussed at the beginning of this paper.

The size of the R groups in R′COR″ influences the degree of stereospecificity of these reactions, as the results of runs 5, 10, and 14 show when

(Continued)

Product	$[\alpha]_{589}^{25}$	Optical Purity %
$C_6H_5\overset{\overset{\displaystyle OH}{\|}}{\underset{\underset{\displaystyle H}{\|}}{C}}CH_2OH$	+4.91° (C, 4.03, E^h)	8.3
$HOCH_2CH_2\overset{\overset{\displaystyle OH}{\|}}{\underset{\underset{\displaystyle H}{\|}}{C}}CH_3$	+3.34° (C, 4.03, E^h)	~30
$HO(CH_2)_3\overset{\overset{\displaystyle OH}{\|}}{\underset{\underset{\displaystyle H}{\|}}{C}}CH_3$	+0.257° (Neat)	
$C_6H_5\text{-}\overset{\overset{\displaystyle OH}{\|}}{\underset{\underset{\displaystyle H}{\|}}{C}}CH_3$	+2.94° (C, 13.14, B^a)	7.4

^{f} C = chloroform.
^{g} Reaction run at room temperature.
^{h} E = 95% ethanol.
^{i} Molar ratio of chelate to substrate = 3.2.
^{j} Runs 1 and 2 were in pentane; all others were in toluene.

R″ is varied from H to CH$_3$ to n-C$_4$H$_9$. From these limited results it seems that the greater the difference in size between R′ and R″, the greater will be the stereospecificity. However, other variables—such as the structure of the asymmetric chelating agent—also have an important influence on reaction stereospecificity. This variable is under study.

The difference in the optical yield upon reduction of α-tetralone vs. β-tetralone with (−)-TMCHD · LiAl$_4$ indicates that the stereochemical outcome of a given reaction may be very sensitive to small changes in the steric environment around the prochiral center. Noteworthy is the result of run 12, where 30% optical purity was achieved, which is considerably higher than that of all the other runs (with the possible exception of run 13). The result of run 12 suggests that when other functional groups capable of reacting with Chel* · LiR are present in the substrate, they can have a strong influence on the overall stereochemical outcome.

In the reduction of 1-hydroxy-3-butanone the reaction can be envisioned as proceeding intramolecularly *via* a six-membered ring intermediate formed by an earlier reaction of the hydroxyl group with AlH_4^-, giving a H_3AlOCH_2 species. The activation energy difference between the two diastereomeric transition states for intramolecular carbonyl reduction might then be greater than that for direct attack on the carbonyl in an intermolecular reduction.

An attempt to obtain evidence for this hypothesis was made. Reduction of 1-hydroxyl-4-pentanone might proceed *via* a stereochemically less favorable, seven-membered ring intermediate, and the product (1,4-pentanediol) of much less than 30% optically purity might result. Although optically active diol was obtained, no assignment of optical purity could be made since the diol could not be transformed stereospecifically, despite several attempts, into 2-methyl-tetrahydrothiophene-1-dioxide, whose maximum rotation is known (18). The validity of the above hypothesis thus remains moot.

One reaction in the literature with which the TMCHD chelates can be directly compared in terms of optical yield is that studied by Nozaki (19), in which sparteine · Li-n-C_4H_9 reacted with benzaldehyde. 1-Phenyl-1-pentanol was obtained in 6% optical purity. The optical yields obtained in the present study were generally higher. In addition, sparteine is a natural product occurring in a plant called "broom tops" and is available in only one absolute configuration, thereby limiting its utility.

Summary

The results of this study suggest that optically active chelated lithium reagents may be used generally for asymmetric synthesis according to the scheme:

$$\text{Chel*}\cdot\text{LiR} + \text{R}'\text{-}\overset{\overset{\text{E}}{\|}}{\text{C}}\text{-R}'' \xrightarrow{\text{H}^+} \text{R-}\overset{\overset{\text{R}'}{\|}}{\underset{\underset{\text{R}''}{|}}{\text{C}^*}}\text{-EH} + \text{Chel*}$$

The chelating agent may then be recovered unchanged and reused, as was done many times during this work. Either of the absolute configuartions of the chiral product may be obtained at will, either by varying the absolute configuration of Chel* or by varying the mode of synthesis. As additional results are accumulated on a variety of substrates and types of reactions, it may be possible to predict with confidence the stereochemical outcome of a particular reaction. As additional insight is gained into the factors critical to stereospecificity, perhaps optical yields approaching enantiomeric purity will be realized.

Experimental

Resolution of *trans*-1,2-Diaminocyclohexane (DACH). A total of 1000 grams (8.76 moles) DACH (Adams Chemical Co.), 1323 grams (8.76 moles) (+)-tartaric acid, and 6 liters water were used with the Asperger procedure (*8*). Crop 1 tartrate separated, 542 grams, upon cooling to 0°C. The mother liquor was concentrated to about 4.5 liters, and crop 2 separated, 267 grams. Further concentration of the mother liquor to about 2.5 liters gave crop 3, 180 grams.

Optically active (−)-DACH was recovered from the tartrate salt by adding the latter to an excess of aqueous NaOH and continuously extracting the mixture with benzene under nitrogen. Crop 1 gave 216 grams distilled (−)-DACH, bp 71°–73°C/8 mm, $[\alpha]_{589}^{25}$ −40.3° (C, 5.23, benzene), corresponding to 97% optical purity as determined from a sample of optically pure (−)-DACH · 2HCl having $[\alpha]_{589}^{25}$ −15.6° (C, 0.20 gram per ml H_2O) (*8*).

The mother liquor remaining after crop 3 (−)-DACH tartrate separated was treated as described for crop 1 tartrate, and 540 grams of distilled (+)-DACH were recovered $[\alpha]_{589}^{25}$ + 20.3° (C, 5.05, benzene). The material was placed in a Schlenk tube, which was then placed in a constant temperature bath at 20°C. The temperature of the bath was lowered slowly to 9°C over 19 days as crystals grew. The tube was inverted, and the solids were filtered from the mother liquor. The arm of the Schlenk tube containing the solids was heated, and the molten (+)-DACH was removed from the tube with a pipette. It displayed $[\alpha]_{589}^{25}$ + 38.7° (C, 5.32, benzene), which is 94% optically pure; 137.9 grams were obtained. The mother liquor, $[\alpha]_{589}^{25}$ + 13.4° (C, 5.03), 331 grams, was charged into a new Schlenk tube and put back into the bath at 9°C. The bath temperature was lowered over 18 days to −3°C, as a second crop of crystals formed; these crystals were recovered and melted. The material displayed $[\alpha]_{589}^{25}$ + 36.2° (C, 5.23, benzene) or 87.5% optical purity, wt 56.2 grams. The mother liquor displayed $[\alpha]_{589}^{25}$ + 8.22° (C, 5.09, benzene).

Preparation of (+)- and (−)-*N,N,N′,N′*-Tetramethyl-1,2-cyclohexanediamine ((+)- and (−)-TMCHD). The Eschweiler-Clarke (*11*) procedure was used with formaldehyde and formic acid. A 90% yield of (+) and (−)-TMCHD was obtained, having $[\alpha]_{589}^{25}$ ± 17.2° (neat), d = 0.888; $[\alpha]_{589}^{25}$ ± 20.0 (C, 5.06, benzene).

Asymmetric Syntheses (Run 14). A charge of 0.19 gram (5 mmoles) LiAlH₄, 25 ml toluene, and 0.85 gram (5 mmoles) (−)-TMCHD, $[\alpha]_{589}^{25}$ − 17.2° (neat) (100% optically pure) was stirred in a beaker for one hour at room temperature. The turbid gray mixture was cooled to −80°C, and a solution of 1.20 grams (10 mmoles) acetophenone in 10 ml of toluene was added dropwise while the reaction mixture was maintained at −70° to −80°C. When acetophenone addition was complete, the reaction mixture was maintained at −70° to −80°C for about 30 minutes, then allowed to warm to 0°C. Water, 5 ml, was added, followed by 30 ml of 1N HCl. The liquid phases were separated, and the aqueous phase was extracted with 15 ml pentane. The combined organic phase was then extracted with 15 ml 1N HCl, 15 ml 10% $NaHCO_3$ solution, 15 ml H_2O, dried over Na_2SO_4, and finally concentrated on a rotary evaporator. By

VPC analysis, the product was 92% 1-phenyl-1-ethanol and 7.4% toluene; no (−)-TMCHD was present. The optical activity of the product was measured with a Perkin Elmer model 141 polarimeter: $[\alpha]_{589}^{25} + 2.94°$ (C, 13.14, benzene), corresponding to 7.4% optical purity by direct comparison with an authentic sample of optically pure 1-phenyl-1-ethanol. The other reactions summarized in the table were run similarly, with no attempt made to optimize reaction conditions to obtain maximum stereospecificity.

Literature Cited

1. Fischer, E., *Ber.* (1894) **27**, 3231.
2. Ritchie, P. D., "Asymmetric Synthesis and Asymmetric Induction," Oxford University Press, London, 1933.
3. Morrison, J. D., Mosher, H. S., "Asymmetric Organic Reactions," Prentice-Hall, Englewood Cliffs, N. J., 1971.
4. Marckwald, W., *Ber.* (1904) **37**, 1368.
5. Bentley, R., "Molecular Asymmetry in Biology," Academic, New York, Vol. I, 1969; Vol. II, 1970.
6. Eliel, E. L., "Stereochemistry of Carbon Compounds," McGraw-Hill, New York, 1962.
7. Jaeger, F. M., Bijkerk, L., *Proc. Kon. Ned. Akad. Wetensch.* (1937) **40**, 12.
8. Asperger, R. G., Liu, C. F., *Inorg. Chem.* (1965) **4**, 1492.
9. Woldbye, F., *Rec. Chem. Progr.* (1964) **24**, 197.
10. Smith, A. J., U.S. Patent **3,163,675** (1964).
11. Clarke, H. T., Gillespie, H. B., Weisshaus, S. Z., *J. Amer. Chem. Soc.* (1933) **55**, 4571.
12. Langer, Jr., A. W., *Trans. N.Y. Acad. Sci.* (1965) **27** (7), 741.
13. Langer, Jr., A. W., U.S. Patent **3,451,988** (1969); **3,541,149** (1970).
14. Rausch, M. D., Sarnelli, A. J., Advan. Chem. Ser. (1973) **130**, 248.
15. Langer, Jr., A. W., Whitney, T. A., U.S. Patent **3,734,963** (1973).
16. Althouse, V. E., Feigl, D. M., Sanderson, W. A., Mosher, H. S., *J. Amer. Chem. Soc.* (1966) **88**, 3595.
17. Clark, D. R., Ph.D. Thesis, Stanford University, D. A. No. 71-19,662 (1970).
18. Cram, D. J., Whitney, T. A., *J. Amer. Chem. Soc.* (1967) **89**, 4651.
19. Nozaki, H., Aratani, T., Toraya, T., *Tetrahedron Lett.* (1968) 4097.

Received February 12, 1973.

INDEX

INDEX

A

Acidity, kinetic	192
Acidity, thermodynamic	9
Acids, metalation of weak	7
Active compounds, optically	235, 274
Addition rate, effect of butadiene	208
Addition to double and triple bonds, nucleophilic	264
Agent	
–cation interaction, chelating	120
effect of chelating	133, 204
on ion paring, effect of chelating	125
metalation of the chelating	10
steric effects of the chelating	15
Agents, charge-transfer	138
Agents, skeletal structures of polytertiary amine chelating	114
Alcohols, preparation of oxo	208
Aliphatic amines, chelating ditertiary	249
Aliphatic polylithio compounds	211
Alkali metal	
catalysts, polymerization using N-chelated	163
complexes, stereochemical properties of N-chelated	56
compounds, magnetic resonance studies of polytertiary amine chelated	113
compounds, U.S. patents on N-chelated	18
reactions of	57
Alkylaromatic lithiation, aromatic and	214
Alkyllithium polymerizations	171
Allyl anion	204
Allyllithium	
and crotyllithium from olefins, preparation of	38
and crotyllithium–TMEDA complexes, preparation of	38, 39
–TMEDA and crotyllithium–TMEDA, reaction of	40, 41
–TMEDA solid complex, preparation of	49
Amine(s)	
bridgehead-type	187
chelated alkali metal compounds, magnetic resonance studies of polytertiary	113
-chelated organolithium reagents, stereochemical properties of	58

Amine(s) *(Continued)*	
chelating ditertiary aliphatic	249
mixtures containing primary and secondary	153
structure and concentration, effect of	192
Ammonium salts, quaternary	140
Anion	
allyl	204
generation on the polymer backbone	179
intermediate, allyl	205
Anionic	
graft copolymers	177
initiator	184
techniques, grafting by	177
techniques, metalation by	177
Anthracene dianion, molecular orbitals for the	105
Applications of N-chelated organolithium compounds, synthetic	248
Aromatic	
and alkylaromatic lithiation	214
compounds, polylithiation of	214
hydrocarbons, metalations of	250
hydrocarbons, telomerization of ethylene with	189
polycyclic	217
telomer waxes	194, 196
Aromatics, poly-n-alkyl	196
Arrhenius equation	136
Asymmetric induction	235
Asymmetric synthesis *via* lithium chelates	270
Atomic charge distribution for the benzyl carbanion	87
Atoms, metalations of organic molecules containing hetero	256

B

Backbone, anion generation on the polymer	179
Base(s)	
chelating	1
coordination, metal–	77
interactions, metal–	79
Benzene	
by n-butyllithium–n-TMEDA, lithiation of	46
chelated lithium salts in	159

Benzene *(Continued)*
  with lithium chelates, stereospecific interaction of 126
  metalation 251
  preparation of phenyllithium from 25
  preparation of phenyllithium–TMEDA in 48
  -soluble inorganic complexes .. 158
Benzyl carbanion, atomic charge distribution for the 87
Benzyllithium–TMEDA
  complexes, ring-isomer content in 49
  complexes, isomer content of .. 34, 35
  lithiation of benzene by n- ... 46
  TED, reactions of 36, 37
  TED solid complex, preparation of 49
(Benzyllithium)₂–TMEDA in toluene 34
  preparation of 48
  reactions of 34, 35
Biochemical transformations 271
Birefringence, crystalline 195
Block copolymers, SBS 183
Bonding, covalent 12
Bond(s)
  carbon–halogen 230
  carbon–nitrogen 230
  carbon–oxygen 230
  formation of carbon–carbon ... 228
  lengths, naphthalene 67
  nucleophilic addition to double and triple 264
  stereoselective reactions at the Li–R 272
Bridgehead-type amines 187
BuLi
  hydrogenolysis rate 7
  ortho metalation with 254
  –TMEDA
    in the halogen–metal exchange reaction 266
    lithiation of toluene with n- .. 216
Butadiene
  addition rate, effect of 208
  concentration, effect of temperature and 207
  polymerization, effect of ion pair structure on 12
  polymerization of 173
  telomerization 205
Butyllithium–tertiary diamine complexes, toluene metalation by n- 31

C

Carbanion
  atomic charge distribution for the benzyl 87
  geometries 76
  moiety 193

Carbanion *(Continued)*
  organometallic complexes, structural properties of 60
  system, delocalized 56
Carbon
  –carbon bonds, formation of ... 228
  chains, solid paraffinic 199
  –halogen bonds 230
  –nitrogen bonds 230
  –oxygen bonds 230
Carbonyl compounds, chelated complex metal hydrides for reducing 273
Catalysis, N-chelated organolithium 1
Catalyst(s)
  chelated organosodium 3, 201
  compositions, nonstoichiometric.. 206
  efficiency 178
  polymerizations using N-chelated alkali metal 163
Catalytic behavior of N-chelated organolithium reagents 58
Cation interaction, chelating agent– 120
Chain(s)
  polyethylene waxes, straight ... 165
  solid paraffinic carbon 199
  transfer, factors affecting the ... 202
  transfer mechanisms 15
Charge distribution for the benzyl carbanion, atomic 87
Charge-transfer agents 138
Chel* precursor 272
Chelated
  alkali metal compounds, magnetic resonance studies of polytertiary amine 113
  complex metal hydrides for reducing carbonyl compounds 273
  lithium
    compounds, reactions of optically active 274
    halides 115
    salts in benzene 159
    salts, polytertiary amine 115
  organolithium catalysts 3
  organolithium compounds, synthetic applications of N- .. 248
  organolithium reagents, stereochemical properties of amine 58
  organosodium catalysts 201
  salts, structural features of 120
  sodium naphthalenide 117
Chelates
  asymmetric synthesis *via* lithium 270
  organolithium 5
  properties of 6
  stereospecific interaction of benzene with lithium 126
  structure of lithium 3
Chelating
  agent
    –cation interaction 120
    effect of 133, 204

INDEX

Chelating *(Continued)*
  agent
    on ion pairing, effect of 125
    metalation of the 10
    skeletal structures of polytertiary amine 114
    steric effects of the 15
  bases 1
  ditertiary aliphatic amines 249
  polyamines, tertiary 202
  polyethers 1
  tertiary polyamines 1
  -type diamines 187
Chemical shift, α-methylene proton 7
Chromatography (GPC), gel permeation 181
Cis-trans isomerization, photochemical studies of 73
Complexation of *trans*-TMCHD, selective 145
Complex
  dissociation 156
  metal hydrides for reducing carbonyl compounds, chelated 273
  preparation of
    allyllithium-TMEDA solid ... 49
    benzyllithium-TED solid 49
Complexes
  benzene-soluble inorganic 158
  contact-ion pair 61
  crystalline organolithium 6
  inorganic 142
  isomer content of benzyllithium–TMEDA 34, 35
  metalations of organometallic π 258
  organosodium 267
  preparation of
    allyllithium– and crotyllithium–TMEDA 38, 39
    the tertiary diamine 24
    proton NMR spectra of sodium iodide 151
  solubility of organolithium–tertiary diamine 28
  stability of organolithium–TMEDA 29
  stereochemical properties of N-chelated alkali metal ... 56
  structural properties of carbanion organometallic 60
  synthesis and isolation of 58
  synthetic reactions of organolithium–tertiary diamine 50
  tertiary diamine organolithium .. 23
  toluene metalation by *n*-butyllithium–tertiary diamine complexes 31
Components of a light distillate telomer 197
Compositions, nonstoichiometric, catalyst 206
Compounds
  aliphatic polylithio 211

Compounds *(Continued)*
  chelated complex metal hydrides for reducing carbonyl 273
  N-chelated organoalkali metal .. 7
  conductivities of organolithium.. 133
  optically active 235
  polylithiation of aromatic 214
  reactions of optically active chelated lithium 274
  synthetic applications of N-chelated organolithium 248
  U.S. patents on N-chelated alkali metal 18
Concentration, effect of amine structure and 192
Concentration, effect of temperature and butadiene 207
Conductivities of organolithium compounds 133
Conductivity, effect of solvent on .. 133
Conjugated diolefins, telomerization of 201
Constant(s)
  hyperfine coupling 69
  instability 188
  solvents, low dielectric 137
Contact-ion pair complexes 61
Content of benzyllithium–TMEDA complexes, isomer 34, 35
Content in benzyllithium–TMEDA complexes, ring-isomer 49
Coordination, metal–base 77
Copolymers
  anionic graft 177
  raw graft 182
  SBS block 183
Coupling constants, hyperfine 69
Covalent bonding 12
Crotyllithium
  from olefins, preparation of allyllithium and 38
  –TMEDA
    complexes, preparation of allyllithium– and 38, 39
    preparation of 49
    reaction of allyllithium–TMEDA and 40, 41
Crystalline birefringence 195
Crystalline organolithium complexes 6
Cyclization, thermally induced ... 239

D

Delocalized carbanion system 56
Diamine
  complexes
    preparation of tertiary 24
    solubility of organolithium–tertiary 28
    synthetic reactions of organolithium–tertiary 50
    toluene metalation by *n*-butyllithium–tertiary 31

Diamine *(Continued)*
 mixtures 144
 organolithium complexes, tertiary 23
  preparation by metalation ... 24
Diamines, chelating-type 187
Dianion, molecular orbitals for the anthracene 105
Dielectric constant solvents, low .. 137
Dimers 6
Diolefins, telomerization of conjugated 201
Dipole–dipole interactions 6
Directed metalation 222
 heterocyclic synthesis *via* 239
 in ruthenocene 225
 in thiophene 227
Directing abilities of substituents .. 231
Directing mechanism 232
Directing substituents, sulfonamides as ortho- 240
Displacement reaction, olefin 199
Dissociation, complex 156
Dissociation equilibria 131
Distillate telomer, components of a light 197
Distribution for the benzyl carbanion, atomic charge 87
Ditertiary aliphatic amines, chelating 249
Double and triple bonds, nucleophilic addition to 264

E

Efficiency
 catalyst 178
 exchange 184
 grafting 178
Eigenfunctions 102
Electron spin resonance (ESR) .. 113, 132
Energy, salt lattice 154
Equation, Arrhenius 136
Equilibria, dissociation 131
(ESR), electron spin resonance .. 113, 132
Ethylene
 polymerization 164
 telomerization of 264
  with aromatic hydrocarbons .. 189
  with olefinic telogens 198
  with olefins 197
Exchange efficiency 184

F

Ferrocenes, pseudoasymmetry in .. 238
Formation, isomer 31
Formation of carbon–carbon bonds 228

G

Gel permeation chromatography (GPC) 181
Generation on the polymer backbone, anion 179

Geometries, carbanion 76
Gilman procedure 41
Graft copolymers, anionic 177
Graft copolymers, raw 183
Grafting by anionic techniques .. 177
Grafting efficiency 178
Grignard reagent 228

H

Halides, chelated lithium 115
Halogen bonds, carbon– 230
Hetero atoms, metalations of organic molecules containing 256
Heterocyclic synthesis *via* directed metalation 239
Hexamers 6
Hydrides for reducing carbonyl compounds, chelated complex metal 273
Hydrocarbons
 metalations of aromatic 250
 polylithiation of 211
 telomerization of ethylene with aromatic 189
Hydrogenolysis rate, BuLi 7
Hyperfine coupling constants 69

I

Induced cyclization, thermally ... 239
Induction, asymmetric 235
Initiator, anionic 184
Inorganic complexes 142
 benzene-soluble 158
Instability constant 188
Interaction
 chelating agent–cation 120
 dipole–dipole 6
 metal–base 79
 steric 205
Intermediate, allyl anion 205
Ion pair bonding 12
Ion pair complexes, contact- 61
Ion pair, effect of chelating agent on 125
Ion pair structure on butadiene polymerization, effect of 12
Isolation of complexes, synthesis and 58
Isomer content of benzyllithium–TMEDA complexes 34, 35
 ring- 49
Isomer formation 32
Isomerization, photochemical studies of cis-trans 73
Isomers, tetramethylcyclohexanediamine 144

K

Kinetic acidity 192
Kinetic control *vs.* thermodynamic control 32

Kinetic vs. thermodynamic
  metalations 9

L

Lattice energy, salt 154
Light distillate telomer, components
  of a 197
Lithiated tertiary amines and
  diamines 41
Lithiated N,N,N',N'-tetramethyleth-
  ylenediamine, preparation of.. 49
Lithiated trimethylamine, preparation of 50
Lithiation, aromatic and
  alkylaromatic 214
Lithiation of benzene by n-butyllithium–TMEDA 46
Lithiation of toluene with
  n-BuLi–TMEDA 216
Lithium
  chelates, stereospecific interaction
    of benzene with 126
  chelates, structure of 3
  compounds, reactions of optically
    active chelated 274
  halides, chelated 115
  salts in benzene, chelated 159
  salts, polytertiary amine chelated 115
Liquid polybutadienes 167
Li–R bond, stereoselective reactions
  at the 272
Low dielectric constant solvents .. 137

M

Magnetic resonance studies of polytertiary amine chelated alkali
  metal compounds 113
Mechanism, directing 232
Mechanisms, chain transfer 15
Metal
  –base coordination 77
  –base interactions 79
  catalysts, polymerizations using
    N-chelated alkali 163
  complexes, stereochemical proper-
    ties of N-chelated alkali .. 56
  compounds, N-chelated organo-
    alkali 7
  compounds, polytertiary amine
    chelated alkali 113
  compounds, U.S. patents on
    N-chelated alkali 18
  –halogen interchange 184
  hydrides for reducing carbonyl
    compounds, chelated com-
    plex 273
Metals, reactions of alkali 57
Metalation(s)
  by anionic techniques 177
  of aromatic hydrocarbons 250
  benzene 251

Metalation(s) (Continued)
  with BuLi, ortho 254
  by n-butyllithium–tertiary
    diamine complexes 31
  of the chelating agent 10
  directed 222
    heterocyclic synthesis via ... 239
    kinetic vs. thermodynamic 9
  of organic molecules containing
    hetero atoms 256
  of organometallic π complexes .. 258
  at ring sites 253
  in ruthenocene, directed 225
  solvent for 249
  in thiophene, directed 227
  of toluene 189
  of weak acids 7
α-Methylene proton chemical shift 7
Methyllithium tetramer 4
Mixtures
  containing primary and secondary
    amines 153
  diamine 144
  pentamine 149
  tetramine 146
Moiety, carbanion 193
Molecular orbitals for the
  anthracene dianion 105
Molecular weight, influence of
  pressure on 191
Molecules containing hetero atom,
  metalations of organic 256

N

Naphthalene bond lengths 67
Naphthalenide, chelated sodium .. 117
N-Chelated
  alkali metal catalysts, polymeri-
    zations using 163
  alkali metal complexes, stereo-
    chemical properties of 56
  organoalkali metal compounds .. 7
  organolithium catalysis 1
  organolithium compounds, syn-
    thetic applications of 248
  organolithium reagents, catalytic
    behavior of 58
Nitrogen bonds, carbon– 230
(NMR) nuclear magnetic resonance 113
  spectra of sodium iodide com-
    plexes, proton 151
  spectra of tetramines 147
Nonstoichiometric catalyst
  compositions 206
Nucleophilic addition to double and
  triple bonds 264

O

Olefin displacement reaction 199
Olefins, telomerization of ethylene
  with 197
Oligomerization 142

Optically active chelated lithium
  compounds, reactions of 274
Optically active compounds 235
Orbitals for the anthracene dianion,
  molecular 105
Organic molecules containing hetero
  atom, metalations of 256
Organoalkali metal compounds,
  N-chelated 7
Organolithium
  catalysis, N-chelated 1
  catalysts, chelated 3
  chelates 5
  complexes, tertiary diamine 23
    preparation by metalation ... 24
  compounds, conductivities of .. 133
  compounds, synthetic applications of N-chelated 248
  reagents 56
    catalytic behavior of
      N-chelated 58
    stereochemical properties of
      amine-chelated 58
  –tertiary diamine complexes ... 28
  synthetic reaction of 50
  –TMEDA complexes, stability of 29
Organometallic complexes, structural properties of carbanion 60
Organometallic π complexes,
  metalations of 258
Organometallic syntheses 2
Organosodium catalysts, chelated .. 201
Organosodium complexes 267
Ortho metalation with BuLi 254
Oxo alcohols, preparation of 208
Oxygen bonds, carbon– 230

P

Paraffinic carbon chains, solid ... 199
Patents on N-chelated alkali metal
  compounds, U.S. 18
Pentamine mixtures 149
Permeation chromatography
  (GPC), gel 181
Permethylated tertiary polyamines,
  skeletal structures of n- 144
Permethylated tetramines, separation of N- 148
Phenyllithium from benzene,
  preparation of 25
Phenyllithium–TMEDA in benzene,
  preparation of 48
Photochemical studies of cis-trans
  isomerization 73
Phthalimidine 240
Plastics-range polyethylene 164
Polar-modified alkyllithium
  polymerizations 171
Polyamines, chelating tertiary 1, 202
Polyamines, skeletal structures of
  n-permethylated tertiary 144
Polybutadiene, liquid 167
Polycyclic aromatics 217

Polyethers, chelating 1
Polyethylene, plastic-range 164
Polyethylene waxes, straight-chain 165
Polylithiation of hydrocarbons ... 211
Polylithioanthracene, products from 219
Polylithio compounds, aliphatic ... 211
Polymer backbone, anion generation
  on the 179
Polymerization
  of butadiene 173
  effect of ion pair structure on
    butadiene 12
  ethylene 164
  factors affecting 11
  polar-modified alkyllithium 171
  using N-chelated alkali metal
    catalysts 163
Poly-n-alkyl aromatics 196
Polytertiary amine chelating agents,
  skeletal structures of 114
Polytertiary amine chelated alkali
  metal compounds, magnetic
  resonance studies of 113
Polytertiary amine chelated lithium
  salts 115
Polythiation of aromatic compounds 214
Precursor, Chel* 272
Preparation
  of allyllithium and crotyllithium
    from olefins 38
  of allyllithium– and crotyllithium–
    TMEDA complexes 38, 39, 49
  of benzyllithium from toluene .. 29
  of (benzyllithium)$_2$–TMEDA in
    toluene 48
  of benzyllithium–TED solid
    complex 49
  of crotyllithium–TMEDA 49
  of lithiated N,N,N',N'-tetramethylethylenediamine 49
  of lithiated trimethylamine 50
  of oxo alcohols 208
  of phenyllithium from benzene .. 25
  of phenyllithium–TMEDA in
    benzene 48
  of tertiary diamine organolithium
    complexes by metalation .. 24
Pressure on molecular weight,
  influence of 191
Primary and secondary amines,
  mixtures containing 153
Procedure, Gilman 41
Processes, separation 142
Prochiral substrates, reactions of .. 275
Products from polylithioanthracene 219
Properties
  of amine-chelated organolithium
    reagents 58
  of aromatic telomer wax 196
  of carbanion organometallic complexes, structural 60
  of N-chelated alkali metal complexes, stereochemical 56
  of chelates 6

INDEX

Proton NMR spectra of sodium iodide complexes 151
Pseudoasymmetry in ferrocenes ... 238

Q

Quaternary ammonium salts 140

R

Rate, BuLi hydrogenolysis 7
Rate, effect of butadiene addition 208
Raw graft copolymers 183
Reaction(s)
  of alkali metals 57
  of allyllithium–TMEDA and crotyllithium–TMEDA 40, 41
  of benzyllithium–TED 36, 37
  of (benzyllithium)$_2$–TMEDA in toluene 34, 35
  at the Li–R bond, stereoselective 272
  olefin displacement 199
  of optically active chelated lithium compounds 274
  of organolithium–tertiary diamine complexes, synthetic 50
  of prochiral substrates 275
  telomerization 2
  temperature, effect of 207
  transmetalation 11
Reactivities in proton abstraction and polymerization 204
Reagent, Grignard 228
Reagents, organolithium 56, 58
  stereochemical properties of amine-chelated 58
Reducing carbonyl compounds, chelated complex metal hydrides for 273
Resonance (ESR), electron spin 113, 132
Resonance (NMR), nuclear magnetic 113
  studies of polytertiary amine chelated alkali metal compounds 113
Ring-isomer content in benzyllithium–TMEDA complexes 49
Ring sites, metalation at 253
Ruthenocene, directed metalation in 225

S

Salt lattice energy 154
Salts in benzene, chelated lithium 159
Salts, quaternary ammonium 140
Salts, polytertiary amine chelated lithium 115
Salts, structural features of chelated 120
SBS block copolymers 183
Secondary amines, mixtures containing primary and 153

Selective complexation of *trans*-TMCHD 145
Separation of *N*-permethylated tetramines 148
Separation processes 142
Shift, α-methylene proton chemical 7
Sites, metalation at ring 253
Skeletal structures of *n*-permethylated tertiary polyamines ... 144
Skeletal structures of polytertiary amine chelating agents 114
Sodium iodide complexes, proton NMR spectra of 151
Sodium naphthalenide, chelated .. 117
Solid complex, preparation of allyllithium–TMEDA 49
Solid complex, preparation of benzyllithium–TED 49
Solid paraffinic carbon chains ... 199
Solubility of organolithium–tertiary diamine complexes 28
Solvent on conductivity, effect of .. 133
Solvent for metalation 249
Solvents, low dielectric constant .. 137
Spin resonance (ESR), electron 113, 132
Stability of organolithium–TMEDA complexes 29
Stereochemical properties of amine-chelated organolithium reagents 58
Stereochemical properties of *N*-chelated alkali metal complexes .. 56
Stereoselective reactions at the Li–R bond 272
Stereospecific interaction of benzene with lithium chelates 126
Steric effects of the chelating agent 15
Steric interaction 205
Straight-chain polyethylene waxes 165
Structural features of chelated salts 120
Structural properties of π carbanion organometallic complexes 60
Structure(s)
  on butadiene polymerization, effect of ion pair 12
  effect of telogen on telomer 192
  of lithium chelates 3
  of *n*-permethylated tertiary polyamines, skeletal 144
  of polytertiary amine chelating agents, skeletal 114
Substrates, reactions of prochiral .. 275
Substituents, directing abilities of 231
Substituents, sulfonamides as ortho-directing 240
Syntheses, organometallic 2
Synthesis and isolation of complexes 58
Synthesis *via* directed metalation, heterocyclic 239
Synthesis *via* lithium chelates, asymmetric 270
Synthetic applications of *N*-chelated organolithium compounds ... 248

Synthetic reactions of organolithium–tertiary diamine complexes 50
System, delocalized carbanion ... 56

T

Techniques, metalation of anionic 177
TED, reactions of benzyllithium–..36, 37
TED solid complex, preparation of benzyllithium– 49
Telogen on telomer structure, effect of 192
Telogens, telomerization of ethylene with olefinic 198
Telomer, components of a light distillate 197
Telomer structure, effect of telogen on 192
Telomer waxes, aromatic 194, 196
Telomerization 203
  butadiene 205
  of conjugated diolefins 201
  of ethylene 264
    with aromatic hydrocarbons .. 189
    with olefinic telogens 198
    with olefins 197
  reactions 2
Temperature, effect of reaction .. 207
Tertiary
  chelating polyamines 202
  diamine complexes
    preparation of 24
    solubility of organolithium .. 28
    synthetic reactions of organolithium 50
    toluene metalation by n-butyllithium– 31
  diamine organolithium complexes 23
  polyamines, chelating 1
  polyamines, skeletal structures of n-permethylated 144
Tetramer, methyllithium 4
Tetramethylcyclohexanediamine isomers 144
Tetramethylethylenediamine, preparation of lithiated N,N,N',N'- 49
Tetramines 146
  NMR spectra of 147
  separation of N-permethylated .. 148
Thermally induced cyclization ... 239
Thermodynamic acidity 9
Thermodynamic control, kinetic control vs. 32
Thermodynamic metalations, kinetic vs. 9
Thiophene, directed metalation in 227
(-)-TMCHD-LiAlH$_4$ 274
(-)-TMCHD·LiC$_4$H$_9$ 274
TMCHD, selective complexation of trans- 145
TMED (N,N,N',N'-tetramethylethylenediamine) 1

TMEDA
  in benzene, preparation of phenyllithium– 48
  complexes, isomer content of benzyllithium– 34, 35
  complexes, preparation of allyllithium– and crotyllithium– 38, 39
  complexes, stability of organolithium 29
  and crotyllithium–TMEDA, reaction of allyllithium– 40, 41
  lithiation of benzene by n-butyllithium– 46
  lithiation of toluene with n-BuLi 216
  preparation of crotyllithium– ... 49
  reaction of allyllithium–TMEDA and crotyllithium– 40, 41
  solid complex, preparation of allyllithium– 49
  in toluene, preparation of (benzyllithium)$_2$– 48
  in toluene, reactions of (benzyllithium)$_2$– 34, 35
Toluene
  with n-BuLi–TMEDA, lithiation of 216
  metalation 189
    by n-butyllithium–tertiary diamine complexes 31
  preparation of benzyllithium from 29
  preparation of (benzyllithium)$_2$– TMEDA in 48
  reactions of (benzyllithium)$_2$– TMEDA in 34, 35
Transfer agents, charge 138
Transfer, factors affecting the chain 202
Transfer mechanisms, chain 15
Transformation, biochemical 271
Transmetalation 188
  reactions 11
trans-N,N,N',N'-Tetramethyl-1,2-cyclohexanediamine (TMCHD) 270
Trialkylaluminum simulation 2
Trimethylamine, preparation of lithiated 50
Triple bonds, nucleophilic addition to double end 264

U

U.S. patents on N-chelated alkali metal compounds 18

W

Waxes, aromatic telomer 194
  properties of 196
Waxes, straight-chain polyethylene 165
Weight, influence of pressure on molecular 191

The text of this book is set in 10 point Caledonia with two points of leading. The chapter numerals are set in 30 point Garamond; the chapter titles are set in 18 point Garamond Bold.

The book is printed offset on Danforth 550 Machine Blue White text, 50-pound. The cover is Joanna Book Binding blue linen.

Jacket design by Norman Favin.
Editing and production by Spencer Lockson.

The book was composed by the Mills-Frizell-Evans Co., Baltimore, Md., printed and bound by The Maple Press Co., York, Pa.

QD
1
A355
#130

JUL 29 1975